NATIONAL ENERGY DATA SYSTEMS

NATIONAL ENERGY DATA SYSTEMS

Proceedings of an International Workshop

Edited by

DILIP R. AHUJA
R.K. PACHAURI
N.K. GOPALAKRISHNAN

CONCEPT PUBLISHING COMPANY
NEW DELHI

First Published 1984

Published by

Naurang Rai
Concept Publishing Company
H-13 Bali Nagar
NEW DELHI-110015 (India)

Printed by

Asia Composer
at Sunil Printers, Naraina
NEW DELHI-110028

Preface

DURING the past decade a number of developing countries around the world have instituted mechanisms and methodologies for energy planning as part of overall development planning and economic policy. The importance of formal national policies and plans for the energy sector arises not only out of the price increases of oil that took place in 1973-74 and 1979-80, but also because the demand for energy in the developing countries is expected to grow much faster than in the developed countries, and because indigenous energy resources have not been exploited optimally due to a variety of constraints related to technology, finance and skills. Whereas some of these constraints have received a fair amount of attention in recent years, another major resource which is essential for energy planning, namely, that of suitable data for planning has not received nearly as much attention as other subjects.

There are historical reasons for inadequate development of data systems and information flows for energy planning. Firstly, suitable data in general are often not available in the developing countries for any kind of planning at the national level, much less at disaggregated regional levels. Further, since planning for energy supplies in the past has essentially been industry oriented, there is a notable lack of linkages across different forms of energy in a number of countries in the third world. Consequently, data on power may be available with the electric utility industry, but seldom does this industry have any information available or accessible on substitutes such as oil products or natural gas supplies, even though developments in the latter may have significant effects on the former. (In India, for instance prices of diesel oil can significantly curtail the demand for electric pumpsets in agriculture). Also developments in the energy sector cannot be viewed in isolation of developments in the economy as a whole and in other sectors.

Energy planning must take into account information and data pertaining to non-energy sectors as well.

It was in realisation of the importance of this subject area that the Tata Energy Research Institute decided to organise a workshop to deal with the entire spectrum of data systems for energy planning, in which specialists from the third world as well as those with an interest in the subject from the developed countries were to be invited to participate in a structured exercise. A number of specific theme papers were commissioned for the purpose, so that the discussions that took place would be reasonably pointed and based on adequate research by those actively involved in this field of knowledge. The workshop was organised at New Delhi in December 1982, and the papers presented were carefully edited, revised and in some cases re-written to suit the needs of a volume such as this.

Quite apart from the valuable knowledge and insights provided in the papers themselves, the workshop discussions provided very useful information that has been distilled and presented in this volume. The participants to the workshop consisted of a mix of academics, researchers, very senior policy makers and others drawn from energy organisations, and therefore, the discussions had both a fairly sound academic base as well as an underlying practical bias. This volume, therefore, is designed not only for academics and researchers, but should be of value to those actively involved in energy decision making.

Some useful directions emerged from the workshop, which could be very briefly summarised and mentioned at this stage. Even though the papers and discussions covered a broad canvas, it was found important to maintain specificity with respect to a central model for discussion and analysis, which by choice was that of India. However, the summary presented herein and the papers in later chapters are applicable in varying degrees to different countries of the world. Indeed, it was for this reason that a number of country papers were written and presented, so that the workshop while focusing indepth on the system obtaining in one country, did go into general issues based on the experiences of other countries.

In the case of electric power there was a uniform consensus

that improved methodologies for load forecasting were particularly important, and these in turn would throw up the need for superior data which could be developed over a period of time. In general, it was also felt that substantial data gaps existed at the low tension distribution end of power systems, and these could be made good by extensive instrumentation and metering. Consumer metering is essential for rational pricing of power, even though the costs and problems associated with metering rural and remote customers may be substantial. But it was pointed out that prices are vital signals for customers and provide an effective basis for demand management. An area of research which the developing countries would be well advised to go into on a continuing basis is that of electricity pricing, tariff design and related issues.

The need for better end use data is common to all forms of energy, but since electricity for industry is consumed in large quantities by a generally small number of consumers, there is some importance attached to regular reporting of different information by such consumers. A system of incentives and disincentives to ensure a steady flow of information from such consumers would be particularly useful. It was also found that, in general, there were major deficiencies in the availability of non-technical data such as on financial aspects or non-energy data pertaining to other sectors of the economy. This again is a problem that exists with other sources of energy as well. Indepth discussions and analysis of present data processing systems also revealed the importance of computerisation for effective planning and management of electric power systems in developing countries.

In the case of oil and natural gas, discussions covered the whole range of problems associated with distribution and marketing of oil products as well as going to the other end of the spectrum to cover problems of exploration and drilling. It was agreed that a system of data collection be instituted to accurately quantify physical losses of oil and oil products in refining and distribution activities. Reference to the Indian situation in this context was made to suggest that data be collected extensively at the depot level.

In the area of exploration for and production of hydrocarbons it was found that knowledge on terms pertaining to reserves, such as indicated, total, recoverable, etc., needs to be spread outside the oil industry, so that those concerned with energy issues are able to comprehend developments in the hydrocarbons area. Also, since estimated data on oil and natural gas reserves are probabilistic in nature, it is desirable to provide a range of figures rather than a set of single figures. Reliability of estimates in the oil business is ultimately produced only through actual drilling. Hence reserves allocated for exploration activities are essentially resources meant for creating reliable information, and thereby providing a reliable basis for energy planning.

Since the industrial sector is growing rapidly in a number of developing countries and because it is often the largest consumer of conventional energy, considerable attention was devoted to studying data relating to industrial energy consumption. It was found that as a methodology energy analysis had a very useful role to play in building up valuable information on energy use in various processes and activities. It is particularly important to collect end use information on energy consumed in the industrial sector. Referring to the case of India the workshop found that whereas the Annual Survey of Industries (ASI) published regularly was extremely useful, in some cases it was desirable to supplement this information with data at the plant level. It would also be useful to cut down the lags in the publication of ASI data. A system of incentives and disincentives is essential to bring about better reporting of energy data by industrial units, (if necessary by enforcing legislation that was enacted in 1981 in India). Another area which has been neglected in the past is a study of temperature distribution of energy uses in industrial processes and activities.

The rural energy scene in developing countries represents yet another area where not enough has been done to create information useful for planning purposes. A useful means by which the current gap can be filled up would be through the development of surveys pertaining to different agroclimatic regions using a compatible but not necessarily common design. In developing these surveys it is important to draw a distinction between surveys for short-term and medium-term purposes. A mix of such surveys should be devised and adopted after carefully assessing the needs for such surveys. In the rural sector, in general, it is necessary to establish the "need" for energy by specific end uses so that technologies can be matched with each particular type of application. The important role of animal energy in India and many other developing countries was accepted by the workshop. It was felt that the actual potential of this energy form and the possibility of improvements through better yoke design etc. have not been surveyed. These need to be carried out and disseminated.

Since the workshop was international in character, attention was also paid to the possibilities of international exchange in this field. It was agreed that linkages should be established between National Energy Information Centres wherever they existed in developing countries, such as the one in Thailand. It was found that inter-country comparisons for specific sectors and end-uses was also important. For instance, countries like China and India had areas of strong commonalities wherein comparisons could be made. The case of Brazil where such energy balances have been produced for the different states of the country was found interesting and worthy of emulation in other large countries. The workshop also revealed a consensus on several general issues. For instance, the relevance of using the coal replacement measure in India perhaps did not exist any more and energy statistics in this country could perhaps now be computed in terms of coal equivalent units. There was also acceptance of the importance of training programmes particularly at the field level, resulting from which better measuring devices and instrumentation in the energy field would take place.

The whole area of energy planning and its dependence on good energy data can be developed if each country were to set up a permanent expert group to review and follow up developments from time to time. It is also important for institutions inside and outside the government to develop quantitative modelling techniques for a study of energy-economy interaction. This in turn would throw up weaknesses and gaps in energy data systems which could lead to refinement in the long run. Owing to the effects of energy production and use on the environment, it is also necessary for planning purposes to develop data on the environment effects of energy use.

In essence, the chapters of this book and a narration of the salient points debated and discussed cover the range of questions and important conclusions briefly mentioned above. In preparing this volume the editors and the Tata Energy Research Institute have been motivated by the important need to develop a consciousness on the importance of data in the energy planning process. These efforts, fortunately, have shown some signs of response from organisations that could make a contribution to the field over a period of time. The United Nations University for instance, in its current programme of energy planning has taken up a project on energy data for looking deeper into this area. The Government of India, through its representatives who participated in the workshop, has also been showing some interest in setting up a permanent mechanism for reviewing energy data on a regular basis and ensuring that improvements and refinements take place in keeping with the compulsions of good energy planning. It is hoped that at the very least this volume will shed light on important issues related to data needs for energy planning in the developing countries.

DILIP R. AHUJA
R.K. PACHAURI
N.K. GOPALAKRISHNAN

Acknowledgements

THE editors of this volume would like to express their appreciation to the organisations who provided financial support to the International Workshop on National Energy Data Systems, in which the papers included in this volume were presented. These organisations include the Ministry of Energy of the Government of India, Coal India (Limited), the K.D. Malaviya Institute of Petroleum Exploration of the Oil & Natural Gas Commission, the Global and Inter-regional Projects Division of the UNDP, New York (which provided support through the International Cooperative Programme for Training and Research in Energy Planning and Management) and the United Nations Educational, Scientific and Cultural Organisation, Paris. Thanks are also expressed for the various participants of the Workshop who shared their time and knowledge, and helped in numerous ways while discussing issues dealing with energy data for planning purposes. This ultimately led to a much clearer focusing of important issues that have been discussed and presented in this volume. Finally, the editors would also like to express their thanks to the staff of the Tata Energy Research Institute who helped in the preparation of this volume.

List of Participants

International Workshop on National Energy Data Systems
December 15-18, 1982

S. ADITYA

Director, K.D.M. Institute of Petroleum Exploration,
Basin Studies Division, Oil and Natural Gas Commission,
9, Kaulagarh Road,
DEHRADUN-248195.

S.K. AGARWAL

Chief Engineer (Planning), Central Electricity Authority,
Sewa Bhawan, Ramakrishnapuram,
NEW DELHI-110022.

DILIP AHUJA

Fellow, Tata Energy Research Institute,
Jeevan Tara Building, Parliament Street,
NEW DELHI-110001.

HUGO ALTOMONTE

Instituto de Economia Energetica, Bariloche Fundacion,
24 de Septiembre 178, Casilla de Correo 138,
8400 Bariloche, Provincia de Rio Negro,
ARGENTINA.

JOAO LIZARDO R.H. de ARAUJO

Programme de Engenharia de Distemas
E. Computacao (COPPE),
Federal University of Rio de Janeiro,
Ilha do Fundao-CP 1191, 20-000RJ Rio de Janeiro,
BRAZIL.

List of Participants

A.N. BANERJEE

Chairman, Mine Environment Committee,
M-155, Greater Kailash II,
NEW DELHI-110048.

JERROLD BERKE

Deputy Resident Representative,
United Nations Development Programme,
55, Lodi Estate,
NEW DELHI-110003.

RAMESH BHATIA

Professor, Institute of Economic Growth,
University Enclave,
DELHI-110007.

D. BHATTACHARYA

Senior Chemical Engineer,
National Council of Applied Economic Research,
Parisila Bhawan, IP Estate,
NEW DELHI-110002.

CHARUAY BOONYUBOL

Dean, Faculty of Engineering, Chulalongkorn University,
Phyathai Road, Bangkok 5,
THAILAND.

S.K. BOSE

Advisor, Coal India Ltd., 10, Netaji Subhas Road,
CALCUTTA-700007.

B. BOWONDER

Chairman, Centre for Energy, Environment & Technology,
Administrative Staff College of India, Bella Vista,
HYDERABAD-500475.

ROMIR CHATTERJEE

Coordinator, United Nations Development Programme,
International Cooperative Programme for Training & Research in Energy Planning and Management,
1889 F. St.. N.W.
WASHINGTON D.C.-20006, U.S.A.

RENE CODONI

Coordinator, U.N. Asian & Pacific Development Centre,
Pesiaran Duta, P.O. Box 2224, Kuala Lumpur,
MALAYSIA.

MAHESHWAR DAYAL

Secretary, Commission for Additional Sources of Energy,
Dept. of Science and Technology, Technology Bhawan,
New Mehrauli Road,
NEW DELHI-110016.

ASHOK DESAI

Consultant, National Council for Applied Economic Research, Parisila Bhawan, Indraprastha Estate,
NEW DELHI-110002.

NIKHIL DESAI

World Bank, 55, Lodi Estate,
NEW DELHI-110003.

NITIN DESAI

Advisor (PAD & M & I), Planning Commission,
Yojana Bhawan,
NEW DELHI-110001.

K. DHARMARAJAN

Director, Ministry of Energy, Shram Shakti Bhavan,
Rafi Marg,
NEW DELHI-110001.

MOHAMMAD FAZAL

Member, Planning Commission, Yojana Bhawan,
Parliament Street,
NEW DELHI-110001.

P.B. GHATE

Spl. Asstt. to the Minister of Planning,
Planning Commission, Govt. of India,
67, Vasant Marg, Vasant Vihar,
NEW DELHI-110057.

B.N. GHOSH

Dy. General Manager (Technical), Indian Oil Corporation,
Indian Oil Bhawan, Janpath,
NEW DELHI-110001.

N.K. GOPALAKRISHNAN

Director, Documentation & Information Centre,
Tata Energy Research Institute, Bombay House,
24, Homi Mody Street,
BOMBAY-400023.

AJIT GUPTA

Principal Scientific Officer,
Commission for Additional Sources of Energy,
Department of Science & Technology, Technology Bhavan,
New Mehrauli Road,
NEW DELHI-110016.

C.L. GUPTA

Hon. Director, Tata Energy Research Institute,
Field Research Unit, Care Sri Aurobindo Ashram,
PONDICHERRY-605002.

K.S. HINGE

Member, Governing Council, Tata Energy Research
Institute, Bombay House, 24, Homi Mody Street,
BOMBAY-400023.

C.R. JAGANNATHAN

Chairman, Oil India Ltd., Allahabad Bank Building,
Parliament Street,
NEW DELHI-110001.

S.L. JAIN

Director, Central Electricity Authority, Sewa Bhavan,
R.K. Puram,
NEW DELHI-110022.

R. JAYASHANKAR

Lecturer, University of Delhi,
28, South Patel Nagar
NEW DELHI-110008.

DAVID JHIRAD

Deputy Director, Energy Management Training Program,
Institute for Energy Research,
State University of New York at Stonybrook,
NEW YORK-11974, U.S.A.

J.K. KHURANA

Director (Corporate Planning), Bharat Petroleum Ltd.,
Bharat Petroleum Headquarters, Ballard Estate,
BOMBAY-400038.

MAURICE LEVY

Director, Energy Programme, United Nations University,
4, Place Jussieu, 75005 Paris,
FRANCE.

SUCHAI LIMSUMALEE

Head of Energy Assessment Unit,
Regulatory Divn. in the National Energy Administration,
Pibultham Villa, Kasatsuk Bridge, Bangkok 10500,
THAILAND.

G.S. MARWAH

Director, Indian School of Mines,
DHANBAD-826004.

MARINO MENA

Dean College of Engineering, Melchor Hall,
University of the Philippines, Diliman, Quezon City,
PHILIPPINES.

K. MISHRA

Joint Secretary, Ministry of Energy, Govt. of India,
Shram Shakti Bhawan, Rafi Marg,
NEW DELHI-110001.

T.K. MOULIK

Indian Institute of Management, Vastrapur,
AHMEDABAD-380015.

VINOD MUBAYI

Brookhaven National Laboratory, Upton, Long Island,
NEW YORK 11973, U.S.A.

H.K. MULCHANDANI

Head, Projects Division, Indian Institute of Petroleum,
CSIR Complex, NPL Campus, Pusa,
NEW DELHI-110012.

MOHAN MUNASINGHE

Senior Energy Advisor to the President,
10 De Fonseka Place, Colombo 5,
SRI LANKA.

RITA NANGIA

Consultant, Oil Co-ordinating Committee, Kailash,
Kasturba Gandhi Marg,
NEW DELHI-110001.

K.P. NYATI

Dy. Director, Fuel Efficiency,
National Productivity Council, Lodi Estate,
NEW DELHI-110003.

R.K. PACHAURI

Director, Tata Energy Research Institute,
Jeevan Tara Building, Parliament Street,
NEW DELHI-110001.

K.C. PANT

Member of Parliament, 7, Tyagaraj Marg
NEW DELHI-110001.

K.S. RAJAN

Former Secretary, Ministry of Industry,
S-450, Greater Kailash-II
NEW DELHI-110048.

S. RAMANI

Professor of Humanities, Indian Institute of Technology,
MADRAS-600036.

N.S. RAMASWAMY

Director, Indian Institute of Management,
33, Langford Road,
BANGALORE-560027.

S. RAMESH

Joint Secretary, Ministry of Energy, Department of Power,
Shram Shakti Bhavan, Rafi Marg,
NEW DELHI-110001.

S.V.S. PRAKASA RAO

Deputy Director, Central Statistical Organization,
Ministry of Planning, Govt. of India,
Sardar Patel Bhawan, Sansad Marg,
NEW DELHI-110001.

S. SUBBA RAO

Indian Institute of Management, 33 Langford Road,
BANGALORE-560027.

D. SANKARAGURUSWAMY

Advisor, Planning Commission, Yojana Bhavan,
NEW DELHI-110001.

E.A.S. SARMA

Director, Ministry of Energy, Department of Power,
Shram Shakti Bhavan, Rafi Marg,
NEW DELHI-110001.

K.C. SEAL

Director General, Central Statistical Organization,
Ministry of Planning, Sardar Patel Bhawan, Sansad Marg,
NEW DELHI-110001.

D.P. SENGUPTA

Chairman, Department of Electrical Engineering,
Indian Institute of Science,
BANGALORE-560012.

J.C. SHAH

Energy Consultant & Vice-Chairman,
National Council of Power Utilities, 7, Jai Shefali Park,
Satellite Road,
AHMEDABAD-380015.

K. RAVI SHANKAR

Associate Consultant, Administrative Staff College of India,
Bella Vista,
HYDERABAD-500475.

S. RAVI SHANKAR

Information Officer, Tata Energy Research Institute,
Bombay House, 24, Homi Mody Street,
BOMBAY-400023.

KIRK R. SMITH

Energy Program Leader,
East-West Resource Systems Institute, East-West Center,
1777, East-West Road,
HONOLULU, HAWAII 96848, U.S.A.

P.R. SRINIVASAN

National Productivity Council, Lodi Road,
NEW DELHI-110003.

R.N. SRIVASTAVA

Director, Central Electricity Authority, Sewa Bhavan,
R.K. Puram,
NEW DELHI-110022.

S.S. SRIVASTAVA

Additional Director, Central Statistical Organization,
Ministry of Planning, Sardar Patel Bhawan, Sansad Marg,
NEW DELHI-110001.

LU YINGZHONG

The Institute of Nuclear Energy Technology,
Qinghua University, P.O. Box 1021, Beijing,
PEOPLE'S REPUBLIC OF CHINA.

Contents

PREFACE v
ACKNOWLEDGEMENTS xi
LIST OF PARTICIPANTS xii
ABBREVIATIONS xxiv

INAUGURAL SESSION

Inaugural Speech
—MOHAMMAD FAZAL 3

TECHNICAL SESSION—1

OIL & NATURAL GAS DATA SYSTEMS

1. Demand Forecasting for Petroleum Products in India
—H.K. MULCHANDANI & I.B. GULATI 19

TECHNICAL SESSION—2

ELECTRIC POWER DATA SYSTEMS

2. Data System in the Indian Power Sector
—D.P. SENGUPTA 69

TECHNICAL SESSION—3

COAL DATA SYSTEMS

3. Coal Data System in India
—S.K. BOSE & D. SENGUPTA 157

TECHNICAL SESSION—4

INDUSTRIAL ENERGY DATA SYSTEMS

4. Energy Analyses of Three Energy Intensive Industries in India
—D. BHATTACHARYA 191

TECHNICAL SESSION—5

INTERNATIONAL ENERGY DATA SYSTEMS

5. Thailand's National Energy Data Systems
—CHARUAY BOONYUBOL 227

6. Energy Data Systems in Thailand
—SUCHAI LIMSUMALEE 246

7. An Introduction to the National Energy Data System in the People's Republic of China
—LU YINGZHONG 260

8. Argentine Energy Data Systems
—HUGO A. ALTOMONTE 276

9. Evolution and Perspectives of the Brazilian Energy Information System
—JOAO LIZARDO R.H. DE ARAUJO 310

10. Energy Policy Planning Data Systems in the Dominican Republic
—ANTONIO J. D'ERRICO, ANDRES DOERNBERG, ELISA GARCIA, ROBERT NATHANS & MICHAEL OWENS 333

11. Energy Statistics in India
—K.C. SEAL 361

TECHNICAL SESSION—6

12. Rural Energy Data Systems
—B. BOWONDER & K. RAVI SHANKAR 371

13. Energy Data Systems : The Role of Traditional Sources
—RAMESH BHATIA 384

14. Data for Renewable Energy in India
—C.L. GUPTA 427

TECHNICAL SESSION—7

ENERGY DATA SYSTEMS FOR POLICY AND PLANNING

15. Energy Policy and Data Systems
—ASHOK V. DESAI 455

VALEDICTORY SESSION

Valedictory Address
—K.C. PANT 491

SUBJECT INDEX 503
NAME INDEX 510

Abbreviations

ADB	Asian Development Bank
APSEB	Andhra Pradesh State Electricity Board
ASTRA	Centre for the Application of Science and Technology to Rural Areas
ASI	Annual Survey of Industries
ATF	Aviation Turbine Fuel
BBMP	Bhakra Beas Management Board
CASE	Commission on Additional Sources of Energy
CEA	Central Electricity Authority
CFRI	Central Fuel Research Institute
CMPDIL	Central Mine Planning and Design Institute Limited
CNPE	Commission National Politica Energetica
CSO	Central Statistical Organization
CWPC	Central Water and Power Commission
DAE	Department of Atomic Energy
DGMS	Directorate General of Mines Safety
DFC	Demand Forecasting Cell
DGTD	Directorate General of Technical Development
DMR	Department of Mineral Resources
DST	Department of Science and Technology
DVC	Damodar Valley Corporation
EB	Energy Bureau (Argentina)
ECLA	Economic Commission for Latin America
EDIS	Energy Development Information System
EGAT	Electricity Generation Authority of Thailand
ELECTROBRAS	Central Electricas Brasileiras
ESND	Energy System Network Diagram

FAO	Food and Agricultural Organization
FPC	Fuel Policy Committee
GDP	Gross Domestic Product
GIDCO	Gujarat Industrial Development Corporation
GNP	Gross National Product
GSI	Geological Survey of India
HSD	High Speed Diesel
HT	High Tension
HYV	High Yielding Varieties
ICAR	Indian Council for Agricultural Research
IEA	International Energy Agency
IFAD	International Fund for Agricultural Development
IIP	Indian Institute of Petroleum
ILO	International Labour Organisation
IMF	International Monetary Fund
INDAL	Indian Aluminium Company
KEB	Karnataka Electricity Board
KPC	Karnataka Power Corporation
LDO	Light Diesel Oil
LFC	Load Frequency Control
LPG	Liquified Petroleum Gas
LT	Low Tension
MEC	Minerals Exploration Corporation
NCAER	National Council for Applied Economic Research
NEA	National Energy Administration
NEEPCO	North-Eastern Electric Power Corporation
NEIDCT	National Energy Information and Documentation Centre, Thailand
NHPC	National Hydro-Electric Power Corporation
NIC	National Informatics Centre
NPC	National Productivity Council
NSSO	National Sample Survey Organization
NTPC	National Thermal Power Corporation
OCC	Oil Co-ordination Committee

OLADE	Organization Latino Americana De Energia
ONGC	Oil and Natural Gas Commission
OPEC	Organization of Petroleum Exporting Countries
PCM	Phase Change Material
PCRA	Petroleum Conservation Research Association
PERT	Programme Evaluation Review Technique
PETROBRAS	Petroleo Brasileiro S.A.
PLCC	Power Line Carrier Communication
PRC	People's Republic of China
REB	Regional Electricity Board
REC	Rural Electrification Corporation
RES	Reference Energy System
RLDC	Regional Load Despatch Centre
RR	Railway Receipt
SEB	State Electricity Boards
SKO	Superior Kerosene Oil
SCCL	Singareni Collieries Co. Ltd.
SLDC	State Load Despatch Centre
TCR	Tonnes of Coal Replacement
TD, T&D	Transmission and Distribution
TELCO	Tata Engineering & Locomotive Company
TERI	Tata Energy Research Institute
UNCTAD	United Nations Conference on Trade and Development
UNDP	United Nations Development Programme
UNIDO	United Nations Industrial Development Organization
WGEP	Working Group on Energy Policy

INAUGURAL SESSION

Chairman : R.K. PACHAURI

Inaugural Address : MOHAMMAD FAZAL

Inaugural Speech

MOHAMMAD FAZAL

I AM grateful that I have been asked to be with you this morning. Considering my inadequacies, I should imagine that it is very generous of you to ask me to be at this function today.

Right at the outset, I would like to say a few words about the base of data collection in our country, and it may have some relevance in some other countries, particularly in the near-developed or developing societies. After you have collected the data, the assembly and sophistication in further use of the data becomes a matter of expertise. But the arrangements or methods for the collection of the ground data, and how firm or infirm these are, I should think, have to be very carefully considered and examined on the Indian scene.

Coming as I do from a village, I find, for example, the method of collection of data of agricultural outputs very conjectural. I wish to God, I am wrong. The data at the grassroot level is expressed in terms of whether it is a 4-anna crop or a 8-anna crop. Because 16-annas made a rupee, the yield is expressed as 25 per cent or 50 per cent of expected. And I am not too sure if the base, even today, is better than what has been traditionally the method of collection of the agricultural data. All that has to be very carefully looked by experts. How firm is our base? Because if the foundations are not firm, then the whole citadel may be wrong and no matter what expertise you bring in for manipulating the data, a lot of things will go wrong.

I would also like to bring to your notice, the poverty aspect of the Indian society. We all talk time and again of the percentage population below the poverty line, and it seems that the precision has become so sophisticated that we report this to the second place of decimal. As to how this has been built up, I am not too sure. You would be amused to hear this, coming from me, as I am from an organisation which should know it more than anybody else. And yet I am not terribly convinced of the premises of this poverty line, having had some acquaintance with abstract science, like physics which I did; and economics, which I was forced to read. I am also having in mind the utter poverty which was visible before we threw away the yoke of colonialism, when even an item like egg was a great luxury. I am not telling any secret. The one family in 10,000 in the country, which could give 1/8th of an egg to the eldest child, was considered terribly fortunate. And tea was again a luxury in the 30s. There was a premium paid for drinking tea by the tea promoters. At least for the common women in my village and neighbouring areas of my state, 4 or 5 yards of loin-cloth was unaffordable and quite a few of them were half-naked. And yet today, these three very broad indices of living standard tell a different story. You do not come across naked women or men in this country; there are tea shops in every village of this country, and the eggs glare at your face from the steel wire basket in almost every shop. Similarly, fruit was a luxury in the 40s and before that. If you needed fruits, you had to be either a sick person or a city dweller. You had to go to the city centre of a district to get fruit. And yet today, fruits are being sold almost everywhere, in every town, in every large village—and not just the local ones, but also those which are sophisticated here, as grapes. If there were no purchasing power, or that the people were so poor, then I do not think there would have been any market for these commodities. Therefore, the base for this data right from the grassroot level, has to be examined very seriously by experts, because it is on the base that premises are built and conclusions drawn.

We will be gaining a great deal by your advice in the course of time in the Commission, on how the base data may be improved and the collection methods further refined. Now I said this as an aside thinking it had some relevance. You may be discussing what we in the Commission may be terribly interested to know whether these premises which we are trying to share with you, have any relevance.

And while this Conference may be discussing the various data systems in the energy sector, and what sophistication could be introduced in this system, it appears desirable to consider in what manner optimum use can be made of the existing data in the sector. Energy statistics are necessary for assessing, managing and monitoring national energy development. Since data collection, processing and presentation are costly and time consuming operations, they have to be properly planned and the whole system has to be made cost effective. This is especially so for developing countries where not only the resources are scarce but there is shortage of skilled personnel. I am happy to find that the shortage is not all that acute, seeing such a large assembly at this conference and having been told that lot of our experts abroad are inclined to come back. I think the shortage may not be there for long.

It must be recognized that energy data systems are only means to a larger end and, therefore, they should be placed in a proper perspective in the scheme of things. While it would be useful to compare energy data systems of the different developing and under-developed countries, these systems again must be viewed within the total socio-economic context of the concerned country, instead of being viewed as ends in themselves. The institutional and infrastructural facilities that have, therefore, to be established in an individual country on a priority basis for energy data system have got to be designed as part of the total system for economic planning and implementation in these countries.

This Workshop is going to consider the energy data system under 5 headings, namely, coal data system, power data system oil data system, rural energy data system, and finally total national energy data system. This exercise will evidently be for comparison between different developing countries. There is also going to be a discussion on energy data for policy and planning and at the end, there will be a concluding session to summarise the recommendations of this Seminar. May I suggest that while it will be useful to go into details of the data requirements of the different sub-sectors of the energy system, such as coal, oil, power, rural energy, and so on, it will possibly be a useful exercise to put all these sub-sectors within the perspective of the total energy sector and also the total national economy. This perspective may be kept in mind as the discussions proceed. I may like to emphasise that it will be much more useful if energy data systems are discussed as part of the total economic situation in the country, instead of energy sector being discussed in detail separately.

The Central Electricity Authority (CEA) in India regularly prepares and publishes a voluminous amount of data on generation and distribution of power in the different states of the country as also on the efficiencies of inputs and several other relevant matters connected with the power sector. These data are published in Handbooks which are circulated to the concerned Departments as well as the Central Ministries and the Planning Commission. The CEA and the Ministry of Energy also publish detailed data on the salient features of the electricity supply to industry, installed capacity and generation, on transmission and distribution lines and transformers, utilisation of utilities and non-utilities, details of licences and electrical accidents, financial statistics and programmes of electricity supply in the different states. All these data are prepared on a massive scale and circulated in booklets, which no doubt are very useful, but are not always read by decision makers in their totality, and by others concerned at different levels. An exception might be the recent booklet published by the CEA highlighting the power supply situation in India which gives data on the generation capacity of all the sources of energy, energy supply, per capita consumption, and so on as well as the outlays in the Sixth Plan. This booklet was brief and presented in a much more readable manner. I am sure many people must have read this booklet.

What I want to emphasize is that the presentation of data is very important. With the increasing use of computers, it should be possible to improve the presentation and to disaggregate the data and compile them in formats which would have relevance to different categories of users and decision makers. For example, at the highest level, the data should be presented in a much more aggregated form, giving the salient features of energy demand and supply, shortfalls in the supply and demand and the physical areas where attention is to be given by these top level decision makers. The second level should be provided printouts of detailed data which would give explanation for the data that was submitted to top level decision makers. For example, at power plants where the demand of inputs is not being met, the actual problems such as shortage of coal, quality of coal supply, labour problem, etc. need to be specified. At the lower level, details of each of these critical areas, including the thermal power stations which demand attention, should be provided through computer printouts. The data in itself is not the panacea and data has to be tailored to the need of the person who is going to use the data at different levels.

This is just one example of how the computer could be utilised in presenting data that is so painstakingly collected and compiled by our staff. The data have to be presented in an appealing form to meet the needs of decision makers at different levels. The machinery for collecting the data may be streamlined and the formats in which the data have to be presented would have to be modernised and developed with the help of computers. In this connection top level decision makers as well as the next level have a very decisive role to play. They have to specify the type of energy data for making investment decisions which would be in keeping with the larger interests of the economy.

The role of the experts should be to intelligently to inform the decision makers about the capability of these energy data systems and how best these could help them make better decisions in the larger and overall interest of the economy. In the present set-up, however, one finds that the specialists work in their own world. I mean no disrespect. I am only trying to highlight what is my impression. I may be wrong—and the decision makers have to take decisions with little useful inputs from the work done by the specialists. This is especially true for the new growing field of power and energy. This compartmentalisation of specialist, administrators and the political level decision makers is a typical feature of developing countries which has to be removed. Data system may be the preserve of specialists but they have to be used by administrators and political decision makers. Therefore, there needs to be a more and better communication between these three levels so that the data are utilised more effectively.

Since energy sector of the economy is a critical area, the

energy data systems have a vital role to play in the national progress and development. This workshop is timely and it is hoped that its deliberations will help in providing guidelines on how we can use our energy data system more effectively in reaching our goals of national development, not only in this country but in other countries also. We have to also ensure that whatever is outdated in our system is abolished and we embrace what is productive and progressive. I am making this observation having in mind that no matter how sophisticated the data and its presentation, if our systems and if our norms of behaviour, norms of thinking, norms of operation, remain archaic and primitive, they will have nothing much productive to offer. India has the making of a big society and the will to achieve this in a short span of time has to be further strengthened. In this great national task quick decisions are to be taken from time to time, based on well documented and accurate data. I hope you will make some contribution in these endeavours. I have great pleasure, therefore, in now inaugurating this conference. Thank you.

DISCUSSION

GOPALAKRISHNAN

THE IDEA of organising this Workshop is an outcome of our experience in handling energy information over the last six to seven years. We found that in quite a few situations, when people ask for some particular document, what they really required was certain data which, within our limited resources, we could not provide. We then thought about organising some kind of a data bank, but it is not easy to identify the different data elements that go along with each specific aspect of energy. On the other side, we came across quite a few discussions at the international level about the relevance of energy data systems. I particularly refer to the Workshop held in 1979 under the auspices of the International Energy Agency. The discussions in these kinds of meetings mostly refer to the comparability of energy data for the purposes of funding international projects or for promoting trade. Definitely this is a purpose that should be served by a data system. But we also need data systems at the national level for the purpose of formulating policies or for the design of projects relevant to our needs. I would like to point out one particular instance. For instance, a country decides to double the coal output. It is not just a matter of doubling the quantity, there are implications in several other sectors of economic activity. We have to definitely increase the number of railway wagons, modernize the technology for mining, employ more miners, and things of like manner. We require data on aspects of coal-related technologies. And these type of data elements are probably not covered in the international energy data systems which have been discussed so far.

So with this background the idea of organizing a Workshop on National Energy Data Systems came up. Since we wanted to have definite guidelines for discussion, we decided to get theme papers done by experts. We were lucky in this case. We approached the authors and requested them for papers and within a short time they agreed to contribute.

This Workshop is not going to discuss confidential data. I believe we are primarily concerned with the data that are in the public domain. Now, so far as these data are concerned, there is no single agency in the country responsible for this. Another aspect is that there are quite a few organizations concerned with particular aspects of the energy data but these data are not readily made available. If somebody asked for them, probably they may be provided. But nobody knows where to ask or whom to contact for the particular data. Now one of the objectives of this exercise will be to examine the possibility of having a single source for energy data in the public domain. In other words, some kind of a national effort may be necessary to set up a centre to establish linkages between the various agencies concerned. Of course, we may have also to discuss, if possible, the kind of processing that may have to be done with the primary data but I am not sure that we will have sufficient time to go into that aspect.

PACHAURI

I just wanted to mention that this Workshop is part of the activities of the UNDP's International Cooperative Programme for Training and Research in Energy Planning, of which TERI is one of the seven founding institutions. The Cooperative's main objective is to impart education and conduct research in energy planning. So, therefore, this Workshop is essentially in the realm of developing knowledge in the area of energy planning.

I would like to refer to the inaugural address this morning. I think Mr. Fazal has made a valid point that when we are talking about energy data, we are not talking *only* about energy data and, if we ignore that aspect, we do it at our own peril because energy is an intimate part of economic activity. Therefore, it is not enough to merely consider energy information or energy data *per se*. One has to look at the other linkages with the other activities in the economic system. Hopefully, through the discussions we will have over here, we would be covering those elements of non-energy data that are of direct relevance to energy planning and energy decision making. We are fortunate that we have a spread of disciplines and background among the people participating which I hope will promote not only an intensive look at energy data systems but also broaden our outlooks and highlight the linkages between the energy and non-energy activities. So, I think, this is the point of emphasis we have to keep in mind during the rest of the discussions of the Workshop.

Mr. Gopalakrishnan referred to the genesis of this workshop and told you essentially why TERI has been interested in this. I may also add that this is a subject which is really not being looked at in sufficient depth in developing countries. Typically what has happened is that a particular system has been regarded as the preserve of one department, or one agency, with the result that its usefulness has remained confined to that particular agency only. And in the course of research, one finds out the barriers that exist. It is the purpose of this Workshop to identify those barriers and hopefully arrive at some conclusions or recommendations as to how these barriers can be lowered. So that, not only is research helped by this broader view of energy and non-energy data, but also decision making and planning in the agencies that are responsible for these functions. These are just some remarks which I thought that I should bring to bear on this discussion and perhaps what we could do now, if there are any doubts or questions or comments, is to throw it open. I would be happy to get the views of anybody who is interested.

J.C. SHAH

Since we are starting off on the subject, I wonder if you would like to spend the next few minutes in informing us of the work done, not just in India but elsewhere too, in creating an internationally feasible or active energy data system. I think there has been some work done. I am not sure if all of us are aware of what is happening.

PACHAURI

One person who may be able to give us some information on what the East-West Centre proposes to do is Dr. Kirk Smith. Dr. Mohan Munasinghe from his experience at the World Bank might also be able to tell us something. So may I request Kirk to say something and then we can go to Mohan.

SMITH

We are working in data systems in two areas. One, we are trying to develop a fairly detailed energy data for the Asia-Pacific region which is our region and includes basically the ESCAP region, Asia as far west as Iran, Oceania and South Pacific and does not include Latin America, Africa, Europe or the Middle East. We are going to be relying on you and others and use detailed government and non-government statistics for the other countries. We hope to be able to sell at cost tapes of this system to anybody in this region or anyone else who may want it.

The second area we are working on is perhaps a little bit esoteric. It seems to be more in the purview of physicists than economists. This is what we may call energy indexing, that is, how to add accurately one source of energy to another and include not only the energy quantity and the joule content or the metric ton coal equivalent but also the energy quality. This has not been too much of a problem in the past, when one was adding basically the cost of one fuel to another. But in the future as nuclear power could become more important, as renewable energy sources become more important, the difference in energy quality among the different sources becomes rather dramatic and it can confound national energy statistics substantially. It could mask progress in conservation and mask certainly international comparisons. So this energy indexing area, though somewhat academic, certainly will relate to the accuracy of the national energy systems.

MUNASINGHE

As far as the World Bank is concerned, most of the data gathering activity is directly concerned with, or arising from, the operations of the World Bank. As you know, in the past we have been concerned with lending for electricity, power, coal, oil, gas, and so on. In developing these lending operations we have collected large amounts of data on most of the developing countries, dealing with these energy sub-sectors and the particular government institutions that deal with them and so on.

More recently we have started a scheme of energy assessment which tries to put all of the sub-sector data into a broad framework and to make it consistent within the energy balance. As I said the energy data activity of the World Bank is not an end in itself; it simply is an upshot of our operational activity and concerns. Recently we have also tried to codify this data and put it up on a computer data base. For example, we have power, oil, gas, coal, and other data sheets for each country. We also work very closely with the UN in the preparation of their energy statistics, for example, on the data sheet. Summing up, in the past we have been more concerned with sub-sector and project oriented data. More recently we are working towards a better understanding of energy data system on a national basis particularly to look at trade in different forms of energy.

PACHAURI

I wonder if I could request Professor Maurice Levy to say something on this subject since he has been involved in articulating the UNU's programme on Energy Data Systems. And after him I would request Vinod Mubayi to say something.

LEVY

Just a few months ago, the UNU started a programme on research in energy planning for developing countries, namely, focusing on the specific problems which developing countries have in planning their own energy future. Of course, one of the various problems in developing countries, especially in very poor developing countries, for example, in Africa, is the absence of data. We are interested in methodological problems related to that, namely, what are the most important data, what are the data which are essential, also what is the sensitivity of planning to the data and finally, may be, we may be replacing some data by other ways to assess the problem. So this is a more methodological exercise which we hope to start in the next few months.

Now in the past five years, on the other hand, we were involved in another type of project, for integrated rural energy system. We now have knowledge of about 12 such systems in various parts of the world and these systems are a source of data. We are analysing and evaluating the way in which the system works, not only in terms of quantity, but also in terms of quality, and the socio-economic impact. I think this could be a contribution in the sense that one distinct factor of energy planning in developing countries is the importance of rural energy.

Last week I was in China for a Workshop which the UNO organized on the rural energy system. In China the rural energy sector is probably the most important one because 80 per cent of the population live in rural areas. In industrialized countries, rural energy is only a minor part of energy planning. In fact most of the time it can be just neglected, except in respect of machinery, fertilizers, and things like that. In developing countries the rural energy sector should be taken into account. This is, of course, a very difficult problem because of, I suppose, both commercial and non-commercial aspects. That is why I was interested in this Workshop because I think it is important to have a comprehensive view of the way in which data are arrived at in different parts of the world.

MUBAYI

There are three things about data which strike the researchers working in this field as obviously important which could perhaps be summarised as comparability, quality, and compatibility. I hope that at some place there will be some referenec to the fact of the different units and the reporting structure of data for different countries. For example, in India, there is a unit called 'coal replacement' for reporting energy statistics. Now this unit is not used anywhere else in the world. For researchers doing national and international comparisons, sometimes the comparability information is not immediately available.

Secondly, a reference has been made to going beyond what energy data mean in terms of heat flow alone or what is implied by the first law of thermodynamics, to the quality of energy. This is important for planning, especially related to new technology assessment. The third aspect refers to compatibility. We have been talking about different sectoral and sub-sectoral energy data systems. What is perhaps important also, is to look at the overall data system, from the stage of flow of energy from extraction of resources to its final use in different sectors like industry, household, agriculture, transport and so on. There are a number of data gaps which are necessary for policy and for analytical planning studies. When you try to visualize what the flow of energy is, right from the extraction of the basic source in the case of fossil fuel, or in the conversion of natural resources like flow in water, or in the extraction of energy from sources like uranium, to its final use in various end-use or demand sectors of the economy, there are a number of areas where there aıe very large gaps in the existing data. For policy and planning and intervention at secondary or tertiary stages of energy conversion, data on device efficiency may be more important than very detailed information at the primary stage of conversion.

TECHNICAL SESSION 1

OIL AND NATURAL GAS DATA SYSTEMS

Chairman	: C.R. JAGANNATHAN
Theme Paper	: N.K. MULCHANDANI and I.B. GULATI
Discussants	: S. ADITYA J.C. KHURANA
Appendix I	: The Organization of Oil & Natural Gas Data System in India, R. JAYASHANKAR

ONE

DEMAND FORECASTING FOR PETROLEUM PRODUCTS IN INDIA

H.K. MULCHANDANI and I.B. GULATI

FOR planning the growth of the petroleum refining industry, it is essential to prepare reliable estimates of the future demand of the various petroleum products. The 'demand forecasts' form the basis for creating additional refining capacity and for adopting suitable refinery processing schemes, so as to match the production with the required product slate.

Since its inception in 1961, the Indian Institute of Petroleum has been engaged in the task of preparing demand forecasts for the industry and the government. The Institute has been doing this task, with the involvement of the oil companies, major users and concerned government agencies. Periodic reports, based on the projections are issued. The forecasts are reviewed from time to time and updated.

Over the years, the public sector oil industry has grown. The oil companies and the government have more co-ordination and monitoring of the sales, and the task of preparing demand forecasts has become more scientific, as reliable input data started becoming available.

The product slate of petroleum products is large and the products range from a wide variety of fuels (LPG, MS, ATF, HSD, LDO, Fuel Oil and so on) to lubricants, greases, bitumen, waxes and other specialities. Each of these has several diverse end-uses. Some of the products form feed stocks for various petrochemicals and fertilizers. Therefore, preparing demand forecasts for petroleum products becomes a different type of exercise than in the case of coal or electricity. In the case of petroleum products, it becomes essential to forecast the demand of the major products individually in addition to the total. The forecasting technique for giving reliable demand data may differ from product to product. Some products, because of the end-use pattern, may be more amenable to a particular methodology of projections than the others.

Some of the exercises done in the country during the last six years to prepare the demand forecasts for petroleum products are reviewed as follows.

1. In mid-1976, when some stability was observed in the petroleum prices in the international market, following the turbulent period of 1974-75, the IIP, at the request of the Ministry of Petroleum once again took up the task of preparing demand forecasts. The efforts resulted in the Rolling Report of November 1976, which after modifications was adopted in May 1977, by the industry/government group. This covered the period 1976-77 to 1980-81 and 1983-84. The demand projections for these seven years were based on 1976-77 actuals, and the revised Fifth Five Year Plan targets for macro-economic indicators and growth of the main petroleum consuming sectors.

In this report, both trend extrapolation and regression analyses were used for all major products. For selected products such as MS, HSD, naphtha, fuel oil, petroleum coke and wax, demand was also worked out on end-use basis. Table 1.1 shows two of these projections and actuals realized for total of all products and for HSD.

TABLE 1.1 : **The projections and actuals realized for total products and HSD**

(in million tonnes)

	Total		HSD	
	Projected	Actual	Projected	Actual
1976-77	24.07	—	7.08	—
1977-78	25.88	25.54	7.65	7.73
1978-79	28.34	28.24	8.26	8.64
1979-80	30.78	29.88	8.93	9.80
1980-81	32.86	30.90	9.64	10.33
1983-84	37.97	—	12.14	—

While in 1977-78 and 1978-79, the actual consumption of total products and of HSD alone was very close to the projected figures, in the subsequent two years, the actual consumption was less by one MT in 1979-80 and by about 2 MT in 1980-81. On the other hand, the consumption of diesel (HSD) was more by about one million tonne than the forecasts. The forecasts of HSD demand was done mainly on end-use analysis. The road transport had to take a share of rail traffic in these few years.

2. The revision of the 1977 report was done in September 1978. The 1977-78 actual consumption figures were used as the 'base' and demand projections were built on this. In this exercise the demand for the perspective period 1983-84 to 1987-88 was also estimated.

This relatively long range exercise was done for each product. The projections were again mainly based on end-use analysis.

For the long range period from 1983-84 two scenarios, one of moderate toning down of middle distillate demand and the other of more severe policy prescriptions, were developed (Table 1.2).

TABLE 1.2 : **IIP Projections : September 1978**

	IIP Projections September 1978 (million tonnes)	Actuals (million tonnes)
1977-78	Base	25.49
1978-79	28.0	28.24
1979-80	30.5	29.88
1980-81	33.0	30.90
1981-82	35.2	32.32
1982-83	37.9	—
1983-84	39.8	—
1984-85	41.9	—
1985-86	43.9	—
1986-87	45.8	—
1987-88	48.0	—

In the near range period the forecasts given in this report were quite close to those made in 1977. As pointed above, the actual total consumption was less by 2 MT in 1980-81 and by 3 million tonnes in 1981-82 than these projections. However, both 1979-80 and 1980-81 were years of shortages combined with greater pressure on the road transport system and consequent high requirement of diesel for the transport sector. There was some unfulfilled diesel demand resulting in local shortages. The production from the two eastern sector refineries was also disturbed for a long period.

The end-use analysis of HSD based on norms of consumption, growth of vehicle population and obsolence rate, is limited by the specific assumptions made and the projections yield to greater forecasting error if some major additional factors come into play.

The 1978 report which covered projections up to 1987-88, was adopted by the *6th Plan Working Group for Petroleum Sector* also and incorporated in full in its report.

3. The next revision of the demand forecasts was done in 1980 taking 1979-80 actuals as base. The Ministry's groups decided that the projections were excessively influenced by the shortages in 1979-80 particularly of HSD, and felt that a large unfulfilled demand exists. Based on the indications at that time, it adopted for 1980-81, a growth rate of 16 per cent both for kerosene and for diesel; lowering it to 12 per cent for 1981-82 and 10 per cent for 1982-83. Such a high growth rate for diesel raised the estimates for 1980-81 to a very high level increasing the overall growth rate for all products to 12.2 per cent as against 6.0 per cent of the previous year. Thus the projections made in 1980, resulted in uniformly high figures all along. The actual consumption in 1980-81 fell short of these projections by 2.4 MT and in 1981-82 by 4.5 MT (Table 1.3). This brings forth the pitfalls in being led by 'short range' aberrations in long-term forecasting exercises.

From the point of railway movement and monsoons, 1981-82 was a normal year. The economic activity due to various reasons was, however, slow. Thus the 1981-82 consumption represents the actual demand fairly well.

The Expert Committee on Aditional Grassroot Refining Capacity, which finalized its report in 1980, also worked out the demand estimates for 1980-81 to 1983-84 and further. These were very close to the IIP 1980 Report figures. For the further period up to 1987-88, the growth rates were toned down, feeling that various measures to contain the demand of petroleum products will by then show their impact (Table 1.3).

TABLE 1.3 : **The predicted and actual consumptions for total petroleum products**

Year	IIP Report June 1980	Additional Refining Capacity Report	Actuals (milllon tonnes)
1978-79	—	Base	28.24
1979-80	Base	30.9	29.88
1980-81	33.3	33.7	30.90
1981-82	36.8	36.5	32.32
1982-83	39.6	40.0	—
1983-84	42.7	42.7	—
1984-85	45.8	44.9	—
1985-86	49.0	47.2	—
1986-87	52.1	49.7	—
1987-88	55.3	52.4	—
1988-89	58.8	—	—
1989-90	62.8	—	—

4. In 1982, a fresh exercise was done for preparing the demand forecasts. The Oil Co-ordination Committee (OCC), working through a number of sub-groups and relying mainly on the end-use method has worked out the projections up to 1989-90, and has prepared a detailed report. Simultaneously, IIP which by this time had developed a Data Bank for Petroleum Statistics and had computerized all the historical data, used some of the statistical forecasting techniques as applied to all petroleum products and with the help of computer programmes, prepared forecasts up to 1990-91, and also for the perspective terminal year 2000-01. Three different methods were used. The data from the 'Time Series Decomposition' technique which gave the highest of the three values is given below (Table 1.4) along with the OCC Report Figures.

TABLE 1.4 : **Time series decomposition methods**

	OCC Report-'82	IIP Report-'82 (Time series decom-position)	IIP Report-'78	Energy Policy Committee	
				R	O
1981-82	Base	32.4	35.2	—	—
1982-83	34.4	33.9	37.9	35.5	35.4
1983-84	36.4	35.5	39.8	—	—
1984-85	39.0	37.1	41.9	—	—
1985-86	41.9	38.6	43.9	—	—
1986-87	44.8	40.2	45.8	—	—
1987-88	47.9	41.8	48.0	44.7	40.96
1988-89	51.1	43.4	—	—	—
1989-90	54.2	44.9	—	—	—

R=Reference Level Forecasts.
O=Optimal Level Forecasts.

For comparison the Energy Policy Working Group forecasts figures and IIP 1978 Report Figures are also given in Table 1.4. It is observed that :

(i) the Time Series Decomposition technique gives much lower figures than the earlier IIP projections and even the 1982 OCC projections; but the figures are of the same order as worked out by the Energy Policy Committee for its Optimal Level Scenario.

(ii) the predominantly end-use analysis based figures of the IIP 1978 Report compare well with the OCC 1982 Report Projections towards the terminal period.

The 1987-88 demand forecast figures of both these reports, are lower by about 4 to 5 million tonnes than the estimates of the Addition Refining Capacity Expert Committee. The Optimal Level figures of the Energy Policy Committee are still lower by about seven million tonnes. This indicates that the measures and policy prescriptions which the Energy Policy Committee envisaged for efficiency improvement, inter-fuel substitution, and containing the demand of the petroleum products are not likely to have the expected impact.

BASIC INPUT DATA

Demand forecasting work requires a fair amount of authentic basic data. The data on product-wise, region-wise sales are pooled and well recorded.

For all products, month-wise, area-wise figures are available for the last 20 years or more. We have put all this historical sales data in our computer data files.

Data on macro-economic parameters, growth targets for various industries and sectors, production targets for various types of vehicles and on other related economic activities are all well recorded and available from the Planning Commission. The Economic and Statistics Division of the Department of Petroleum also maintains the production and consumption data of petroleum products and brings out a fairly comprehensive statistics report annually. For OCC, Bharat Petroleum collects and maintains detailed records of sales of petroleum products.

However, sufficiently reliable information on end-use pattern of some of the products is not documented. In recent years, along with the oil companies, the IIP has carried out a field survey for the assessment of diesel oils use in the agriculture sector. In this extensive study, the first of its kind, diesel oil requirement was assessed separately for all the states on the basis of population of agricultural machinery including pump sets, average annual operating hours, specific fuel consumption, effective utilization, load factors, etc. The need for keeping proper records of agriculture machinery and diesel consumption for various blocks was recommended.

For the transport sector, the Norms of Consumption (specific fuel consumption per vehicle per year) for various types of vehicles were first developed by the IIP in 1963-64. A need to revise and update these was felt.

With the participation of the OCC and oil companies, the IIP took up an extensive study for this purpose in 1980. This covered a very wide field survey, organised by the regional co-ordinators of oil companies, on the basis of a questionnaire designed by the IIP. The average fuel consumption norms were developed for each type of gasoline and diesel operated vehicles. Average Fleet Utilization Factors were also developed.

For lubricants, greases, and specialities, sufficient field data on end-use applications is not available. This is true particularly for industrial oils.

For waxes a comprehensive field survey for end-use pattern and future growth rate for major applications was conducted in 1976. Currently, this study is being updated.

In summary, a comprehensive computerised data base is the first need of any meaningful exercise in forecasting. Comprehensive data base will enable to evaluate various forecasting techniques such as end-use, statistical and econometric models. This Data Base Management System will be a good foundation for the MIS required by policy makers for future planning and forecasting. Requirements of such a data base is shown in Table 1.5. It will be desirable if a central agency is authorised to maintain and update the Data Base, retrieve and bring out the necessary output as per the objectives of the energy planners.

TABLE 1.5 : **Requirements of petroleum data base**

Category	Frequency	Coverage
1. Production	Yearly	By Refinery, By-Product, By Type of Crude
2. LPG Supply	,,	By Source
3. Consumption	Monthly	All-India, By Region, By State, By Company, By Depot, By-Product, (especially Naphtha, HSD, LDO, FO, LSHS, HHS, etc.)
4. Sales	,,	By Company.

Acknowledgements

The authors acknowledge with thanks the assistance rendered by their colleagues D.C. Dhall and S.C. Rastogi in the computation and compilation of the tables given in this paper.

DISCUSSION

PACHAURI

Mr. Mulchandani, could I request you to say something about the requirements of basic input data part for reliable forecasting.

MULCHANDANI

As I told you earlier, there are various forecasting techniques and different techniques require different types of data. Also, the requirements for various agencies differ. For example, IIP is just interested in estimating the demand factor, whereas the Ministry may look at it from a different, perhaps holistic angle, and provide some cushion for contingencies. So there are various requirements for various people. Those who make demand forecasts, normally employ different techniques; OCC has employed end-use methods. We also used to do it earlier, sometimes on a plant-to-plant basis, as for naphtha; and sometimes source restricted, as for LPG. I know from personal experience that when we wanted to do some multiple regressions based on economic indicators, we had to run to from pillar to post in various agencies to collect the information. The Planning Commission diverted us to some other source. I think this is not desirable. Those people who are in the petroleum sector must sit down and plan what sort of data are required, and where and how it should be kept.

MUNASINGHE

Surely, in a country as large as India you must have many different centres of demand and refinery locations. Do you use some kind of spatial distribution model or month by month consumption figures?

MULCHANDANI

Yes, I forgot to tell you that apart from demand forecasting on an all India basis we also do demand forecasting on a regional basis. There are two methods: either from data for various depots in the country, (then you know which region takes how much), or from the historical ratios, if depot data are not available. And as for refinery locations, we choose them after examining which regions have supply demand gaps in middle distillates. Then we choose a few locations based on transportation models.

AHUJA

It seems like a lot of agencies are producing a lot of forecasts. Do you know how they are put together ? To decide how much oil to import, for example? The second question is in your Table 1.1; you have either systematic under projection or over projection. Did you analyse what were the causes of this bias ?

MULCHANDANI

In answer to your first question, we just identify the gaps. The policy makers decide about import matters. I have already told you there are various considerations—building up buffer stocks, etc. And as for your second question regarding analysis, this we do. There used to be a group in the Ministry called the Demand Forecasting Group. This was represented by the Ministry of Finance, the Planning Commission, Railways and Agriculture, Transport, Oil Corporations, and various other users. We sit together and see why these things are not tallying, and what are the reasons.

NANGIA

A comment related to the distribution model. I am working in the OCC as an Economist and I can say we do have a large transport distribution model and we are using it for location, time phasing, and process selection decisions.

NIKHIL DESAI

I have three questions. One, when you say consumption, are they basically sales data and in the absence of not having stock change data, do monthly sales data convey the meaning you would like to get ? Second, when you speak of end-uses, how disaggregated do you get—is it transportation as a whole, or rural, urban, goods, passenger, public, private transportation? Third, what do you think is perhaps the most difficult data to obtain and which you would like to have?

MULCHANDANI

We get monthly data, from the Oil Corporations. They record it in the monthly proformas. Or we get it from the year books compiled by the Ministry of Petroleum.

NIKHIL DESAI

How disaggregated do you get?

MULCHANDANI

We don't go very far, as far as rural and urban disaggregation is concerned. Many years back we tried for kerosene.

NIKHIL DESAI

What are the most difficult data to obtain in your opinion, the most necessary and yet the most difficult to get?

MULCHANDANI

We would like to have all that I have listed. At different times we use different techniques. But the most important data is at the very micro level, that is, the depot consumption. If we have it, we can compile many things out of that.

JAYASHANKAR

I have a feeling that emphasis is being given to yearly data in the forecasting. Seasonal variations in demand for petroleum products are very wide indeed. For example, for HSD the difference between peak demand and the lean demand varies from 30 to 35,000 (units). So the forecast should be month-wise instead of yearly.

MULCHANDANI

You are perfectly right. I fully agree. That is why of late we have started relying on monthly data and are working on monthly data.

SEN GUPTA

I have a very simple question. You have just been talking about monthly forecasts and monthly projections. Now on all an India basis, what is the time lag, approximately, between when you try to collect information from the depots and collate them?

MULCHANDANI

We get the monthly data after a lag of two months.

KHURANA

I can add a few comments to what Mr. Mulchandani has just said. In fact, since April 1978, the petroleum consumption is computerised with one of the oil companies. All the data is available month-wise in considerable detail, such as, supply region, supply zone, consumption zone-wise, revenue district, state, monthly, quarterly and annual summaries, and bunker sales. For the last 2 or 3 years, some end-use information is also on tape now, that is the retail and direct sale of motor spirit and HSD, industry sale by category, industry sale of ATF by location and category, own-use summary of the industry, etc. A difficult exercise which is now somewhat on the rails is the industry end-use analysis for four major products: ATF, HSD, LSD and furnace oil. This analysis goes into great detail, such as agriculture, textile, coal, steel, transport, government, DGT&D, etc. Now just to answer the question that you have raised about how this is accomplished. All the oil companies have their own respective customer master files and one end-use code is attached to the customer master file and all these data are processed with the help of that code. So the IIP is welcome to use this data base any time they want for their analysis.

For demand forecasting this depot-wise data is inadequate. For distribution planning, data are not yet available. The channel of this data is quite different and independent from the consumption. The major problem here is that of communication of such data, as in India the computerised transmission lines do not function as yet. Transmission is by mail at the moment, or by tapes sent by air.

I do support Mr. Mulchandani that an effort should be made to centralise this. There are certain bits of data which were included in the second systems report such as production of machinery, product-wise. The whole objective was to prepare data from the points of view of crude allocation, production planning, distribution planning or inventory control. If some efforts are made to centralise all these, I think it will be a step in the right direction.

JHIRAD

Do you keep track of losses from the crude oil point of view like the end-use? That would be very helpful in making some evaluation of industrial energy substitution and efficiency.

MULCHANDANI

We only keep track of refinery losses, that is all, nothing beyond that.

JAGANNATHAN

I think the marketing companies would keep track of losses, distribution, etc.

MULCHANDANI

There is the National Productivity Council, Petroleum Conservation Research Association (PCRA) and a few other agencies who keep a track of these.

JAGANNATHAN

Gentlemen, I will just close the discussion on this particular topic. We have one or two other aspects of the petroleum industry to discuss. You have heard of the tribulations in forecasting the demand of what we need. It is even more so when it comes to exploration. Quite often we are asked when are you going to be self-sufficient, when are you going to produce so and so. That is a difficult question to answer. Because we do not know what itself is sufficient because the demand is going up and we have as many projections as there are committees appointed for the purpose. However, in an oil industry, decisions are made in an environment of uncertainty, speaking purely from the explorationist's and the geo-scientist's side. Unfortunately we do not have as yet anywhere in the world a direct oil finding tool to let us know how much oil is there under the ground. Therefore, forecasting relies very heavily on the interpretative skill borne on the data that one collects. We can easily identify the sedimentary basins. We straightaway run into a problem. There are a lot of basins which have not been explored as well as they should have been. In areas where oil has been produced historically, we have a lot of data. Using that as a model, even though it is a very dangerous thing to project it to other basins, we tend to project various estimates.

Various expressions are used in these estimates, like prognosticated reserves, proved reserves, indicated reserves, positive reserves, probable reserves—nothing wrong in this jargon, provided the end-users are aware of what they mean. There have been a lot of such figures floating round. They have willy-nilly to give some idea to the planners as to what production one can achieve. What is important to realise that a lot of it is in the unexplored area, and the high degree of probabilistic nature of these projections. These are constantly being updated, as we survey and drill the area. Even after drilling, there is a lot of doubt as to how much oil there is, and as to the optimum rate of production the field can sustain given a reserve and how quickly you want to exploit. The question of economies comes in. Do you wait till Doomsday to exploit it or do you want it now. There are certain yardsticks available here. If you exploit the known reserves very fast, you are liable to leave behind irretrievably a lot of oil. Therefore, at least that is realised. I am particularly emphasising the problems of geo-scientists and the data that needs to be fed to the powers that be, that is the planners. They have to take into account multifarious regions in allocating resources or allocating the targets for various activity.

Having discovered a field, then again the mathematician or the reservoir engineer takes over. He puts in a lot of guess work, for example, how much of this oil is recoverable by primary means and by secondary recovery means. This again is constantly updated. There is nothing final about it. Even a proved reservoir can change. Because when we estimate the proved reserves, we allocate a certain recovery factor based on certain historical data, how it is produced, how the pressures behave, etc. One can realise that one has been very conservative, that the field has not been behaving as it should, therefore, we should reduce the recovery from this. This is not often being realised by a lot of people who are not familiar with this. I want to emphasize that the data in this region has to be understood properly. Nevertheless, one has to make an estimate of what rate of production the field can comfortably sustain. This changes every year. The important thing is to realise these things and constantly update findings to see whether we have been conservative or over-optimistic. There has been a lot of reluctance on the part of oil companies to come out with this figure, one of them was my own company in the past. But it will very soon be realised that in order to plan the down-stream facilities some numbers must be quoted. The refineries or the pipelines or the petro-chemical plants, whatever is based on this hydrocarbon resource, cannot wait till you get very reliable data.

As an illustration, we discovered a field in the north-eastern part of India in Assam in 1953; by 1956 with only about a dozen or so wells, we had to tell the government how much we were going to produce from there. We stuck our necks out and said 3 million tonnes. We did not have resources for 3 million tonnes, and what we did there was a bit of guessing. We needed to drill further and the geology was complicated but we took a calculated risk and said 3 million tonnes. They had to set up refineries for this and the refineries were chosen within a distance of 700 miles from the centre of production. Our refinery was set up 250 miles away from the field. Therefore, we had to build a pipeline. The nature of crude was such we could not pump it in winter. The pour point was low and it was highly waxy. Therefore, we had to solve the problem of how to pump it out in the winter months. We came out with a solution. It is a physico-chemical process that we patented. We had to design the plant in Assam from the drawing table. The refineries were set up. It took some time for all these things. We were happy to find more reserves, and therefore, our prediction came true. From 1967 till today we are producing the same amount. Let me give you some numbers. We told them that we had estimated about 30 million tonnes discovered, proved, and indicated. Today we have still that amount, in the reserves, having produced some 54 million tonnes. That means, we have been discovering every year a little bit more than what we have produced. One takes risks and sometimes one can go wrong but nevertheless such decisions have to be taken in these things.

Coming back to the sort of data we need, as the objective of this Workshop has quite rightly laid down, we want accurate data and in good time. It bears repetition again. The data has to be sifted. For example, in a completely unexplored basin, say, Bengal, we have drilled so many wells yet we haven't found any oil there. But we still feel it is there somewhere. One has to make a guess to estimate how much is there. If you find, how much will it be. In order to prove that lot of effort is necessary. It has been an area where some data are available. There are other areas, for example along the foothills of the Himalayas, where we just don't have very much information. We see only the very surface and nothing at depth. It is a complicated area. We can't get the information from deep down. Even the technology to get deep information to predict the size of containers and so forth is going to be difficult. It is the same about a lot of other areas off-shore and on-shore. But this task goes on and data is very painfully, slowly collected. Simultaneously we keep informing the government any way that it is likely to generate so much production, it is likely to contain so much—this is all based on certain norms. There is no hard and fast rule as to how to do this. One uses one's own experience or analogy from other areas of the world. The important thing to do is to converse in the same language. The assumptions must be somewhat similar. My plea is that organisations like Oil India, the ONGC, should get together in these matters (and they are) and cooperate more closely in defining what we mean by proved and prognosticated reserves, how we derive these, so that everybody knows what we mean. At least in this country let there be no confusion. Prognosticated reserves in China, Russia or elsewhere mean something else. But at least here we should know what we are talking about.

The data in some cases are based on insufficient information. The realisation of that is important. Information has to be constantly updated. I am saying nothing profound. This is very obvious to those who are in the game. I am merely saying that the different departments, the Planning Commission, the end-users, refineries and so forth, must realise what we mean by these things. We have put out certain figures that by 1990 we must produce this much and these must be taken in the light in which they have been mentioned. They are probabilistic, one can do a lot of statistical analysis with these and say what is the degree of reliability and so forth.

The same applies to natural gas. Well, when we discover

oil, we also have a certain amount of gas dissolved in it, which comes out at a certain rate. The prediction of that is important. The deliverability of the associated gas changes with time. One must realise this. We have used a lot of mathematical modelling in predicting what will happen if we produce at a particular rate, what will happen if we increase the rate, what will happen if we inject water into the void it created, what will happen if we inject gas. This is done all over the world. There is nothing new. But what I want to convey to you is that we do apply these modern techniques in predicting the deliverability rates of gas, oil, and so forth.

Of the large gap in our understanding of the sedimentary basins in this country, there is no doubt. In some areas we are well aware of this because we have been working there for a longer time; in some areas we are just scratching the surface. That is why it is important, as government policy, to throw open some of the blocks to foreign companies. Because we need to discover this oil, if it is there, quickly. Otherwise our economies, sort of, will be shot to hell and gone because we have to import oil at a very high price. We need to get quick data, quick action on the ground by drilling to find the oil. This is why they have chosen this route to quickly find the resources. May be my colleagues from the ONGC and some other distinguished visitors here may want to react to some of these thoughts.

ADITYA

Well, the Chairman of this session, being from the same profession as I, has left very little for me to discuss. I will just confine myself to what efforts have been made towards the creation of a Mineral Resources Data Bank. Since 1975, the Department of Science and Technology, Natural Resources Division, took up this work on setting up a Mineral Resources Data Bank, called National Information System for Science and Technology, NISST. The idea obviously was that a lot of planning requires information about the mineral resources. As the Chairman of the session has already mentioned, resource assessment has got a lot of uncertainty involved in it; a large range of probability can be attached to the quoted figures. When the formats for data collection came to the ONGC, we found that there was really very little in them concerning the exploration and production aspects of oil and natural gas.

There was a second committee on storage, automatic processing and retrieval of geological data. This was a well-known programme of the International Union of Geological Society. This also was taken up by the Science and Technology Department and all participating agencies engaged in mineral resources, exploration, development and production. There was a format again worked out to make an inventory of oil gas fields, and to collect all relevant data on oil gas exploration. It had identification, locale and all the break-up, even methodology and technology used for making assessment of resources. Then someone who wanted information for a particular purpose found that this inventory or this format was not satisfying, and not able to give the information. This essentially brings us back to the important point that any design of a data base, particularly a computerised data base, has to be tuned to meeting the valid requirements, or as many valid requirements (you can't meet all the requirements) as possible.

Between 1978 and 1979, there have been quite a few joint working groups, even an Indo-US Joint Working Group on exchange of information on energy, environment and resources. The work on the designing of proper computer base data bank is still going on. We, on our part in the ONGC, have started a computerised Geo-Data Base. We are essentially designing it for helping us in our exploration work. We are feeding in all the information that we get from the sub-surface by drilling the well, or about the formation, or whatever parameters we use for exploration, and some amount of the oil-field information which is normally required by others. We are designing it to have certain general information required by our management and all the users who need it for planning purposes.

NITIN DESAI

I just wanted to talk about the question of reserve data. When energy planners, energy economists, sit to work, what they are concerned with is the data on reserves which is available to them. I think it is true that there are no clear norms for defining reserves as proven, indicated, prognostic, etc. I think this is the point which this Workshop should consider to see whether it is possible to get reasonably clear and uniform principles for categorising reserves. In fact, I personally feel that one of the problems is that the whole exploration effort is thought in terms of whether it is going to prove a certain quantity of reserves. Actually, we should think of the exploration effort as a means of gaining information. In fact, we have been arguing from time to time with some of the people involved in oil exploration in India, that they should categorise sedimentary basins by state of information, which is really a state of data, and then plan exploration efforts so as to upgrade the quality of data and the information for each sedimentary basin. We should not conceptualize exploration efforts in terms of the quantity of reserves which will be proven. I feel we could refine the concept of prognosticed reserves a little bit. At the moment what does prognosticed reserves mean? It means, as Mr. Jagannathan was explaining to me, given a sedimentary basin of such and such size, oil could have been created in it, so much could have been entrapped, and this is the percentage which is likely to be recoverable. It is a standard calculation. You go ahead, dig a well, find some oil there, and it gets into the indicated category, when you delineate the field and it gets into the proven category.

There is a great deal of variation in the confidence with which you have a reserve estimate in the prognostic category. There are some sedimentary basins in which you have done practically no work, but you have some geophysical indication of its (the reservoir's) existence. In some you may have done seismic work, or finer seismic work, and I feel that one of the things that we should try is to attach some sort of discount factor for the quality of information on prognostic reserves that are available. If we do that, it will be easier for us operationally to conceptualize the exploration programme, as something which will move us from a lower category of information to a higher category of information, rather than in terms of proven quantity of reserves, which is done usually on the basis of some very crude norms. I think this is the concrete task which this workshop can engage itself in. I should be very happy to find out what is the practice in other countries with regard to this.

There are certain questions beyond the reserve stage, many technical issues on reserve estimation, etc. I am sure this will be very difficult to discuss in a seminar of this nature and in any case I have grave doubts as to whether reservoir studies are really standardized. There are also many other issues like monitoring of exploration efforts, operational questions of a programme, etc. Again I do not think this is a place to go into these. But I do believe that this conceptual question on the categorisation of reserves and on the quality of information, the degree of certainty with which you have a particular prognostic reserve estimate, is a matter which affects not just the oil sector, but all energy sectors, and which I think this Workshop can address itself to.

Again beyond the reserves stage, I personally do not see too many problems as far as the production, processing, and marketing data are concerned. Oil companies are fairly well organized for this. They have been improving their data base and the retrievability of the data. On the consumption side, Mr. Mulchandani had mentioned the manner in which they do the work. My main problem with that data is that it is all product-oriented and not use-oriented. Energy is used for certain purposes, for motive power, for heat, for light, as a feed stock, for electrolysis. We have practically no information giving us for different classes of users, the manner in which energy is being used. This is very important for planning purposes and analytical purposes. Take kerosene, for instance. We do not know how much kerosene is used in India for lighting kerosene lamps, where replacement is electricity for electric light; and how much is used for cooking, where replacement is totally different. Take the example of motor spirits. I was very surprised to find out the other day that now we have reached a stage in India where the major part of motor spirit consumption is by two-wheelers and not four-wheelers. This is why I feel that the problem of data base on the consumption side is too product oriented. It is not organised to allow us to ask questions about inter-fuel substitution or about short-term planning. If I have a kerosene shortage, I do need to know whether it is an area where kerosene is used predominantly for cooking or for lighting. Because what I have to do to meet this shortage will vary with the nature of the use of kerosene. This is my main difficulty with the consumption data and I am very keen on finding out whether any country has managed to organize the data base not on a product basis but on a user basis.

JAGANNATHAN

Thank you, Dr. Desai. Let me clarify a couple of things on the first point you raised about the classification of reserves. What I want to convey is not that nobody here or in other parts of the world has a definition for all these things. What I was emphasizing is that when we say something, we mean either the ONGC or ourselves in Oil India, we must mean the same thing. We have definitions of these things, but the true meaning and significance of these must be understood. This is the point I want to make. But you have a valid point, you as the end-user of all these data, must realise what we mean by these things.

As regards the exploration drilling, again that is a good point. We do sometimes drill, not with the sole intention of finding oil, but what we call parametric wells or information borings, just to get the information from sub-surface structures. This brings me to another point which I should have made earlier on. It is not always when we get the seismic data and the geological data can we say that we are going to find oil. It is not always that we find oil with the first well. Quite often we come across these doubts in a lot of people's minds. Are we wasting public funds in drilling to satisfy somebody's whims? Exploration is an expensive business. It is a gamble. We must have not only the stamina but the courage to drill these wells. Unless you drill, you neither get information nor are you going to find oil. We have been lucky, we have been more than lucky, in finding oil with the first well in Bombay High, the first well in Naharkatia, with the first well in Ankleshwar, with the first well in Cambay. But there are other areas where we have been drilling and have not found it. For example, before we did Naharkatia, our predecessor company had drilled 200 wells in Assam alone, and the wealth of information that we gained resulted in the discovery of Naharkatia. But that was the first well. What I am trying to say is that we do not just drill there to satisfy somebody's crazy ideas. We have to justify why we should drill in a basin. It must satisfy certain basic conditions like, are we drilling in an area where if hydrocarbons existed, could they be trapped, is it a reasonably large enough structure to spend that kind of money? Has it satisfied the other basic conditions, does a source rock exist in that area? If it does, is there a reservoir rock? If oil is generated, can it be contained in a reservoir rock? Is there a cap rock, something to seal it off from escaping from there? All these conditions have to be satisfied. All this is very fundamental and nothing new. Nevertheless, Dr. Desai, you may rest assured that we and the ONGC, at least as leading organizations on the exploration side in this country, will try to arrange and define the jargon we use. Mr. Aditya, any comments?

S. ADITYA

I can only add to what you have already explained. Particularly I would like to clarify the point Dr. Desai has raised about prognostic and proved reserves. Actually these definitions, as accepted by almost all of us, are that we do not call it a reserve till we really have seen the existence of it. We would rather call it a resource, may be even a sub-economic one. But in years to come, it may become economic and you can really count it as a resource. You can go even to the extent of making an estimate of what is really there that you can expect. First you can start with sub-economic resource, then you can come to economic resource, and then you come to a reserve, when you know something exists. The uncertainty is about the extent of the quantity that exists. So obviously there will be always more certainty established with more information coming. So, we have categorization of reserves also.

In the organization I work, we have accepted the categorization as A, B, C-1, C-2 and if we go by the standard oil companies' practices, the same things are considered as proved, what is in hand about which you are, may be, 90 per cent certain, and possible and probable, depending on the degree of confidence you have. These days, the reserve, resource assessment is not done by using the adjectives possible and probable. You can even have the whole range of figures from may be 95 per cent probability to 5 per cent probability and in certain cases where you do have a less explored basin area, you can start with a 60 per cent probability figure. Obviously the lower estimate that you give is with a 60 per cent probability and the higher estimate will have 5 per cent probability. Depending on the amount of risk the exploration investment is willing to take, you can count what is the asset that is accepted and plan accordingly. The whole resource appraisal is also to be taken for the purpose of planning and there will be different degrees of risk taking involved. If the decision is to accept a higher risk, we can take a higher resource appraisal with a lower probability attached to it. We are applying these things even for the areas where we have proved reserves but with a degree of uncertainty. So in such an area we start with 100 per cent or 90 per cent certainty of reserves around a producing well and go to 5 per cent probability around the tested wells and in other parts of the field determined from seismic structure mapping or any other geological information that we may have. This system exists.

To the next question of categorization of basins, we have tried a very broad classification. In the basins where we are already producing, extension of exploration can be done with much greater confidence. As a matter of fact, very recently when we had discussions with the World Bank about getting loans from them, they were willing to advance loans on such basins where we are producing oil so that there is perhaps a trend of discovery which can be inferred, but not necessarily established. We have basins which we classify as perhaps frontier areas of exploration in times to come. These are what you call geologically very ancient basins, may be anything between 150 to 500 million years old sediments. So older the sediments, more complex the sediments, less the chances of finding oil. These are some of the general rules. Such basins come last in the priority of exploration because expectation is less. We have categorised this and have tried to identify the objectives and indicate which deserve priority of exploration and also indicate the expectation. As Dr. Desai mentioned, certainly, till now we have been giving point estimates. This is rather misleading. One has to give a range of figures with confidence values attached to them.

ASHOK DESAI

As another Desai, may I ask a supplementary? The published figures of reserves for Bombay High that I have seen are about 280 million tonnes, and an unpublished figure I have seen, which nevertheless comes from a very high source, is about 1700 million tonnes. What I would like to know is whether this enormous difference is entirely due to these degrees of probability that Mr. Jagannathan has been talking about or whether there has been a fresh discovery? What is the process by which these figures are decided and put out?

ADITYA

I am not sure if you are referring to just the Bombay High Region or including the satellite structures for which we are still working out the development plans.

NITIN DESAI

One figure 1700 is the figure for *in situ* oil, that is oil in place. The other figure is the oil recoverable without any secondary recovery other than water injection.

ADITYA

We calculate reserves for an area what we call in place reserve, or geological reserve. It is there in the sub-surface but when you drill a well the formation has got certain petrophysical characteristics, the pressure and other factors or the porosity or formability of the rock, the nature of oil. All these will determine how much really will come from the oil that is entrapped in the sub-surface. This is what Mr. Jagannathan has already referred to as the recovery factor asset. If you produce under a particular physical condition you may get 18 to 20 per cent recovery. If you have what we call pressure maintenance system or a fluid injection system, or gas, or even some exotic mix of polymer and visible compound or may be even using hot fluid thermal recovery having burnt the whole thing and recovering the resultant fluid, you can get, may be, a higher recovery. This is what we call enhanced recovery process. You first make an estimate on what is your regime of development and accordingly you can make an assessment of certain recovery factors. That is what you are going to get positively. You can aim at increasing the recovery factors also. Every country does it. Even average recovery is sometimes taken, if you do not know any details about the reservoir condition. When you have drilled a few wells, and measured the parameters down hole, then only you can make a better judgment even though you have obviously sampled only a small part of the reservoir. As you go on developing, you can be more sure of how much oil you are going to really take. So 1700 is in place, 200 may be at the moment taken as recoverable reserve. It may turn out to be more than 200 also as you go on developing the field. I will quote about Ankleshwar. We had taken the decision to have water injection right from the start of development. We took 50 per cent as recovery factor, not for all the producing gas, because depending again on the character we have different recovery factors. Now the actual development of the field has been done since 1963-64, for the last 20 years. We know now it has produced much more than what we expected and we have reassessed and made a final plan, so that we can get the maximum of that is in place, and the maximum ratio of recovery that we can have. This is a continuous process. That perhaps clarifies the different figures quoted and may be in place obviously will be much higher.

KHURANA

Can one expect 50 per cent in the end?

ADITYA

Not in all cases.

NITIN DESAI

Chairman, on reserve figures, oil in place is reasonably well defined. But it is not a very meaningful figure, because you cannot get all that oil for use. Meaningful figure is recoverable reserves and it is not uniform. It is not as if you can say generally that it is 18 to 20 per cent or so. It will vary on the characteristics of the field, the pressure there, quality of the oil, and so on and not only that, it will vary with the type of development you contemplate for that reserve. So this introduces a grey area. I can give my personal opinion, which is that, my friends in the ONGC, and so on have generally tended to be fairly conservative on this aspect both on the assessment of the oil in place as well as their assessment of the recovery part. I would like everybody to keep this in mind when they look at the figures for India. The 1700 that he is talking of, is actually an estimate which I may mention is higher than the oil in place estimate on which they are working. But the big difference is not due to that but due to the fact that one is oil in place and the other is oil recoverable. The difficulty is that this figure of recoverable oil which is meaningful is not a well defined figure, when you want to do international comparison, or aggregate different resources.

JAGANNATHAN

You cannot do that. It will be more confusing. However to answer Dr. Ashok Desai's question, the 1700 tonnes, may refer to, as Mr. Aditya was trying to explain, the census restricted to Bombay High or the summation of all satellite fields, or it could refer to oil in place or some of it. Therefore, if you could tell us who gave you this figure, we may be able to tell you. That is, at what point of time. . . .

ASHOK DESAI

Well, it was your secretary who gave me this figure.

JAGANNATHAN

When?

ASHOK DESAI

Two months ago. How much was the oil in place then?

JAGANNATHAN

The recovery factor varies from field to field. Mr. Aditya was talking about the Ankleshwar field. It is one of the textbook like structures, very favourable thermobility and so forth. That is why the recovery factor is very high. It is even more than 50 per cent. The average expectation around the world, taking the good with the bad, taking heavy oil with light, is about 30 per cent. If you get 30 per cent you are doing well with a primary recovery. To give you another idea, in Assam we started with 25 per cent. Now we have some of them at about 30 per cent. The average expectation now is in the region of 35 per cent.

ASHOK DESAI

How much do you get for the secondary?

JAGANNATHAN

Another 5 per cent, depending on the type and the timing. If you start very early in the life of the field, you are likely to get more. This is again to be a very slow process, over a period of 30, 40, 50 years.

ASHOK DESAI

This figure of 200 vs. 1700, is it based on initial experimental drilling or some simulation study?

C.R. JAGANNATHAN

It is a volumetric estimate and it is also cross-checked with other reservoir studies. Basically what we do is to find the volume of a container. You know the thickness of the oil layer, and the areal extent. You multiply these two, and apply the process correction. Oil is contained in the inner rock, in limestone or in sandstone. We find out how much it is possible to contain by measuring the porosity of the rock. We get the theoretical volume and then you have got to deduct the quantity of water. There is a certain amount of water in it. It is not produced. If you really allow for all these, you get the reserve figure. Then you add and do a little bit of cosmetics, to get the final figures.

PACHAURI

Chairman, I would like to address myself to the second part of what Dr. Nitin Desai said, that is with respect to data being totally product oriented. This is a very serious problem for all sources of energy. I am sure you would agree with me that the same problem exists with power, and the same problem exists with coal. I remember that in the discussion in the Working Group of Energy Policy, there seems to be an area of darkness of about 10 million tonnes of coal, which are consumed in various industries and really we do not have a handle on where they are being consumed. But I think the problem here really is that perhaps our policies in the energy sector have been much too supply oriented. We just have not looked at the demand aspect of it, whether it was power or oil we just have somehow projected a demand without necessarily seeing how the demand was generated. I think time has now come with energy prices being so high, that it is essential to look at the structure of demand. It is only when the gap between the policy maker and the planner and the data collection agencies is bridged and there is more effective communication, then perhaps you will see an organization of data, along end-uses, which would permit some better and deep analysis into the structure and the nature of demand itself. I do not know how this can be done. One way, of course, is getting together like this but at the same time one must realize that organizationally perhaps a lot has to be done at the levels of various government departments, etc., dealing with energy to see that this kind of orientation trickles down to those organizations that are involved in data collection.

NITIN DESAI

One needs to know not just end-use by sector, but also by the type. . . . For instance, I want to know how much electricity is used in industry, for running motors, how much for furnaces, heat oriented applications, and motive power oriented applications.

PACHAURI

Yes, for inter-fuel substitution this knowledge becomes absolutely essential.

NITIN DESAI

Yes, it is not just who is using it, but for what purpose it is being used. Basically there are only four or five things that you can do with it.

JAGANNATHAN

What you have said is unexceptionable. There are no two opinions about it. We need this data. Perhaps TERI might give a start to this one.

NIKHIL DESAI

The points made by Dr. Nitin Desai and Dr. Pachauri are well taken. There are, on the other hand, problems even with the product consumption data in the country for both energy and non-energy consumption, e.g., whether one takes stock changes into account or not. Having some experience in dealing with cross-country data, I have found that for a given country or set of countries, it is very difficult to get consistent figures for oil consumption which is about the easiest data to get among all the energy fuels. The UN, at some point, introduced energy consumption of products only, it has not included non-energy consumption. Whereas when you look at domestic oil consumption in that country, then you are including non-energy consumption, refinery as well as other non-energy consumption. That is a problem which I don't know how we can solve. The only way is to begin with certain definitions which are not standardised yet.

The other thing in addition to the use data is we don't know what device, what equipment, what are the efficiencies, what are the changes in technical efficiencies over a length of time. We also do not know how economic efficiency changes. What energy is being used for, and how economic cost of production and energy uses have changed over a period. Petroleum products, once again, are one energy form, other than power, where price data should be easily available. But those data are not available for more than 50 or 60 countries in the world. How reliable are reserve and production data for other countries, who keeps them and who gives it to whom? I was wondering if people from other countries could give any idea in particular if there is any problem created by the oil companies on what the reserves are in certain fields and what the production offtakes are.

JAGANNATHAN

As far as I can see, there are some data published but it is very difficult to say whether it includes everything. Oil companies seems to be very variant and for various business reasons they do not give it by fields. But for well developed fields these data are available.

NIKHIL DESAI

My experience has been that even for oil production numbers across the countries, and not just backward ones, data published by the UN and IEA vary.

JAGANNATHAN

That is because some people don't respond. They send out various circulars to all countries. They do not ask for field-wise data, just the total. Some countries respond, some don't. These are the difficulties. I agree with you it is very difficult to get absolute answers to these. Perhaps whatever is published is about the nearest we can get.

AHUJA

How much of a gamble is oil exploration? What indications tell you where to drill in the first place?

JAGANNATHAN

Well, when I said gamble, I meant there are a lot of uncertainties. I will again emphasise, that there is no way whatever you have as of today in the world to tell you, if you drill there, you will find oil. We haven't got that tool. What it tells you is place *X* is a better area to drill than *Y*. First of all, the geologists eliminate all areas other than sedimentary rocks. They don't look in volcanic rocks or deep seated igneous rocks, or rocks that originated from high temperatures, for they do not contain oil. Oil is essentially organic remains going through a very complex chemical reaction and becoming either liquid or gaseous hydrocarbon. These things are only in the sedimentary area. In a country like ours, there are vast sedimentary areas, so also in other countries. The gamble is only because we don't know unless we drill. So one asks next, could that basin have generated hydrocarbons? Does the geological history indicate that it has gone through certain processes in which oil could have been generated? Whether the right conditions of temperature, of generation of organic matters are there? How do we know that? We go and see the rocks. If they are exposed on the surface, collect them, do a lot of geochemical analysis of these things and find out if they contain a lot of organic matter, whether they are oil prone or gas prone or hydrocarbon prone. That is one indication. If you find this, you ask where did it go? Is it trapped somewhere? Then look for the trap. The seismic and other surveys are only to tell you whether any traps exist. Seismic survey is primarily done when you don't see a rock covered by recent soil like the Indo-Gangetic alluvium, or Brahmaputra alluvium. There are structures which are open. You can see the anticlinal classical structures and so forth. But then oil is contained in a variety of traps, subtle and some not so subtle. The job of the geologists and geo-physicists is to find out where it is trapped and if it is, is there a reservoir rock, a cap rock to seal it off? When all these conditions are satisfied, then you sink your money and pray like hell.

KHURANA

I have a basic question. One finds different tables for energy in terms of coal replacement or its equivalent. We seem to follow replacement whereas equivalent seems to be a more international standard because the efficiency of the appliances used keep varying with the evolution of technology. Could we not standardise in India on some basis which is in line with the international standard? It is something which puzzles me a bit, why we should go on replacement basis all the time, when everywhere else the equivalent is used?

PACHAURI

Chairman, I think since the question has been raised, perhaps we could make a few comments on this. I personally felt that there was some merit in the coal replacement measure some years ago. But I fail to see any merit any longer because essentially we have been through a period where the railways have changed over from the use of steam locomotives to diesel oil and electrification. Now at the stage when we had a large number of steam locomotives and traction was in the form of that type or power, I think coal replacement measure had some relevance. I am just taking the example of railways. But it applies to various other industries. But Dr. Nitin Desai may like to comment on this. I personally feel that there is really not much benefit in preserving the coal replacement measure. It certainly does make international comparison very difficult because the rest of the world is following coal equivalent or tonnes of oil equivalent.

NITIN DESAI

If you see the conversion factor for coal equivalent and coal replacement, the big difference is in two sectors: in the transport sector and in the household sector. As far as the other sectors are concerned, frankly there is hardly any difference between these. Each tonne of oil is equivalent to two tonnes of coal in both measures, roughly speaking. It is really in the transport sector and the household sector that the big difference arises.

Dr. Pachauri is right that in a situation where we are using both steam coal as well as diesel in the railway system, it would have been unrealistic to give an energy consumption figure in coal equivalent terms because it would have understated the extent to which we would have to produce coal really to replace that amount of fuel oil. And that is why we adopted that system; similarly in the household sector. But I do believe there is the need to have a second look at this because we are confusing two issues. The quantification and measurement of the amount of energy used in the economy, with questions of interfuel substitution, and so on. We have now in government publications started reporting both the coal replacement as well as the coal equivalent. Who knows one of these days coal replacement will be dropped. Right now we are doing both.

PACHAURI

The argument was that in the case of the transport sector, that is misleading because we are going to replace diesel oil for coal.

ASHOK DESAI

It is not just two tonnes of coal per tonne of diesel, but. . . . about 6.7 or something like that. This distinction government has been trying to make between coal equivalent and coal replacement figures is really a distinction between final energy consumption and useful energy consumption and this difference arises not simply in these two sectors but in every sector, and this difference will change over time as the relative efficiency of use of different fuels change. Therefore, it is better if one has to have two sets of figures, to have clearly, figures of final energy consumption and useful energy consumption both in the same unit, rather than use two different units for these two things.

SMITH

Just a follow-up on this discussion. This morning I saw that India is probably in the forefront in the coal replacement figure. Indeed as a wider range of energy sources are considered, other countries may have to turn to something like the coal replacement figure that includes inter-fuel substitution parameters in it. Indeed, India 20 years ago started on this, no other country has taken it up. May be other countries are at fault, not India.

NIKHIL DESAI

I have a comment on that. If the choice of unit, coal replacement or whatever, depends on deciding what the marginal fuel is. If so, with increasing electrification in almost all the economy, would the coal replacement or any such unit be of any meaning, if electricity is one single fuel?

SMITH

You still have to know how much coal it will take.

NIKHIL DESAI

Still, not the marginal fuel.

SMITH

There is some in many circumstances, certainly in power generation.

NITIN DESAI

I will give you an example. In LPG and kerosene too there is a conversion efficiency problem. There you do not need this correction. The total figures are only in terms of simple calorific equivalents. There is a convenient correction, there is efficiency correction also. The point Dr. Ashok Desai was making is if you want to do this, do it systematically for all fuels. Do not do it for one or the other. Then whatever unit you use for final energy and useful energy it does not matter. You use the megajoule, if you want. It can be absolutely anything.

APPENDIX I

THE ORGANIZATION OF OIL AND NATURAL GAS DATA SYSTEMS IN INDIA

R. JAYASHANKAR

INTRODUCTION

Organization structure and respective functions

It is essential to have a birds eye view of the organization structure of the Indian Petroleum Sector and the functions of each one of its constituents before one can begin to understand the data system.

The Ministry of Petroleum and Chemicals came into being in November 1963, namely the Department of Petroleum and the Department of Chemicals and Fertilisers. The Department of Petroleum is divided into the following divisions :

(a) Exploration division.
(b) Marketing, distribution and co-ordination division.
(c) Refineries division.
(d) Pricing division.
(e) Petrochemicals division.
(f) Finance and Accounts division.
(g) General Administration division.

The broad functions of the various divisions of the Department are indicated below :

EXPLORATION DIVISION

This division deals with all matters relating to the exploration and production of crude oil and natural gas, both on shore and off shore. Matters relating to public sector enterprises engaged in exploration and production of crude oil and natural gas, viz., Oil and Natural Gas Commission; and (ii) Oil India Limited, are dealt with in this division.

MARKETING, DISTRIBUTION AND COORDINATION DIVISION

The functions assigned to this division include assessment of requirement, availability and distribution in the country of petroleum products, allocation to states and union territories, demand forecasting and matters relating to retail outlets.

The following undertakings are controlled by this division :

(a) Indian Oil Corporation (Marketing Division).
(b) Bharat Petroleum Corporation Ltd. (Marketing Division).
(c) Hindustan Petroleum Corporation Ltd. (Marketing Division).
(d) Indo-Burma Petroleum Co.

REFINERIES DIVISION

This division is looking after all matters relating to refineries and pipelines including operations and expansion of existing refineries and new projects for refineries and pipelines. The following public sector undertakings are working under the administrative control of this division :

1. Indian Oil Corporation Limited (Refineries and Pipelines Division).
2. Madras Refineries Limited.
3. Cochin Refineries Limited.
4. Bharat Petroleum Corporation Limited (Refineries Division).
5. Hindustan Petroleum Corporation Limited (Refineries Division).
6. Engineers India Limited.
7. Lubrizol India Limited.

PRICING DIVISION

This division deals with the pricing policy in respect of crude oil and bulk refined petroleum products. The Economics and Statistics Cell which is concerned with the collections, scrutiny compilation, interpretation and analysis of economic and statistical data relating to exploration and production of crude oil/gas, refinery production etc., and presentation of data, is also functioning in this division.

Functions of other organizations (other than the public sector undertakings) under the administrative control of the Department of Petroleum.

Oil Industry Development Board

The functions of the Board involve rendering of financial and other assistance for the promotion of all such measures as are in its opinion, conducive to the development of the oil industry in all its facets.

Oil Coordination Committee

The Oil Coordination Committee coordinates the planning of supplies, demand projections, and so on, for the distribution of petroleum products and works closely with the oil industry and the Petroleum Ministry. The Committee reviews the operations of the oil industry and initiates special studies and projects for meeting the future requirements of petroleum products in an optimal manner. The Committee also administers the industry pool accounts on behalf of the Department of Petroleum.

Petroleum Conservation Research Association

The main task of this association is to contain the consumption of petroleum products by curbing wasteful practices and by improving efficiency in fuel utilisation.

EXPLORATION AND PRODUCTION DATA

The Ministry of Petroleum monitors and reviews the progress of the oil exploration programmes on a regular basis through a number of reports, review meetings and through the involvement of its officers in the Boards of the Undertakings, namely, Oil and Natural Gas Commission and Oil India Limited. A list of various fortnightly, quarterly and yearly reports being furnished by ONGC and Oil India Limited to the Ministry is as follows :

Fortnightly Report

1. Situation report giving status and progress of drilling and production.

Monthly Reports

1. Information for Cabinet Committee on industry and trade.
2. Drilling report.
3. Programme for increasing productivity.
4. Royalty statements.
5. Structure-wise—project-wise, production and despatches of oil and gas.
6. Information required in connection with preparation of monthly summary for the Cabinet.
7. Monthly returns relating to production of LPG from natural gas.
8. Monitoring of monthly production performance on crude oil production, gas supplies and LPG production.

Quarterly Reports

1. Progress report of the public sector undertakings regarding geological, geophysical, drilling and production.
2. Progress report on implementation of the items of the

Agreement between India and U.S.S.R. (Agreement-wise).
3. Review of performance.

Yearly Reports

1. Annual Report and Accounts (to be tabled in Parliament).
2. Status of wells, metreage drilled and number of deep drilling rigs.
3. Chronology of important events.
4. Review of plan performance.

Types of data

(1) Reserves of crude oil (million tonnes) and natural gas (billion cubic metres) on-shore data for Gujarat, Assam and Rajasthan, and off-shore data for Bombay High.
(2) Number of deep drilling rigs, owned by Assam Oil Company, Oil India Limited, and Oil and Natural Gas Commission.
(3) Status of wells, (numbers) company-wise in terms of oil wells, gas wells, dry wells, under test, injection.
(4) State-wise achievement of geological field parties in square kilometres.
(5) State-wise achievement of geophysical field parties in line kilometres.
(6) State-wise achievement of drilling in terms of number of wells and meterage.
(7) Company-wise production of crude oil, on shore and off shore in tonnes.
(8) Production of natural gas in million cubic metres, on shore and off shore.

REFINERY DATA

In the Oil Coordination Committee (OCC) an Industry Committee has been set up for deciding on the allocation of crude oil and monthly production pattern based on oil imports and exports, and national and regional demands, logistics of transportation and other allied matters.

Types of data

1. Petroleum refining capacity.
2. Refinery crude throughput and production of petroleum products on calendar/financial years.
3. Refinery-wise crude charged and total production of petroleum products.
4. Refinery production and pattern of refinery production in India on calendar/financial years.

CONSUMPTION AND DISTRIBUTION DATA

The oil economy budget is the starting point for the distribution planning process. The oil economy budget projects production and availability of indigenous crude oil, planned maximum/achievable crude throughputs of different refineries, the total indigenous production of finished petroleum products, exports and the demand and requirements of the various petroleum products for the period in question. The Oil Coordination Committee closely monitors on a continuous basis, the supply and distribution of all petroleum products in the country as a whole and in the different states. Weekly returns are submitted to the Ministry and the Oil Coordination Committee covering significant aspects of sales, stock levels by the oil companies.

Types of data

1. All India total consumption of petroleum products on yearly basis.
2. Product-wise consumption of petroleum products on yearly basis.

3. Seasonal indices of the sales/consumption of major petroleum products.
4. Sales/consumption and market participation of oil companies.
5. Deliveries made to international/coastal bunkers on yearly basis.
6. State-wise sales of selected petroleum products on yearly basis including international bunkers.
7. Estimated state-wise per capita sales/consumption of major petroleum products excluding international bunkers.
8. Estimated per capita consumption of selected petroleum products.
9. Estimated sector-wise consumption of selected petroleum products.
10. Tonnage of Indian tanker fleet.
11. Share of traffic in petroleum products in the total traffic handled at major ports.
12. Coastal movement of petroleum products.
13. Average freight rates from Ras Tanura to Indian ports.
14. Rail freight rates in respect of petroleum products (dangerous and non-dangerous).
15. Retail outlets in India as on 1st of January each year.
16. State-wise distribution of retail outlets, multipurpose/ farm fuel centres, LDO/SKO dealers and LPG distributors.
17. State-wise number of LPG domestic consumers.

DEMAND DATA

In July 1974, the Ministry of Petroleum and Chemicals constituted a Demand Forecasting Cell comprising representatives of the Ministry of Petroleum and Chemicals, the Railway Board, the Planning Commission, the Indian Institute of Petroleum and the Oil Companies. The IIP coordinated the work of this Cell. The last report of the Demand Forecasting Cell was prepared by IIP and published in June 1980. Demand Forecasting Cell (DFC) estimates short, medium and long term demands of petroleum products. Demand estimates are made both for the country and for each of the 12 regions in the country into which it was divided for the purpose. Short-term estimates cover the revised estimates for the current year and estimates for the next year. Medium range estimates cover a period of 2 to 3 years beyond the short range period and long range about 8 to 10 years beyond the medium range.

In 1981, the forecasting task was entrusted to the OCC. The OCC sought the participation of all the oil companies and constituted working groups for each of the petroleum products. The OCC has very recently finalised a report on demand for petroleum products up to 1991-92.

PRICE DATA

The pricing of petroleum products of India has evolved primarily on the basis of the Report of the Oil Prices Committee accepted by the Government in two stages. The refining, pipeline transportation and marketing and distribution costs form the main constituents in the price built-up of petroleum products. The OPC formulated a pricing structure on the "Retention Price Concept" duly assessing the costs and also a reasonable rate of return on the capital employed separately for the above activities.

Types of data

1. Index number of wholesale prices of petroleum products, petro-based industries and other energy items in India.
2. Prices of indigenous on shore and off shore crude oil in India.
3. Basic ceiling selling prices (ex-storage) of petroleum products.
4. Retail prices of selected petroleum products in major cities.
5. Trend in retail selling prices of petroleum products in metropolitan towns in India.

6. Rates of dealer's commission for MS, MSD and LPG in India.
7. Realisation of excise and custom duties from crude oil and petroleum products.
8. Composite incidence of excise/custom duties on petroleum products.
9. Central excise duty on petroleum products.
10. Sales tax rates applicable on crude oil and petroleum products in different states.

MISCELLANEOUS

Foreign trade

1. Imports of crude oil and petroleum products from 1960 onwards.
2. Total exports of crude oil and petroleum products from 1955 onwards.
3. Trend in India's overall trade balance and trade balance with OPEC countries.

CONCLUSION

Thus it will be evident that the amount of data that is generated at all stages of Indian Petroleum sector is enormous.

Computerisation of data processing will eliminate the enormous amount of clerical work of data compilation, duplication and accumulation of unnecessary information. A beginning has been made in this direction. Petroleum consumption data are computerised with one of the oil companies. All the data is available month-wise in considerable detail such as supply region, supply zone, consumption zone-wise, etc. ONGC has a computerised geo-data base to assist them in their exploration work.

There is a great paucity of data for distribution planning. The task of forecasting is mainly centred on demand for petroleum products. Transportation capacity from oil depots to the consuming centres can often prove to be a constraint when sudden surges of demand occur requiring the induction of additional tanklorries into the system. It is essential that forecasts of tanklorries, wagon requirements should be carried out and data pertaining to the same should be disseminated to the oil depots.

To conclude, the objective should be to construct data systems from the points of view of the end-objective—production planning, distribution planning or inventory control.

TECHNICAL SESSION 2

ELECTRIC POWER DATA SYSTEMS

Chairman : J.C. SHAH

Theme Paper : D.P. SEN GUPTA

Discussants : S. RAMESH

MOHAN MUNASINGHE

LU YINGZHONG

TWO

DATA SYSTEM IN THE INDIAN POWER SECTOR

D.P. SEN GUPTA

INTRODUCTION

This is a brief survey of the data system in the Indian Power Sector with particular emphasis on technical data rather than on data concerning finance or management. The paper initially describes the organisational structure in the power sector and the functions of its major constituents. It describes various types of data that are collected and the modes of data acquisition. The processing and utilization of these data constitute the next part of this paper—followed by a brief description of data banking and data retrieval that exists. Some recommendations are offered in conclusion.

Organization structure and respective functions

It is absolutely essential to have a bird's eye view of the organisation structure of the Indian Power Sector and the functions of each constituents before one can begin to understand the data system. The organisation structure is shown in Fig 2.1.

The diagram shows that the power industry has a fairly complex organisational structure. By and large, the major responsibility of power generation and supply rests with the State Electricity Boards (SEB). (In a few states the responsibility of power generation lies with other organisations also.

Appendix 1 provides the ownership-wise share of installed generating capacity). Firm lines (Fig. 2.1) indicate the administrative links amongst the central, state, and private sectors and the dotted lines indicate the interlinks.

The Central Electricity Authority (CEA), the Regional Electricity Boards (REB) and the State Electricity Boards (SEB) constitute the core of the organisational structure. The data system primarily interlinks these three and as such the structure of each and its functions will be dealt with in some detail in the sections that follow.

The Central Electricity Authority

The Central Electricity Authority (CEA) plays the most vital role in the entire data system in the power sector. The CEA was constituted in 1951 and became a full time body when the CWPC was bifurcated.

The functions of the CEA are best described in their own words.[1] These are to:

(i)* develop a sound, adequate and uniform national policy, formulate short-term and perspective plans for power development and coordinate the activities of the planning agencies in relation to the control and utilisation of national power resources;

(ii) act as arbitrators in matters arising between the State Government and the Board and licensee or other person as provided in the Act;

(iii)* collect and record the data concerning generation, distribution and utilisation of power and carry out studies relating to cost, efficiency losses, benefits and such matters;

(iv)* make public from time to time information secured under the Act and to provide for the publication of reports and investigations;

(v) advise any State Government, Board, Generating Company or any other agency engaged in generation or supply of electricity on such matters as will enable

The asteriks (*) represent functions that are particularly relevant in the present context.

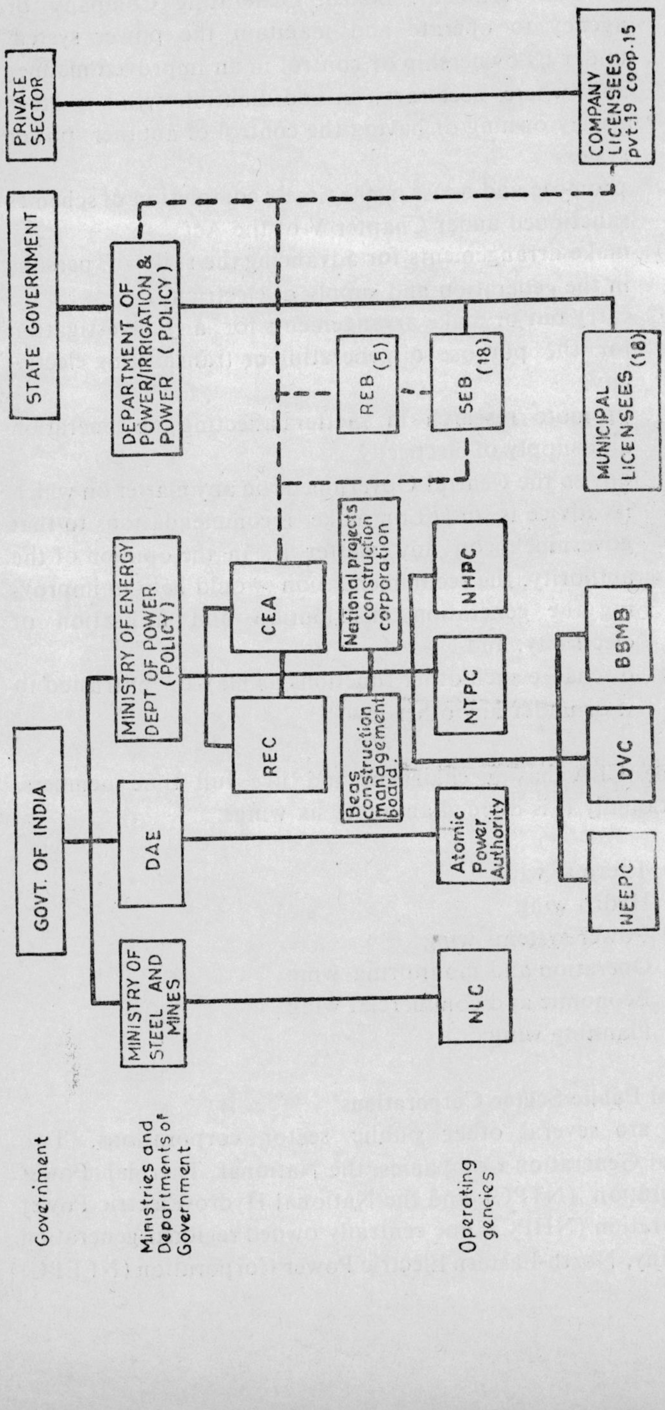

FIG. 2.1 : *Organisation structure of electricity supply industry in India*

such Government, Board, Generating Company or agency to operate and maintain the power system under its ownership or control in an improved manner and where necessary, in coordination with any other agency owning or having the control of another power system;

(vi)* promote and assist in the timely completion of schemes sanctioned under Chapter V of the Act;

(vii) make arrangements for advancing the skill of persons in the generation and supply of electricity;

(viii)* carry out or make arrangements for any investigation for the purpose of generating or transmitting electricity;

(ix) promote research in matters affecting the generation and supply of electricity;

(x) advise the Central Government on any matter on which its advice is sought or make recommendations to that government on any matter if, in the opinion of the authority, the recommendation would help in improving the generation, distribution and utilisation of electricity; and

(xi) discharge such other functions as may be entrusted to it or under any other law.

The CEA has a chairman and five full time members. Functionally it is divided amongst six wings:

(i) Thermal wing
(ii) Hydro wing
(iii) Power systems wing
(iv) Operation and monitoring wing
(v) Economic and commercial wing
(vi) Planning wing.

Central Public Sector Corporations

There are several other public sector corporations. Two Central Generation Companies, the National Thermal Power Corporation (NTPC) and the National Hydro-Electric Power Corporation (NHPC), one centrally owned regional generation company, North-Eastern Electric Power Corporation (NEEPC) and one state generation company, Karnataka Power Corporation (KPC) have been established so far.

The Department of Atomic Energy

The Department of Atomic Energy (DAE) comes directly under the Prime Minister and enjoys considerably greater autonomy than most public sector corporations. This department is responsible for designing, erecting, commissioning and operating of nuclear power plants. Currently the DAE has two plants in operation—one in Tarapore and the other in Rajasthan.

The Rural Electrification Corporation

The Rural Electrification Corporation (REC) is primarily a financing organization which monitors rural electrification programmes.

The Regional Electricity Boards

The Regional Electricity Boards (REB) were created in 1964. The country has been divided into five regions of contiguous states (Fig. 2.2). The five REB's are.

		States/Union Territories
1. Northern	:	Haryana, Himachal Pradesh, Jammu and Kashmir, Punjab, Rajasthan, Uttar Pradesh, Chandigarh and Delhi
2. Western	:	Gujarat, Madhya Pradesh, Maharashtra, Goa, Daman and Diu, Dadra and Nagar Haveli
3. Southern	:	Andhra Pradesh, Karnataka, Kerala, Tamil Nadu and Pondicherry
4. Eastern	:	Bihar, Orissa, Sikkim and West Bengal
5. North-Eastern	:	Assam, Manipur, Meghalaya, Nagaland, Tripura, Arunachal Pradesh and Mizoram.

A list of the functions and responsibilities of the REBs have

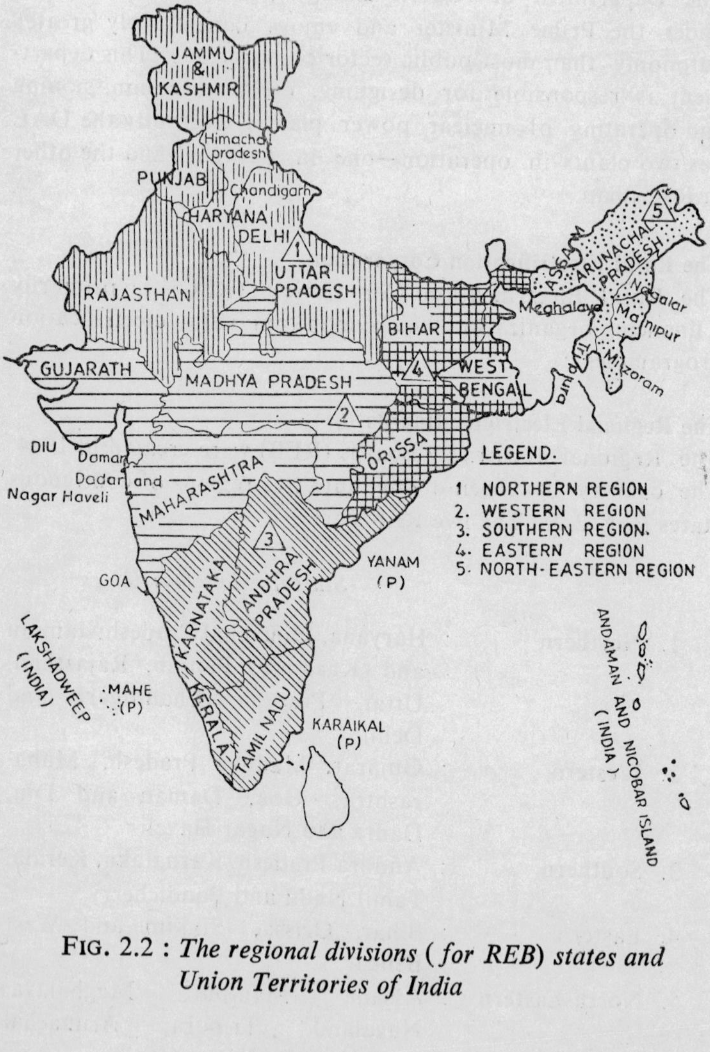

FIG. 2.2 : *The regional divisions (for REB) states and Union Territories of India*

a direct bearing on the data they collect, process and compile. The REBs are expected to[1]:

(i) devise integrated operation of the constituent power systems for achieving economic generation and transmission of power,

(ii) prepare a coordinated overhaul and maintenance schedule for the generating plants,

(iii) draw up generation schedule for enabling utilization of available capacity in each system to the maximum extent,

(iv) formulate the pricing policy for sales or exchange of power from one power system to another,

(v) review the progress of power development schemes in the region.

State Electricity Boards

Most SEBs have a full time chairman and two or three technical members and a secretary. Usually one of the members looks after financial aspects and accounts and the other member (or members) looks after the technical side.

The functions, duties and responsibilities relating to construction of power plants, operation of the power plants and power systems, transmission and distribution and supply of electricity to the ultimate consumers are distributed among several Chief Engineers. Below Chief Engineers are Superintending Engineers in charge of Circles, Executive Engineers looking after Divisions and Assistant Engineers in charge of Subdivisions—units which are in ultimate contact with the consumers. The Accounts/Finance wings of the SEBs also come under a similar hierarchical system.

The SEBs can function as autonomous bodies—the policy matters being directed by the state governments.

Types of data

Many classifications of the data in the Power Sector may be possible. One way of grouping the data broadly—(not that they are independent of each other) is as follows:

(a) Technical data

(b) Commercial data (relating to accounts and finance)

The technical data pertain to all technical details of generation, transmission and utilization and the commercial data pertain to the associated costs, tariff, income, etc. The present paper will be primarily concerned with technical data. The information relating to personnel management and similar aspects are not covered in this paper.

Technical data

The technical data can once again be subdivided into three categories :

(a) Systems data
(b) Operations data (short-term, medium-term, long-term)
(c) Constructions data

Systems data

These data refer mainly to the parametric constants of generators, transformers, circuit breakers, transmission lines and all such details that enable a systems analyst to model the entire system and carry out various analyses like load flow, short circuit studies, transient and dynamic stability analyses, etc.

Operations data

These data pertain to the performance of the power system as a whole. The performance may be monitored instant to instant and *short term data* collected from systems control. The performance may be averaged over a period and utilized for 'post-mortem' projection and planning purposes using *medium term data* and *long term data*.

Constructions data

As the title implies, these data are concerned with the details of construction of various plants and transmission distribution (TD) systems.

Commercial data

Like the technical side, the commercial part of the data system is just as complex and involves intricate tariff structure—operations cost, constructions cost, future projects, and cash flow. In fact the technical projections are evaluated in terms of cost and here the commercial and technical data are almost inseparably intertwined.

However, as has been earlier stated, this paper will mainly concentrate on technical data.

Data system

The data system in the power sector is complex. An attempt is made to represent this structure in Fig. 2.3. The Central Electricity Authority appears to be the heart of the data system, the REBs being major data sources for the CEA. The SEBs feed data to the CEA periodically on standard formats. The SEBs, of course, have data systems for their own use.

Various utilities placed at the bottom of the diagram provide information to the Regional Power Survey Office for future planning, peak load adjustments, etc., and will be dealt with in the section on Load Forecasting—System-Planning.

The CEAs are linked with various central agencies and supply required information to the Planning Commission and to the Department of Power. Some important data, particularly concerning generation, are sent daily. The National Informatics Centre (NIC) receives data mainly from the CEA and processes and banks these data.

The CEA acquires information from various generating stations. So do the SEBs and REBs which are also linked to the transmission system. For this purpose, REBs and SEBs have Load Despatch Centres, RLDCs, (Regional Load Despatch Centres) and SLDCs (State Load Despatch Centres) respectively. Their roles will be discussed in some detail in the sections that follow.

Data acquisition

The short-term (instant to instant) data on generation and power flow are acquired by the RLDC and SLDC.

The RLDCs are linked to the 220 kV lines but the SLDCs up to 66 kV. The modes of communication are:

1. Power Line Carrier Communication (PLCC) or microwave link
2. Hot Line Communication (P & T)
3. Telex Services if (1) and (2) fail
4. Regular P & T, STD service.

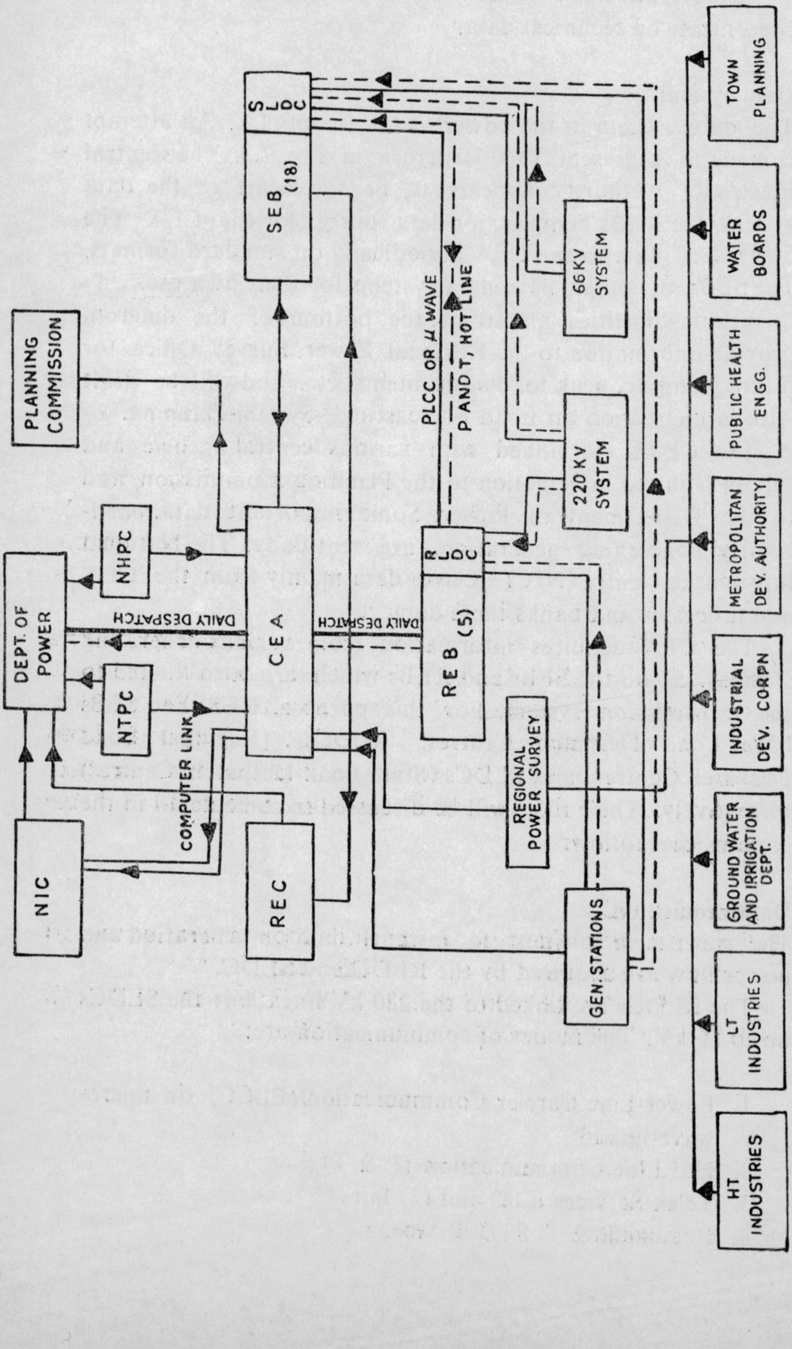

FIG. 2.3 : *Data system in the power sector*

Fig. 2.4 indicates how the RLDC in Bangalore is linked through the PLCC and P & T lines to various generating stations and the SLDCs in the Southern Region.

The information that are collected are Active Power Flow (MW),* Reactive Power Flow (MVAR), and Frequency.

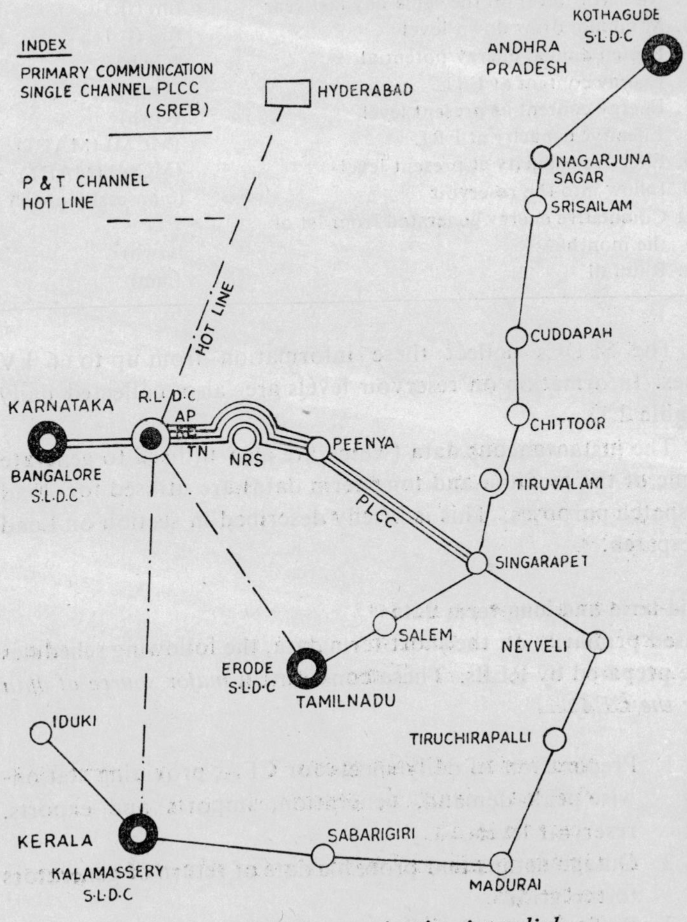

FIG. 2.4 : *PLC Communication inter-link*

*Active Power (MW) is that part of the total power which does work. Reactive Power (MVAR) does not provide work—but it is extremely important in an AC system. For example a capacitor draws reactive power. Reactive power plays an important role in the maintenance of systems voltage,

TABLE 2.1 : **Daily report of hydro reservoirs**

Description	Units
1. Full reservoir level (FRL)	(m) (ft.)
2. Present reservoir level	(m) (ft.)
3. Reservoir level on the same day last year	(m) (ft.)
4. Minimum draw down level	(m) (ft.)
5. Design annual energy potential	(Gwh)
6. Energy content at FRL	(Gwh)
7. Energy content at present level	(Gwh)
8. Effective capacity at FRL	(MCM) (MAFT)
9. Effective capacity at present level	(MCM) (MAFT)
10. Inflow into the reservoir	(cumsecs) (cusecs)
11. Cumulative energy generated from 1st of the month	(Gwh)
12. Rainfall	(mm)

The SLDCs collect these information from up to 66 kV lines. Information on reservoir levels are also collected daily (Table 2.1).

The instantaneous data (which are also utilised to generate some of the medium and long term data) are utilised for load despatch purposes. This is briefly described in section on Load Despatch.

Mid-term and long-term data

Based primarily on the short-term data, the following schedules are prepared by REBs. These constitute a *major source of data for the CEA*.

1. Preparation of daily sheets for CEA, providing station-wise peak demand, generation, imports and exports, reservoir levels.
2. Outage details and probable date of return of generators to the grid.
3. Performance of thermal stations.
4. Preparation of Load Generation Balance Report (LGBR) giving the annual picture of monthly generation, requirement, deficit/surplus and drawing up of maintenance scheduling of generating sets.
5. Drawing up monthly load flow studies and comparing

with actual conditions [operating committee meets once in a month to review the performance of the grid and analyse the problems encountered. Preparation of tie line schedules (assessment) and maintenance programme].

Data collection by SEBs

The SEBs try to collect various information with regard to their respective sectors.

The most important information is a daily account of energy generation and availability and the consumption and peakload of the major consumers.

The SEBs also collect various other details (technical and commercial) for their own use and compiled data are also passed on to CEA.

CEA publishes a General Review which is a comprehensive document providing fairly detailed information of the entire power sector.[2] A set of questionnaire is sent to various SEBs. The statistics division of the SEBs collect the information from sub-divisions and divisions. Information is collected under the following headings :

1. Energy balance.
2. Details of energy sales and purchase.
3. Installed capacity.
4. Details of capacity added.
5. Details of generators, steam turbines added (for thermal power houses).
6. Details of boilers added (for thermal power houses).
7. Details of generator, water turbine added (for hydro power houses).
8. Details of sets retired from service.
9. Details of derating.
10. Details of energy generation.
11. Details of fuel consumption.
12. Details of EHT transformers (above 33 kV) newly commissioned.
13. Details of EHT transformers retired.
14. Details of step-up transformers in service at the generation stations and substations.
15. Details of step-down transformers in service.

16. Details of distribution transformers in service.
17. Details of transmission and distribution lines.
18. Details of EHT lines commissioned during the year.
19. Details of circuit breakers installed.
20. Details of EHT circuit breakers commissioned and deleted during the year (above 33 kV).
21. Details of capacitors installed.
22. Details regarding consumers, connected load and energy consumed.
23. Energy consumption patterns.
24. Details of captive generation.
25. Details of establishment (personnel).
26. Monthly sale and revenue data.
27. Details of theft.
28. Details of accidents.

[There are no queries about parametric details necessary for system analysis. The voltage and ratings are, however, asked for.] It is a formidable task for the SEBs to collate all these data and an equally formidable task for the Commercial Directorate of the CEA to compile all this information into a comprehensive document without the assistance of a computer. Understandably the CEA finds it difficult to bring out an up-to-date version of this General Review. The latest issue now available is for 1979-80.

The SEBs, however, collect information—going down to greater details for their own use and also for providing bulk information to the CEA. Appendices (2 and 3) are sample sheets indicating some of the details that are collated by the Karnataka Electricity Board.

The data collected for financial purposes are represented by a typical data sheet (Appendix 3).

Most of the data described in the previous sections are routine data and mostly concerned with operations. As has been mentioned earlier, each region has a Regional Power Survey Office which collects data from various utilities (Fig. 2.3) for the so-called 'Micro approach' to load forecasting. This will be discussed in further detail in section 'local forecasting—system planning'.

Information regarding constructions are collected by NTPC,

NHPC, DAE by the Electricity Boards and by CEA to monitor progress of various projects.

In addition to all these, CEA compiles data on the performance review of various generating stations by sending what they call 'Roving Monitoring Team' which furnishes considerable amount of technical information regarding the causes of failure of various plants.

The present section has described the acquisition of some of the data, mostly technical. The next section describes the utilization of some of these data that are compiled.

Data utilization

Load Despatch (using short-term data)
The Indian Power System is expected to operate at 50 Hz (cycles/sec.). If generation exceeds the load, the frequency goes up. If generation fails to meet the load, the system dynamically adjusts itself and the frequency falls below 50 Hz.

Frequency is the best indicator of the health of the interconnected system in which, ideally,

$$\Sigma P_{\text{Generated}} = \Sigma P_{\text{Load}} + \Sigma \text{Losses}_{\text{at 50 Hz.}}$$

P being Active Power in MW.

[Reactive Power—control which is called VAr control are not very widely used. This helps to maintain busbar voltages at their stipulated values.] Whereas, RLDC helps in generation scheduling and advises inter-area power exchanges (based on short-term data) the responsibility of load control (when generating capacity falls short) by load shedding is the responsibility of the SEBs.

The RLDC in the Southern Region has a reasonably automatic system that enables increase/decrease of generation of each unit by governor control—(Load frequency control)—but this can be hardly utilized because the generating stations seldom have any reserve to draw upon.

The RLDC goes through the following routine based on the data that it collects. Obviously SLDCs fully participate in these functions.

1. Checking station-wise outage position (that is, which generators are inoperative) every morning at 6 A.M. *to assess the generation availability.*
2. Load forecasting for a typical working day and holiday—based on sustained half hourly demand of the previous day accounting for weather conditions, generation availability and load shedding.
3. Recording frequency continuously.
4. Recording tie line flows half hourly.
5. Generation scheduling; having thermal generation almost flat throughout the day and varying hydro-generation.
6. Drawing up hourly tie line schedule.
7. Assessing surplus power in a region and making it available to other deficit areas or otherwise 'banking' it (during monsoon season for example).
8. Monitoring of station-wise peak demand, station-wise generation, imports and exports between constituent system and other regions.
9. Recording trippings of transmission lines (later analysis of the causes is to be made by the Protection Committee).
10. Monitoring under frequency relay settings and their operation.

Load forecasting—System planning

The Section on Local Despatch has dealt with systems control—day-to-day and hour to hour generation control and inter-area power transfer.

A considerable amount of the data that are compiled are utilized for planning purposes which call for a reasonably accurate projection of load and energy forecasting. Load forecasting should also include a reasonable assessment of the peak load demand.

Load forecast could be classified under :

1. Immediate forecast (day or week).
2. Short term forecast (12—24 months).
3. Medium term forecast (4—8 years).
4. Long term forecast (6—10 years).

Immediate forecast has been referred to in the section on Load Despatch.

Short-term forecast is carried out under the aegis of a Power Survey Committee (Annual Power Survey, APS) set up by the Department of Power. The Power Survey Directorate acts as the Secretariat for the Committee. Since 1962, ten surveys have been carried out and the Eleventh Survey has now been published.[3]

The 9th, 10th, 11th surveys also provide medium/long-term forecasts.

Methodology used by the APS Committee is based on a combination of past trends and 'end-use method' (also called micro approach). Projection for consumptions in domestic, commercial, public lighting, public water works, LT industries and HT industries less than 1 MW are assessed on the basis of past trends[4],[5].

HT industries with consumption greater than 1 MW are identified individually. Detailed data are collected from these industries regarding their production capacity, expansion programme, power and energy requirements (Fig. 2.3). The information is cross checked with the DGTD and concerned Ministries of the Government of India.

The forecast concerning irrigation and dewatering is based on the targets of different states for energising pumpsets and assuming an average rating and energy consumption by each pumpset. The energy requirement by licensees for bulk power supply is also collected. The railway board provides the requirement for electric traction. Having compiled the total load requirement APS adds to the total, the likely TD losses based on past trends.

The method of assessment of load requirement, briefly described above, is a reasonably correct assessment—but usually higher than the projections that are made independently by the Planning Commission, based on input-output model, and detailed material balance studies and capital output ratios.[6]

The State Electricity Boards also make their own power forecasts which are normally on the high side. One of the basic reasons for this is that the industrial requirement (it constitutes the bulk of our electrical energy requirement, national average being 60 per cent) is based on the state's own forecasts of industrial growth and this may not be related to the national plan.

The peakload requirement is obtained by applying suitable load factors obtained from past trends. Account is taken of the load mix and the impact of high load factor in industrial loads in future.

Table 2.2 is from the 11th annual power survey. Tables 2.3 and 2.4 are from Sixth Five Year Plan. These are provided here for immediate reference.

Generation losses or outages

Two types of outages are defined—planned outages and forced outages. There is also what is known as partial loss of generation when the generator cannot deliver power at its full capacity. Planned outages are for maintenance purposes. The SEBs and REBs prepare detailed schedules for individual units. The information is conveyed to CEA as a routine. Forced outages or breakdowns are conveyed to CEA from SEBs, REBs and from generating stations.

Planned maintenance for thermal units

Planned maintenance comprise the following :

1. Complete *annual* boiler overhaul (average duration should be about 30 days but was 52 days during 1980-81).
2. Complete capital maintenance of turbine generators every 3 years.
3. Planned work on new units; shutdown immediately after synchronising.
4. Incomplete boiler overhaul and capital maintenance of turbine generators.
5. Preventive/curative and similar other planned work of short duration.

TABLE 2.2: **Installed capacity, peak availability, peak load, energy availability and energy requirement**

	Actual				Provisional		Estimated		
	1975-76	1976-77	1977-78	1978-79	1979-80	1980-81	1981-82	1982-83	1983-84
Installed capacity MW	20106	21463	23673	26680	28458	30294	33489	38008	43348
Peak availability MW	13452	14197	15097	17027	17402	19188	21142	23976	27401
Peak load MW	13452*	14197*	15097*	17027*	17402*	19188*	25190	27712	30372
Surplus (Deficit) MW	—	—	—	—	—	—	(4048)	(3736)	(2971)
Energy availability Mkwh	74777	83061	85789	96654	98837	104134	127522	142167	161411
Energy requirement Mkwh	74777*	83061*	85789*	96654*	98837*	104134*	139873	153921	168415
Surplus (Deficit) Mkwh	—	—	—	—	—	—	(12351)	(11754)	(7004)

*Restricted.

TABLE 2.3: **Expected generating capacity by 1984-85 (MW)**

Utilities	Hydel	Thermal	Nuclear	Total
Northern region	5240	7743	440	13423
Western region	2247	11104	420	13771
Southern region	6798	4504	470	11772
Eastern region	1409	6779	—	8188
N.E. region	457	546	—	1003
All India (Utilities)	16151	30676	1330	48157
Non-utilities	3	3032	—	3035
All India (including non-utilities)	16154	33708	1330	51192

TABLE 2.4: **Sixth Plan outlay on power (crores)**

Item	States	U.Ts	Centre's	Total
Generation	8002	59	3791	11852
Transmission and distribution	4548	159	705	5412
Rural electrification	1559	17	—	1576*
Miscellaneous	184	12	229	425
Total	14293	247	4725	19265

*Exclusive of the outlay included in Rs. 285 crores for Special Project Agriculture (SPA), Rural Cooperatives, System Improvement Schemes for Harijan Basties.

The CEA obtained data on 181 generating units and 227 boilers during 1980-81 and studied the actual time taken for each unit for planned maintenance. CEA has also closely studied the forced outages and the reasons for this. These studies have been extended to hydel units also (which generally need less maintenance).

It is not always an easy task for CEA to collate information from generating stations regarding the details of maintenance and overhaul scheduling. Appendix 4 provides a list that indicates outages (planned and forced) for different thermal plants (during 1980-81) and also indicates partial non-availability (See also Fig. 2.5).

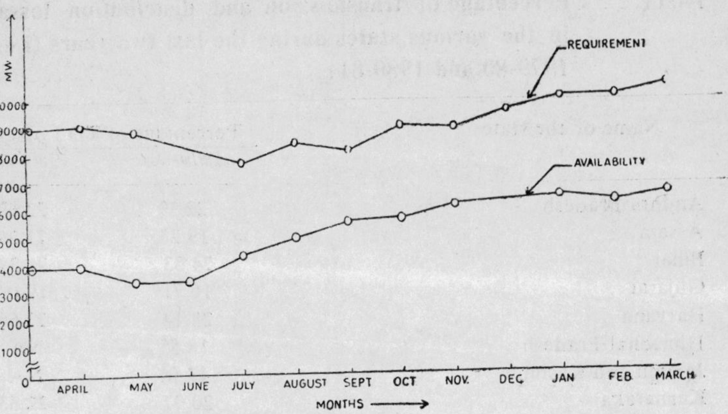

FIG. 2.5 : *Month-wise MW requirement and availability for 1980-81*

Losses in transmission and distribution (TD losses)

It is only during the recent past that increased efforts are being made to collect accurate data on TD losses which happen to be unusually large in India. [Due to lack of data, the performance of the transmission and distribution (T and D) system is hard to measure.] Table 2.5 is provided (for quick reference) to indicate the recent assessments for different states. According to Rajadhyaksha Committee Report, one of the most uncertain features in this data is the percentage of actual loss* in TD (Technical loss) and the percentage loss due to theft (euphemistically called commercial loss). Inadequate metering facilities make it nearly impossible to assess the extent of theft which could easily be 5 per cent of the national average of 20 per cent TD losses.

The details that are sought by the State Electricity Boards and CEA to prevent pilferage of energy (and also materials)

*When current I flows through a conductor, heating takes place, being equivalent to I^2R, (copper loss) where R is the resistance of the conductor. Additional losses also take place in transformers. These losses are called iron loss. TD losses are mainly due to copper loss and partly due to iron loss. High TD losses are the result of (i) inadequate planning, (ii) using wrong size of conductors, (iii) voltage levels not being maintained properly, (iv) long distance transmission of power at low voltage, etc.

TABLE 2.5 : **Percentage of transmission and distribution losses in the various states during the last two years (i.e., 1979-80 and 1980-81)**

Name of the state	Percentage of TD Losses	
	1979-80	1980-81
Andhra Pradesh	22.39	22.69
Assam	19.23	21.38
Bihar	24.53	21.38
Gujarat	16.71	19.68
Haryana	28.14	23.66
Himachal Pradesh	18.55	NA
Jammu and Kashmir	47.01	NA
Karnataka	20.97	22.33
Kerala	15.90	14.57
Madhya Pradesh	22.30	22.40
Maharashtra	16.10	16.39
Manipur	47.24	45.76
Meghalaya	4.32	9.93
Nagaland	32.42	26.59
Orissa	18.29	19.81
Punjab	22.50	19.47
Rajasthan	22.40	25.97
Sikkim	20.51	NA
Tamil Nadu	19.58	18.92
Tripura	31.95	33.97
Uttar Pradesh	19.00	16.19
West Bengal	12.26	13.49

Note : The figures for 1980-81 are provisional
NA=Not available

are given below. If one has to go by trends TD losses are on the increase (Fig. 2.6) and commercial losses are certainly not declining.

MONITORING OF PROGRESS IN PREVENTING THEFT/PILFERAGE OF ENERGY AND MATERIALS, BELONGING TO STATE ELECTRICITY BOARDS

Theft of energy

1. No. of surprise inspections carried out
2. No. of cases in which theft, pilferage or misuse are detected

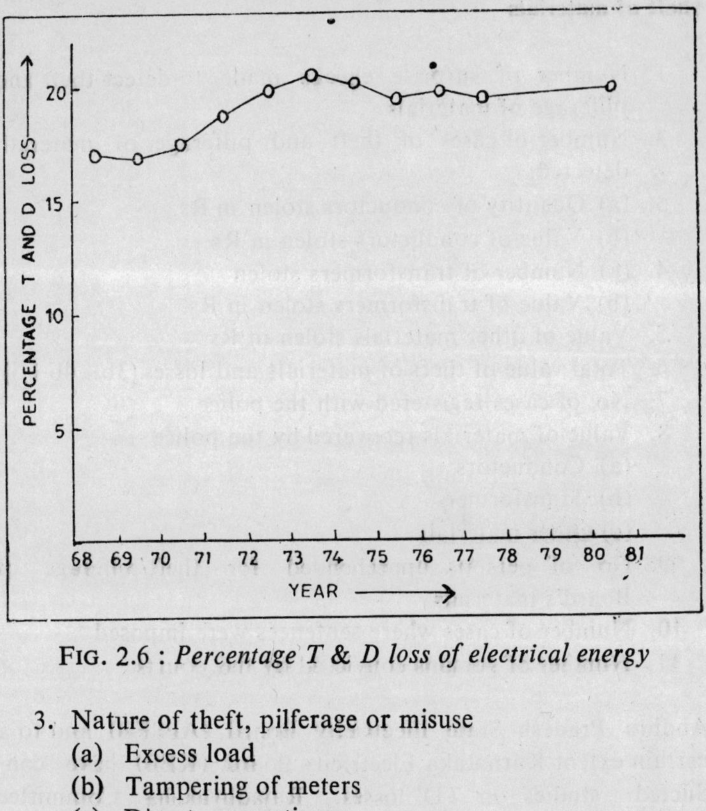

FIG. 2.6 : *Percentage T & D loss of electrical energy*

3. Nature of theft, pilferage or misuse
   (a) Excess load
   (b) Tampering of meters
   (c) Direct connection
   (d) Under billing
4. (a) Total energy estimated as stolen, pilfered or misused (against item No. 3)
   (b) Total cost of energy estimated as stolen/pilfered or misused.
5. (a) Quantum of energy for which it was possible for the Board to effect recoveries
   (b) Total amount recovered
6. Number of persons apprehended for theft of power
7. Number of persons convicted by the courts
8. Steps being taken to prevent theft, pilferage and misuse of energy

Theft of materials

1. Number of surprise checks made to detect theft and pilferage of materials
2. Number of cases of theft and pilferage of materials detected.
3. (a) Quantity of conductors stolen in Rs.
   (b) Value of conductors stolen in Rs.
4. (a) Number of transformers stolen
   (b) Value of transformers stolen in Rs.
5. Value of other materials stolen in Rs.
6. Total value of theft of materials and losses (3b+4b+5).
7. No. of cases registered with the police
8. Value of materials recovered by the police
   (a) Conductors
   (b) Transformers
   (c) Other materials
9. No. of persons apprehended for theft/pilferage of Board's materials
10. Number of cases where sentences were imposed
11. Number of persons convicted by the courts

Andhra Pradesh State Electricity Board (APSEB) and to a certain extent Karnataka Electricity Board (KEB) have conducted studies on TD losses. Rajadhyaksha Committee justifiably recommends that APSEB type of studies should be undertaken in all the states.

Constructions data (Project monitoring)
Inordinate delay in the construction of various thermal and hydel projects has been one of the major factors that has resulted in the power crisis in India. Delay in the execution of projects has resulted in escalation of costs. Appendix 5, provides a list of various projects that have been delayed and the escalation in the cost as a result.

CEA has been trying to provide help in monitoring the work schedule of some of the 62 hydel projects (10974 MW) and 83 thermal projects (14208 MW). With the creation of two Central Generation Companies, namely, NTPC and

NHPC in 1976 project monitoring has become reasonably formalised.

The super thermal station at Singrauli was commissioned in 4 years and NTPC's success is very largely due to a rigid data monitoring system.

NTPC presents monthly progress reports to external agencies like CEA and the Department of Power mainly identifying reasons of slippage in critical areas.

The internal monthly progress reporting is fairly thorough and exhaustive.

A typical set of formats used by NTPC can be classified as below :

1. Pre-award report (contract)
2. Post-award report (manufacturing and despatch)
3. Post-award report (manufacturing and despatch : quantitative report)
4. Construction progress report
5. Weekly execution schedule and progress report for site, internal detail monitoring
6. Engineering progress report

The data obtained enables NTPC to keep a close watch on the progress of a project.

Construction of TD lines is usually the responsibility of the SEBs and its monitoring is not quite as close as the NTPC example just quoted.

Rural electrification

Rural electrification as a planned programme started in 1950s when the role of rural electrification for irrigation purposes was recognised.

The primary burden of carrying out the RE programme rests on the State Electricity Boards. In order to finance the construction of rural feeders and energisation of pump sets the Rural Electrification Corporation (REC) was set up in 1969. The role of the REC is to help SEBs to formulate viable schemes for rural electrification, appraise them, sanction funds for their implementation, monitor and evaluate their progress.

The REC calls for some information regarding the backwardness of a particular area (typically a bunch of villages in a taluk) for which funds are sought for electrification. The chart is shown in Appendix 6. The data provided are not always very authentic.

The SEBs collect information on electrification and progress from Executive Engineers.

The REC compiles a very large amount of data and processes these data in a computer. It calls for quarterly progress report on the achievements of the REC schemes. On T and D it seeks information on different types of feeders, conductors capacitors (number and kVA rating) transformers (number and kVA rating) and the number and the kW connections for Agricultural services, LT/Agro industry, domestic,

TABLE 2.6 : **Village electrification in India**

Month and Year	Number electrified
March 1974	1,56,720
March 1975	1,73,533
March 1976	1,85,808
March 1977	2,02,350
March 1978	2,16,863
March 1979	2,32,770
March 1980	2,49,842
December 1980	2,60,978

TABLE 2.7 : **I.P. sets/tubewells energised**

Month and Year	Number energised
March 1974	24,26,133
March 1975	26,13,630
March 1976	27,92,339
March 1977	30,32,990
March 1978	32,99,901
March 1979	35,98,775
March 1980	39,49,120
December 1980	41,82,313

commercial street light points and the costs incurred for each installation. This is for monitoring the physical achievements.

Commercial data

As has been mentioned earlier, the commercial data system is a complex system on its own. Every link shown in Fig. 2.3 may not be operative so far as the commercial data flow is concerned.

SEBs are primarily responsible for management of their finances and state government policies dictate Tariff rates. The states present proposal for new generating stations to the centre—and projects are assessed by CEA before they are sanctioned by the ministry.

SEBs keep detail account of energy used in various sectors and collect necessary data for the purpose.

Appendices 2 and 3 are examples of typical questionnaires that are sent out by the SEBs to compute the use of electrical energy in different sectors. The response to such questionnaires is meticulously compiled. The bills paid by consumers are also recorded at subdivisional level and finally assorted at the main SEB office. These constitute an important source of data for the SEBs.

The Commercial Directorate at CEA documents various commercial data arriving from different departments and sectors. It collects the electricity tariff schedules from all the SEBs and the electricity supply undertakings/corporations. Details of tariff, utilisation of power and energy by captive plants are also collected from power intensive industries. "Similarly data is collected on inter-state/system exchanges of power and tariff applicable for such exchanges. Power market data such as energy sales and revenue billed, etc., is also collected on monthly basis. The schedules and other data thus collected are analysed and tabulated".[1]

CEA publishes an 'All India Electric Rate Book' summarising various details regarding tariff. There are many other facets of commercial data which are difficult to incorporate in the limited scope of this paper.

Interactions with other energy sectors

The two most important inputs into the power sector are coal and fuel oil and these two are the two major sources of commercial energy. Only a close cooperation of the respective ministries and secretariats and boards can ensure necessary allocation and supply for the power sector. CEA has "been engaged in quarterly and monthly allocation of coal and fuel oil to thermal power stations. CEA has also been monitoring quality and quantity of coal supplied to thermal power stations".

1980-81 allocation, receipt and consumption of *coal* is given below:

Allocation	48.85 Million tonnes
Receipt	36.97 ,, ,,
Consumption	36.72 ,, ,,

A large number of thermal stations run on low stocks. "Daily monitoring of coal stocks of thermal power stations was undertaken and reports furnished to the Cabinet Secretariat, Ministry of Energy, Railway, Department of Coal etc".[1]

In a similar manner monthly allocations of furnace oil, LDO and HSD required by thermal power stations as secondary fuel were made and the Ministry of Petroleum, Chemicals and Fertilizers and the oil companies were advised to supply the allocated quantities to power stations.

The allocation, receipt and consumption of fuel oil for thermal power stations during 1980-81 are given below :

Allocation	1.4 Million KL
Receipt	1.1 ,, ,,
Consumption	1.1 ,, ,,

Data processing and banking

The system of processing the data that are acquired in the entire power sector leaves a lot to be desired. In this respect the major part of the data system is in the pre-computer era. Data that are collected are tediously compiled and stored everywhere in the Electricity Boards, REBs, CEA and all departments involved.

The National Informatics Centre (NIC) has made a beginning having acquired a large computer and the CEA has also acquired a computer and established a data link with NIC. Some of the REBs are planning computerization.

CEA brings out a large number of publications most of which are for internal use. Reference 2 is a comprehensive document but usually about 2-3 years behind time since all SEBs do not respond to CEAs questionnaire in time.

The Department of Power is kept informed about the daily record (see Table 2.8) and this information is the processed version of all the bits and pieces that come from REBs and generating stations. For project monitoring CEA and Department of Power receive abbreviated version of the details that are received from NTPC, NHPC and various Electricity Boards.

It is briefly indicated in the conclusion that follows that complete computerisation of data processing (if not data acquisition itself, to the extent possible) is the only way to eliminate the enormous amount of useless clerical work of data compilation, duplication and accumulation of unnecessary information. Data retrieval which is the hardest of all tasks in the present set-up could be made extremely efficient.

Conclusion and recommendations

The amount of data that is generated at all stages of the Indian Power Sector is enormous. The present paper could provide only a glimpse at a part of this incredibly large and complex data system. As was clearly pointed out, the paper deals mainly with technical data and those concerning finance were hardly touched upon.

The technical data that are available to the public are not even the tip of the iceberg. There is a very large amount of information which are for 'office use' only (as they should be). The method of acquisition of these data is similar to if not the same as what has been described in this paper.

In going through the complex maze of these data structure one is faced with four immediate questions.

TABLE 2.8 : **Daily generation report**

Date : Day :

ALL INDIA POWER POSITION

Type of generation	Region	Generation (Gwh) Today Pro-gramme	Today Actual	Average In this month	Average Last year	Total capacity (MW)Σ	Capacity on-line (MW)*	Aggregate state peak (MW)**	% of peak to total capacity	% of peak to capacity on-line
1	2	3	4	5	6	7	8	9	10	11
A. Thermal	Northern									
	Western									
	Southern									
	Eastern									
	N-Eastern									
	Sub-Total									
B. Nuclear	Northern									
	Western									
	Sub-Total									
C. Hydro	Northern									
	Western									
	Southern									
	Eastern									
	N-Eastern									
	Sub-Total									

(*Contd.*) Table 2.8

1	2	3	4	5	6	7	8	9	10	11
A+B+C Total	Northern									
	Western									
	Southern									
	Eastern									
	N-Eastern									
	All India									

									Actual	
						Cumulative total from 1st April (Gwh)	Type of Gen.	Programme	This year	Last year
							Thermal			
							Nuclear			
							Hydro			
							Total			

Data Missing *Remarks*

* Outages of units of less than 20 MW not considered.

** Covers estimated generation in respect of missing data.

Σ This is the aggregate name plate rating of all the units/or accepted capacity.

1. How authentic are all these data that are being generated?
2. Are all the data useful?
3. Are all the data adequate?
4. Are the methods of acquisition and compilation satisfactory?

There could hardly be a clear cut answer to these questions.

By and large at a gross level the data that are compiled are good and dependable and considering the limitations to be highlighted. Later, CEA, REC and a few other organizations are doing a formidable job of data compilation.

One of the greatest weaknesses in the entire system is inadequate or faulty metering. A large number of meters particularly those obtained from indigenous sources are either inoperative or defective.[4] This not only creates problems for operation and maintenance staff but can legitimately give rise to doubts about the accuracy of some of the data that may be collected at micro level.

The RLDC and SLDC more often than not depend on telephone messages rather than on the sophisticated PLCC or microwave data transmission to ascertain the details of generation, outage etc. and automatic LFC is hardly operative since there is no spinning reserve.

As has been mentioned earlier (losses in transmission and distribution) the information on TD losses can hardly be considered authentic. At best these are gross estimates. In fact Rajadhyaksha Committee[4] strongly makes this point:

> It is unfortunate that such an energy audit (like the one conducted by APSEB) cannot be undertaken for most of the Boards since adequate instrumentation, metering and other sources of data are not available to disaggregate and analyse the losses in the various system elements. It is important that the Electricity Boards instal metering arrangements to monitor losses throughout the network and review these losses periodically to plan system improvement measures.

The data compiled for load forecasting has many gaps and inaccuracies that are difficult to fill or to improve upon. Many projects are proposed and even accepted by the Centre that finally fail or do not come up at all. These are reflected in the inaccuracies of the projections that are made (example: Kudremukh Iron Ore Project).* This is not typically an Indian problem. The CEGB in the UK found itself in an embarrassing position of having a very large surplus installed capacity in the seventies because industries did not come up as they were expected. Surplus power is, however, not a problem for our country.

Inaccuracies mainly occur at microlevel and this is commented upon in the sections that follow. The quotation from Rajadhyaksha Committee Report leads on to the question regarding the adequacy of the data that are compiled.

Obviously the data compiled are not entirely adequate in all respects. This may contradict my earlier statement about the 'massive data' that are compiled. But whether all the data compiled and stored are useful or not is a question that remains to be answered.

One of the most inadequate parts of the data base in the Indian Power Sector is the 'Systems data'. The systems data should comprise:[8]

—technical description of generating units
—technical description of transmission system components
—technical description of distribution system components
—network configuration (switching state) data
—absolute or incremental fuel cost functions attached to a generating unit (unnecessary as long as we do not have surplus power)

No system study to analyse the behaviour of a power system for planning extensions or interconnections is possible without these basic data which seem to be badly lacking in our data

*Even the Working Group on Energy Policy seems over optimistic in its assessment of Electric Power and Energy requirements and both the RLF (Reference Level Forecast and OLF (Optimal Level Forecast) appear to be unrealistic.[7]

base. As soon as a system analysis is to be undertaken there is a rush to dig out some of the relevant data and this unnecessarily delays system analysis for a proposed project.

Although the Roving team of the CEA [and also of the National Productivity Council (NPC)] found out a number of details about the state of health of many generating stations, a standard complaint was regarding the lack of data essential for the maintenance of a plant. Even the pH value of boiler water—an absolutely essential information—is apparently not available at many stations, leave alone other data.

Commenting on TD projects Rajadhyaksha Committee Report points out once again the inadequacies in the systems data and also load forecast as below:

> In order to do this (i.e., preparation of optimised TD projects) detailed data regarding the load at different times of the day and seasonally should be available. Similarly, reasonably dependable long-term forecasts are also required. Currently, macro-level state-wise data on load is available but detailed region and district-wise data required to prepare optimal sub-transmission and distribution systems is not available. This is one reason why the Rural Electrification Programme which plans from LT distribution onwards runs into problems if the distribution system up to the LT stage is inadequate. Collection storage and analysis of data on TD systems is an area in which much greater attention should be devoted.

There are several other areas one could point out where adequate data are not compiled and even if they are compiled they are gross, in the absence of adequate metering and cannot be fully relied upon.

On the other hand volumes of data are stored in files in every office and one wonders whether these data are ever used.

A lot of the details that are collected (samples presented in the appendix constitute an insignificant fraction of the total questionnaires that are sent out as a routine) may at one stage be required to obtain a *total view* with regard to generation,

TD, operations and so on but probably most of these data cease to be of any use afterwards.

In the context of project monitoring the details that are collected by NTPC, for example, are essential, as long as they help the construction of a power station on the basis of a strict time target. But as against the unique example of the Singrauli Superthermal Power Station, one can quote a hundred other examples where projects are years behind schedule which makes all PERT charts or data collection on the progress of the projects and all the intricate exercises in costing, totally meaningless. This of course does not imply that these data should not be collected.

What is absolutely imperative is that all the agencies in the Power Sector realize the usefulness of computers. It seems to be the senseless opposition from trade unions on the one hand and the apathy of the people at the top, that computers are very little used in the Power Sector. One of the results of this is that massive data that are collected and laboriously compiled turnout to be not so useful because manually processing and storing this voluminous information is indeed impossible whereas a computer can do the same job infinitely more efficiently.

The data that flow in, down PLCC or microwave links can be accepted by a digital computer to process the data under command of suitable software.* Essential information are compiled and forwarded to CEA down P and T lines. Data that are likely to be useful in future are stored away and the rest erased. Information that have to be collected by questionnaires from sub-divisions or divisions of Electricity Boards or the worksite of projects are fed into the computer as and when they come. There is no reason why every Circle and Division of a SEB should not be provided with a terminal in the future.

Facilities for cross checking data should be built into the programmes since considerable duplication of data inevitably takes place. CEA has acquired a computer which is linked to NIC but the Commercial Directorate of

*A fairly modern and sophisticated instrumentation and control system has been proposed for 200/210 MW Thermal units.[10]

CEA which handles enormous amount of data does not have a computer and the data are laboriously compiled and manually sent to the NIC. It is not surprising that NIC detects discrepancies in the data that comes from OMI and the Commercial Directorate which are both under CEA. All the wings and Directorates in the CEA should make use of computers and feed into the CEA Central computer or to the one with NIC, for processing and banking.

The Department of Power, Planning Commission and the Prime Minister's Secretariat could all be linked with NIC. Some of the essential data should be routinely processed for display on video terminals. Other data could be recalled 'on request'. Details of progress in construction of various hydel and thermal stations and major T and D lines and the costs can be computerized, updated as the work proceeds.

As things are now, the Department of power, CEA, REBs and State Electricity Boards get into a state of frenzy when the Parliament is in session. Even an idle query from a member sends a shockwave down the line—people working furiously to collect information to answer the query. (Appendix 5). A computer would sort this out in a matter of seconds.

Two aspects should be borne in mind.

1. Computers and computer links are really inexpensive compared to the costs involved in breakdowns, loss of production and delays in construction.
2. Computers do eliminate unproductive labour but do not create unemployment if the management can plan redeployment of labour properly.

In conclusion, three recommendations could be made for the data system in the Indian Power Sector:

1. To build up a comprehensive data base for systems data;
2. To have extensive instrumentation and metering up to LT distribution level;
3. Complete computerization and interlinking of concerned

departments, with suitable software back-up, with CEA or NIC housing the central processor.

Obviously, everything cannot be done immediately and just for the asking of it. But a masterplan may be made and the implementation made in suitable phases. Lack of funds is often the reason given for our inability to undertake such tasks and increasing generation naturally being the topmost priority, other essential areas are neglected. However, if nearly Rs. 20,000 crores are allotted to the Power Sector in the Sixth Plan (See Table 2.4), a small part of this amount could be invested in modernising the data system of the Power Sector. A modest beginning has been made by CEA and NIC, but a lot more remains to be done.

Acknowledgement

This paper has been prepared after extensive discussions with various people involved in the Power Sector and based on some documents that were made available by them. The author gratefully acknowledges the help rendered by:

1. Shri S.N. Roy, Chairman, CEA, New Delhi.
2. Shri M.K. Sambamurti, Member, CEA.
3. Shri Ishkumar, Director, Commercial Directorate, CEA.
4. Shri Thurabi, REB, Bangalore.
5. Shri Narayan, REB, Bangalore.
6. Shri K.V.K. Premkumara, Regional Power Survey of the Southern Region.
7. Shri A.K. Sah, Chairman, NTPC.
8. Shri P.K. Varma, NTPC.
9. Shri N. Sriram, REC.
10. Shri M. Muralidharan, REC.
11. Dr. S. Ramanathan, Planning Commission.
12. Shri Ramchandran, Planning Commission.
13. Dr. Vasudevan, NIC.
14. Shri B.G. Rudrappa, C.E., KEB, Bangalore.
15. Shri H.V. Dayal, Member, KEB, Bangalore.

Special thanks are due to Dr. R.K. Pachauri for his suggestions and encouragement and to Mr. N.K. Gopalakrishnan for providing some documents and other facilities. And finally to Mrs. Sushmita Ganguly and Mr. Samuel Sargunaraj; Project Assistants, IISc., for their help in some data collection, to Mr. Y.S. Ramaswamiah and Mr. R. Manjunath, for typing the manuscript and to Mr. C.A. Kannan for the drawings.

REFERENCES

1. Seventh Annual Report of CEA, 1980-81.
2. Annual General Review CEA, 1979-80.
3. Eleventh Annual Power Survey Committee Report, CEA, 1982.
4. Rajadhyaksha Committee Report, GOI, Publication, 1980.
5. K.V.K. Premakumar, 'Load Forecasting' Lecture Notes.
6. Sixth Five Year Plan, Planning Commission.
7. Report of the Working Group on Energy Policy, GOI Publication.
8. R. Baumann and R. Bayer, 'Data bases for Power System Calculation', *Electrical Power and Energy Systems*, Vol. 4, January 1982.

ADDITIONAL REFERENCES

Standard Design Memoranda on Instrumentation and Control System of 200/210 MW Thermal Units, Data Acquisition System, CEA, March 1982.

KEB Recommendations and Summary of the Tariff Committee, 1981.

Rural Electrification Corporation Limited, 12th Annual Report, 1980-81.

KEB Annual Financial Statement, 1980-81.

Circular Letters from the Federation of Electricity Undertakings of India.

'The First 200 MW Unit of the Singrauli STPS', Urja, *Indian Journal of Energy*, March 1982.

Manual on Basic Electric Design Features for Thermal Power Stations, May 1980, CBIP.

APPENDIX 1

OWNERSHIP-WISE SHARE OF INSTALLED GENERATING CAPACITY

Owner	Share of installed generating capacity		Typical projects
	MW	Percentage	
A. AUTONOMOUS CORPORATIONS/ BOARDS			
(1) Individual States/UTs	19445	72.8	Projects run by State Electricity Boards and Power Corporations and Delhi Electric Supply Undertaking.
(2) Groups of States	2261	8.5	Bhakra, Beas-Sutlej Link, Tungabhadra, Machkund, Rihand.
(3) Centres and States	1421	5.3	DVC, Regional Load Despatch Stations.
(4) Centre	1216	4.5	NTPC, NHPC, NLC
(5) Municipal Corporations	276	1.1	Bombay Electric Supply Undertaking (BEST), Transmission and Distribution only.
(6) Public Limited Companies	1389	5.2	Tata Power, Calcutta Electric Supply, Ahmedabad Electricity Supply and a few Transmission and Distribution Companies.
B. DEPARTMENTAL UNDERTAKINGS			
(1) States	44	0.2	Projects run by Govt. Department of Nagaland, Arunachal Pradesh.
(2) Centre	640	2.4	Nuclear, Thermal Plants.

APPENDIX 2

To be sent to C.E.E. (G) Office by
the 10th of the succeeding month

KARNATAKA ELECTRICITY BOARD

Stat. Form No. MIS IIIA
Division________________

Item		Domestic Ltd.	A.E.H.	Commercial Ltd.	L.T. Power		Street Lighting				I.P. Sets		H.T. Consumers
					Water works	Others	Instns			Con-trol points	L.I.S.	Others	
							F	M	I				
1		2	3	4	5	6	7	8	9	10	11	12	13
1. Total number of live installations at the end of the previous month	Urban												
	Rural												
Current Month	Urban												
	Rural												
Cumulative since April of current month	Urban												
	Rural												
2. No. of installations serviced	Urban												
	Rural												
3. No. of applications pending	Urban												
	Rural												

APPENDIX 2--*contd.*

1			2	3	4	5	6	7	8	9	10	11	12	13
4. Connected load in kW of new installations during the month (Mention contact contract demand for H.T)		Urban												
		Rural												
5. No. serviced under direct connections	Current month	Urban												
		Rural												
	Cumulative	Urban												
		Rural												
6. Total No. of Live installations at the end of the current month		Urban												
		Rural												

Ex. Engineer (El.)

APPENDIX 3

To be completed from the ledgers and furnished to C.E.E. (El.) within 45 days.

KARNATAKA ELECTRICITY BOARD

Stat. Form No. MIS. IV

Division__________ Energy Sales, Revenue and Arrears for the month of__________

Category sales		No. of instal-lations at the end of the month	Energy sold during the month (Units)	Energy sold previous month (Units)	Variation %	Revenue demand Rs.		Variation %	Collection Rs.	Ratio of collection	——×100 demand	Total arrears Rs.	Ratio of arrears	——×100 demand
						Current month	Last month							
1		2	3	4	5	6	7	8	9	10	11	12	13	14
1. Domestic lighting	LT1													
2. A.E.H.	LT2													
3. Commercial Ltg	LT3													
4. Irrigation Pumpsets	LT4													

APPENDIX 3—*contd.*

1	2	3	4	5	6	7	8	9	10	11	12	13	14
5. Industrial power	LT5												
6. Public Lighting	LT6												
7. H.T.													
8. Temp. Supply	LT7												
	LT8												
	LT9												
9. Free lighting													
10. Others													
Total													

Ex. Engineer (El.)

APPENDIX 4

DATA ON LOSS OF GENERATION FORCED AND PLANNED OUTAGES—PLANT LOAD FACTOR (PLF) 1980-81

(*Expressed in percentage*)

	P.L.F.	Partial loss and low load	Planned outage	Forced outage
	1	2	3	4
Northern Region				
Renu Sagar	86	7	7	—
I.P. Station	60	15	10	15
Panki	55	10	22	8
Panki extension	47	21	10	22
Badarpur	46	24	7	23
Obra (T)	46	15	11	28
Obra extension I	40	21	11	20
Bhatinda	38	21	11	30
Obra extension II	38	22	15	25
Panipat	33	28	2	37
Hardua Ganj	34	13	53	—
Hardua Ganj B	32	17	13	38
Faridabad	29	12	16	43
Hardua Ganj A	3	2	95	—
Western Region				
Amarkantak-I	73	18	7	2
Duuvaran	72	22	4	1
Kobra III	69	21	10	—
Trombay	69	9	22	—
Paras	67	29	2	2
Korabi	59	19	12	10
A.E. Co	59	24	16	1
Sabarmati	59	22	13	6
Amarkantak Exten.	57	16	15	12
Gandhi Nagar	57	17	13	13
Korba I	52	24	13	11
Khaperkheba	52	36	4	8
Korba II	51	22	14	13
Nasik	47	21	16	16
Satpura	47	17	3	33
Bhusawal	45	55	—	—
Parli	45	8	20	27
Ukai	36	18	16	30

APPENDIX 4—*contd.*

	1	2	3	4
Southern Region				
Ramagundem B	69	23	4	4
Neyveli	60	14	18	8
Kothagudem A	49	83	14	4
Basin Bridge	47	34	3	16
Vijayawada	41	17	12	30
Ennore	36	21	28	15
Tuticorin	30	20	6	44
Kothagudem C	28	19	9	44
Kothagudem B	21	21	32	26
Nellore	9	2	89	—
Eastern Region				
Mulajore	66	29	3	2
Bandel	60	27	10	3
New Cossipore	58	24	15	3
Bokaro	43	38	19	—
Chandrapura	40	26	10	24
Santaldih	35	26	19	20
Talcher	34	24	28	14
Patratu	33	19	16	32
D.P.L.	29	26	32	13
Barauni	28	15	23	34
Durgapur	25	13	3	59
N.E. Region				
Chandrapur	45	20	3	32
Namrup	20	30	23	27

APPENDIX 5

STATEMENT REFERRED TO IN REPLY TO PART (C) OF UNSTARRED QUESTION No. 5882 ANSWERED IN THE LOK SABHA ON 30-3-1982

Statement indicating thermal projects under construction in the country showing original and revised estimated cost and original and anticipated date of commissioning.

Sl. No.	Name of the project	Capacity (MW)	Estimated cost (Rs. in lakhs)		Original commissioning programme		Anticipated date of commissioning
			Original	Revised			
1	2	3	4	5	6		7
NORTHERN REGION							
I.	*Haryana*						
	1. Panipat St. II (Units 3 and 4)	2×110	7293	11850	Unit-3	9/82	4/84
					Unit-4	3/83	12/84
	2. Panipat St. III (Unit 5)	1×210	11110	Not yet revised	Unit-5	12/84	12/85
II.	*Punjab*						
	1. Ropar (Units 1 and 2)	2×210	16800	32738	Unit-1	5/84	9/84
					Unit-2	11/84	6/85
III.	*Rajasthan*						
	1. Kota (Units 1 and 2)	2×110	8160	14304	Unit-1	3/81	6/82
					Unit-2	9/81	12/82

2. Kota St. II (Units 3 and 4)	2×210	13378	23000	Unit-3	6/85	1/86
				Unit-4	12/85	7/86
IV. *Uttar Pradesh*						
1. Obra St. III (Unit 13)	2×200	15790 (for Units 12, 13)	37600 (for Units 12, 13)	Unit-13	6/78	3/82
2. Parichha (Units 1 and 2)	2×110	8372	15000	Unit-1	3/82	11/82
				Unit-2	9/82	6/83
3. Anpara (Units 1, 2 and 3)	3×210	22710	47120	Unit-1	4/83	12/83
				Unit-2	10/83	6/84
				Unit-3	4/84	12/84
4. Tanda (Units 1, 2, 3 and 4)	4×11	15925	27091	Unit-1	9/84	7/84
				Unit-2	4/84	12/84
				Unit-3	10/84	5/85
				Unit-4	3/85	9/85
5. Unchahar (Units 1 and 2)	2×210	19305	21958	Unit-1	12/85	12/85
				Unit-2	6/86	6/86
6. Anpara 'B'	2×500	41610	57348	Unit-1	1986-87	1986-87
				Unit-2	1987-88	1987-88
V. *N.T.P.C.*						
1. Singrauli STPS [St. I (Units 1, 2 and 3)]	3×210	25566 (for units 1, 2 and 3)	31853 (for units 1, 2 and 3)	Unit-2	6/82	12/82
				Unit-3	12/82	6/83

APPENDIX 5—*contd.*

1	2	3	4	5		6	7
	2. Singrauli STPS [St. II (Units 4, 5, 6 and 7)]	2×210+ 2×500	49437	73217	Unit-4	6/83	12/83
					Unit-5	12/83	6/84
					Unit-6	12/84	1986-87
					Unit-7	12/85	1987-88
WESTERN REGION							
I. *Gujarat*							
	1. Wankabori (Units 1, 2 and 3)	3×210	17214 (for units 1, 2 and 3)	24200 (for units 2 and 3)	Unit-2	3/81	1/82
					Unit-3	9/81	3/83
	2. Wankabori Extn. (Units 4, 5 and 6)	3×210	20679	35060	Unit-4	12/82	3/85
					Unit-5	6/83	3/85
					Unit-6	12/83	3/86
	3. Ukai Units 5	1×210	6683	8430	Unit-5	3/82	1/84
	4. Kutch Lignite (Units 1 and 2)	2×60	7127	7127	Unit-1	10/84	12/85
					Unit-2	4/85	6/86
	5. Sikka (Replacement)	1×120	5478	5478	Unit-2	12/84	6/85
II. *Madhya Pradesh*							
	1. Satpura (Units 8 and 9)	2×210	12968	17524	Unit-8	6/81	9/82
					Unit-9	12/81	3/83
	2. Korba West (Units 1 and 2)	2×210	15099	23557	Unit-1	3/81	11/82
					Unit-2	9/81	5/83
	3. Korba West Extn. (Units 3 and 4)	2×210	12941	18160	Unit-3	9/83	4/84
					Unit-4	3/84	10/84

4. Birsinghpur (Units 1 and 2)	2×210	20032	28658	Unit-1	12/85	12/85
				Unit-2	6/86	6/86
III. *Maharashtra*						
1. Koradi St. III (Units 5, 6 and 7)	1×200+ 2×210	11240 (for units 5, 6 and 7)	24276 (for units 5, 6 and 7)	Unit-6	4/81	3/82
				Unit-7	11/81	9/82
2. Bhusawal Extn. (Unit 3)	1×210	6437	9602		7/81	5/82
3. Chandrapur St. I (Units 1 and 2)	2×210	14003	24696	Unit-1	12/81	11/82
				Unit-2	2/82	3/83
4. Chandrapur St. II (Units 3 and 4)	2×210	12856	19000	Unit-3	8/83	12/84
				Unit-4	2/84	6/85
5. Gas Turbine Generating sets at Uran near Bombay (Units 1, 2, 3 and 4)	4×60	5280 (for units 1, 2, 3, 4)	7250 (for units 1, 2, 3, 4)	Unit-2	1/81	3/82
				Unit-3	3/81	6/82
				Unit-4	5/81	6/82
6. Trombay	1×500	18000	22000		12/82	6/83
IV. *N.T.P.C.*						
1. Korba STPS (Units 1, 2, 3 and 4)	3×210+ 1×500	45080	67348	Unit-1	9/82	12/82
				Unit-2	3/83	2/83
				Unit-3	9/83	1/84
				Unit-4	9/84	1986-87

APPENDIX 5—*contd.*

1	2	3	4	5	6	7
2. Korba STPS St. II (Units 5 and 6)	2×500	42405	61312	Unit-5	1987-88	1987-88
				Unit-6	1988-89	1988-89
I. *Karnataka*						
1. Raichur (Units 1 and 2)	2×210	15925	23948	Unit-1	1/81	6/84
				Unit-2	4/83	12/84
II. *Tamil Nadu*						
1. Tuticorin Extn. (Unit 3)	1×210	7121	8875		12/80	3/82
2. Mettur (Units 1 and 2)	2×210	18970	21670	Unit-1	10/85	1985-86
				Unit-2	4/86	1986-87
3. Neyveli Second Mine cut (Units 1, 2 and 3)	3×210	21398	47958	Unit-1	10/83	9/84
				Unit-2	4/84	3/85
				Unit-3	10/84	9/85
III. *N.T.P.C.*						
1. Ramagundam STPS (Units 1, 2, 3 and 4)	3×210+ 1×500	45914	79471	Unit-1	12/82	2/84
				Unit-2	6/83	3/84
				Unit-3	12/83	2/85
				Unit-4	12/84	1987-88
EASTERN REGION						
I. *Bihar*						
1. Barauni (Unit 6)	1×110	2000+	9820		12/77	12/82
2. Barauni (Unit 7)	1×110	3764			9/81	6/83

3. Patratu (Units 9 and 10)	2×110	4200	11500	Unit-9	9/77	3/83
				Unit-10	3/78	12/83
4. Muzaffarpur (Units 1 and 2)	2×110	8435	14602	Unit-1	7/82	1/84
				Unit-2	1/83	7/84
5. Tnughat (Units 1 and 2)	2×210	17513	28133	Unit-1	6/85	6/87
				Unit-2	6/86	6/88
II. *D.V.C.*						
1. Bokaro 'B'	1×210	6976	15559		11/82	8/84
2. Bokaro 'B' Extn. (Units 1 and 2)	2×210	18693	26576	Unit-1	1986-87	1986-87
				Unit-2	1986-87	1986-87
III. *Orissa*						
1. Talcher St. III (Units 5 and 6)	2×110	3840	8874	Unit-5	9/78	Commis-sioned in 3/82
				Unit-6	6/79	12/82
IV. *West Bengal*						
1. Bandel Extn. (Unit 5)	1×210	3331	8343	Unit-5	3/78	9/82
2. Kelaghat (Units 1, 2 and 3)	3×210	11559	28205	Unit-1	9/78	6/83
				Unit-2	3/79	3/84
				Unit-3	9/79	12/84
3. Durgapur (DPL) (Unit 5)	1×110	4842	7024	Unit-5	6/79	9/83

APPENDIX 5—*contd.*

1	2	3	4	5		6	7
V.	*N.T.P.C.*						
	1. Farakka STPS (Units 1, 2 and 3)	3×21	26476	39730	Unit-1	9/84	2/85
					Unit-2	3/85	8/85
					Unit-3	9/85	2/86
NORTH-EASTERN REGION							
I.	*Assam*						
	1. Bangaigaon Extn. (Units 3 and 4)	2×60	4717	5383	Unit-3	7/82	Commis-sioned
					Unit-4	1/83	9/84
	2. Lakwa Gas Turbines (Units 1, 2 and 3)	3×15	1564 (for units 1, 2, 3)	1811 (for units 1, 2, 3)	Unit-3	1/80	3/82
	3. Waste Heat Utilisation Plant at Kamrup Gas Turbine Power Station	1×22	902	1489		6/80	1/83
	4. Chandrapur Extn.	1×30	1292	2210		3/83	7/84
	5. Mobile Gas Turbines	4×3	734	764		1981-82	3/82
II.	*N.E C.*						
	1. Garo Hills	2×30	4320	Not revised		12/87	1989-90
					Unit-2	6/88	1989-90

APPENDIX 5—*contd.*

Statement indicating on-going hydro-electric projects in the country showing original and revised estimated cost and original and anticipated date of commissioning

Sl. No.	Project	Installed capacity (No. and rating in MW)	Commissioning date of first unit/last unit on the basis of sanctioned project report	Date of commissioning as now envisaged of first unit/last unit	Original estimate in Rs. crores	Latest estimate in Rs. crores
1	2	3	4	5	6	7
	PROJECTS SCHEDULED IN SIXTH FIVE YEAR PLAN AND BEYOND					
	NORTHERN REGION					
	Haryana					
1.	Western Yamuna	6×8	1984-85	1984-85	45.72	45.72
	Himachal Pradesh					
1.	Binwa	2×3	1980-81	1983-84	4.32	9.84
2.	Andhra	3×5	1980-81	1984-85	9.74	21.43
3.	Rongtong	4×0.5	1980-81	1984-85	2.81	6.79
4.	Bhaba (Sanjay)	3×40	1985-86	1985-86	55.80	75.00
	Punjab					
1.	Shanan Extn.	1×50	1978-79	1981-82	4.72	14.13
2.	Mukerian	(6×15)+(6×19.5)	—	1982-83	115.58	115.58
3.	Anandpur Sahib	4×33.5	—	1982-83	80.73	96.12

APPENDIX 5—*contd.*

1	2	3	4	5	6	7
Rajasthan						
1. Nahi Bajaj Sagar		(1×25)+(1×45) +(1×25)+(1×45)	1982-83 1984-85	1984-85 1985-86	40.67 18.70	73.14
2. Anoopgarh		6×1.5	—	1984-85	6.51	6.51
Uttar Pradesh						
1. Yamuna St. II (Khodri)		4×30	—	1982-83 1983-84	17.96	65.16
2. Manari Bhali St. I		3×30	—	1983-84	17.78	68.20
3. Lakhwar Vyas I		(3×100)+(2×60)	1988-89	1991-92	140.97	174.80
4. Manari Bhali St. II		3×52	1989-90	1987-88	82.63	182.00
5. Tehri		4×250	1980-81 1981-82	1991-92	197.92	827.30
6. Vishnu Prayag		4×65.5	1984-85 1985-86	1993-94	104.51	104.51
Common						
1. Pong Extn.		2×60	1981-82 1982-83	1982-83 1983-84	34.04	42.67
2. Dehar Extn.		2×165	1981-82 1982-83	1983-84	59.1	78.30
Central						
1. Salal		3×115	1974-75 1975-76	1987-88	222.13	340.00

WESTERN REGION					
Gujarat					
1. Ukai I.B.C.	2×2.5	1979-80 1980-81	1984-85	3.05	3.05
2. Kadana	2×60	1978-79	1984-85	24.58 (for 4 units)	65.60 (for 2 units)
Maharashtra					
1. Paithan	1×12	1981-82	1984-85	5.96	9.00
2. Bhira Tailrace	2×40	1982-83	1985-86	22.50	41.50
3. Tillari	1×60	1977-78	1984-85	8.16	38.42
4. Bandhardhara	1×10+1×33.5	1983-84	1985-86 1986-87	17.59	17.50
5. Pawana	1×10	—	1985-86	3.95	6.50
Madhya Pradesh					
1. Bodhghat	4×125	1984-85 1985-86	1987-88 1988-89	209.30	350.00
Madhya Pradesh/Maharashtra					
1. Ponch (Joint Hydro-electric Project of Madhya Pradesh and Maharashtra)	2×80	1978-79 1979-80	1984-85	28.28	84.98
SOUTHERN REGION					
Andhra Pradesh					
1. Nagarjunasagar PSS St. I	4×100	1981-82 1982-83	Units 1, 2 commissioned 1982-83	66.09 (for full project)	71.72 (for full project)

APPENDIX 5—*contd.*

1	2	3	4	5	6	7
2.	Srisailam St. I	4×110	1981-82	1982-83 1983-84	38.47 (1962 estimates)	182.25
3.	Donkerayi	1×25	1981-82	1983-84	7.92	13.00
4.	Belimela Dam P.M.	2×30	1981-82	1985-86	17.77	26.50
5.	Nagarjunasagar R.S.C.	2×30	1982-83	1982-83 1983-84	18.19	30.40
6.	Upper Sileru St. II	2×60	—	1985-86	11.98	29.00
7.	Nagarjunasagar PSS St. II	3×100	1984-85 1985-86	1984-85	55.78	72.15
8.	Srisailam St. II	3×110	1983-84 1984-85	1984-85 1985-86	39.38	49.80
Karnataka						
1.	Kalinadi St. I	(6×135)+(2×50)	1976-77 1981-82	Units 1, 2, 3, 4 commissioned 1984-85	125.66 (for full project)	39.00 (for full project)
2.	Varahi	(2×115)+(2×4.5)	1984-85 1985-86	1985-86	136.40	136.40
3.	Gangavali	2×105	1985-86 1986-87	Not indicated	135.83	135.83
4.	Kalinadi St. II	(4×32)+(4×25)+(2×25)	1985-86 1987-88	1990-91 1992-93	205.83	205.83
Kerala						
1.	Idamalayar	2×37.5	1978-79 1979-80	1983-84	23.40	50.00

2. Kakkad	2×25	1981-82	1985-86	18.60	20.50
3. Idukki St. II	3×130	1984-85 1985-86	1985-86	31.68 (1973 Est.)	31.68
Tamil Nadu					
1. Kodamparai PSS	4×100	1978-79	1983-84 1984-85	35.12	113.73
2. Sarvalar	1×20	1978-79	1983-84	8.35	21.57
3. Lower Mettur	8×15	1981-82	1984-85 1986-87	83.60	89.72
4. Kundah Ph. V	1×20	—	1985-86	5.03	5.03
EASTERN REGION					
Orissa					
1. Rengali	2×50	1982-83	1984-85	57.92	88.54
2. Upper Kolab	3×80	1980-81	1984-85 1985-86	53.98 (1974 Est.)	132.26
3. Upper Indravati	4×150	1985-86 1986-87	1986-87 1987-88	208.14	305.90
West Bengal					
1. Jaldhaka	2×4	1978-79	1982-83	3.16 (1973 Est.)	11.64
2. Ramman St. II	2×12.5	1982-83	1985-86	24.99	25.88
D.V.C.					
1. Panchat Hill	1×40	1982-83	1985-86	16.03	16.03
Central					
1. Koel Karo	(6×115)+(1×20)	1982-83 1984-85	Not indicated	157.04	391.40

APPENDIX 5—*contd.*

1	2	3	4	5	6	7
NORTH-EASTERN REGION						
Assam						
1.	Lower Borpani	2×50	1984-85	1985-86	36.36 (1979 Est.)	61.00
Nagaland						
1.	Dikhu	(1×0.75)+(1×0.25)	—	1983-84	0.65	1.55
Tripura						
1.	Gumti	1×5	1982-83	1982-83	1.91 (1978 Est.)	4.30
Meghalaya						
1.	Umiam-Umtru with Upper Khri Diversion	2×30	1986-87	1986-87	38.88	Not yet revised
Neepco						
1.	Kopili	(2×25)+(2×50)	1982-83 1983-84	1983-84 1984-85	57.77 (1974 Est.)	118.66
Central						
1.	Loktak	3×35	1975-76 1976-77	1982-83	77.45	93.30

APPENDIX 6

INDICATORS FOR BACKWARDNESS

Taluk : Bhalki *Dist* : Bidar *Area* : Backward

Sl. No.	Particulars	Bhalki taluk	Bidar dist.	Karnataka State	All India
1	2	3	4	5	6
1.	Ratio of net area irrigated to net area sown	1.08%	3.52%	11.22%	20.05%
2.	Intensity of cropping	128.58%	122.19%	105.84%	114.30%
3.	Density of population/sq. km.	142	151	153	182
4.	Literacy as per 1961 Census (%)	18.33%	14.49%	31.50%	29.35%
5.	Road miles/100 sq. kms	21.22	19.44	23.89	—
6.	Extent of village electrification	50.0%	50.4%	31.0%	18.0%
7.	Electrified agricultural pumpsets/100 sq. km	51	136	68	—
8.	Average population served by bank branch	51896	—	21407	—
9.	Ratio of rural population to total population (1971 Census)	91.58%	85.53%	75.69%	82.04%
10.	Percentage of workers engaged in agriculture to total population	NA	67.92%	65.46%	68.63%
11.	Ratio of cultivable area to geographical area	94.24%	89.75%	75.48%	59.63%
12.	Ratio of cultivated area to cultivable area	77.65%	87.38%	82.91%	81.92%
13.	Famine and droughts	1971-72 1972-73 1973-74			

APPENDIX 6—(*contd.*)

1	2	3	4	5	6
14.	Ratio of gross area irrigated to grossed cropped area	0.9%	2.77%	12.09%	20.91%
15.	Ratio of area under commercial crops to total cropped area	17.81%	19.70%	23.72%	25.22%
16.	Per capita consumption of electricity in units per annum	7.2	—	100	57.47
17.	Scheduled Tribes	208	699	192096	—
18.	Scheduled Castes	25246	123963	317232	—
19.	Average yield in Kg/Hectare under main crops				
	K. Jawar	494 kg.	784		
	R. Jawar	568 kg.	535	537	522
	Groundnut	494 kg.	911	—	—
	Tur	390 kg.	584	—	—
	Urad	494	—	—	—
	Mung	272	—	—	—
	Sugarcane	—	54000	—	—
	Wheat	—	338	—	—
	Gram	—	468	—	—
	Rice	—	686	1960	1040

DISCUSSION

SHAH

Thank you very much, Dr. Sengupta. I am sure all of us would agree that the enormous amount of work he has done in collecting material for his paper has given us today a good view of what is being done, where the weaknesses are, what ought to be done to improve the situation that exist.

The impression I have so far is that the subject is a newly evolving one. The kinds of points that were raised yesterday, about methodologies, units to be used, standardisation, codification, etc., indicate the fact that the field of energy data systems is a relatively new science.

There are some fields, even in India, that have had a national data system approach. I can think of at least one, the demographic system, which has led to our Election Commission machinery that is remarkably successful in conducting elections all over the country in a very short time. The data that they have produced as a consequence, has been really very useful. So, there are instances in which we have data system at the national level, which are quite well developed. But as far as energy in concerned, I do not think we can take the same view again. The Central Electricity Authority, in some areas, have done a fair amount of good work. But I think that the systems of collection of data have yet to be properly organized. But having said this, we should look upon our efforts during the workshop perhaps as an exchange of views based upon experience in more than one country.

In any data collection arrangements, there is likely to be some conflict between the data required by an organization for its own management and control purposes and the data needed by external bodies, whether industry-wise or national but outside the organization. Data are needed for very different purposes. For instance, in Dr. Sengupta's paper we find a reference to the number of breakers installed by utility being reported to the Central Electricity Authority. To my mind this is not the kind of data which the Central Electricity Authority, or any organization outside that particular body, would need. There are many instances of this nature. This is one reason why a whole lot of unnecessary data collects at the national or regional level, and at the same time meaningful data escapes collection. We must keep these two factors in mind and make a proper effort to determine what kind of data we would need and for what purpose outside the organization, and leave whatever is needed for that organization to that organization, and not mix up the two.

We had references yesterday to what use the collection of data can be put to. For instance, to aid in policy making, in investment analysis, in various other features, all these objectives are certainly very valid. One objective which I think needs emphasis especially in the power sector, is one of comparison so that it becomes possible to compare performances. The Central Electricity Authority of India is already making this effort. But because of limited use of the computer specially at the data capturing stage, one finds that the results of whatever comparison one is trying to make come quite a bit late; and secondly in some instances at least one finds that the units used are not identical. Now utilities being organizations with very little commercial competition between themselves have very little incentive to conceal data. The second feature is that they have more than the usual degree of commonality both in process as well as their organizational structure. This lends itself to familiar inter-firm comparison. I should hope that one of the objectives of our national data system would be, as far as the power sector is concerned, should be something that will help the state utilities as well as other corporate bodies to undertake comparisons which ultimately will help their performance.

I would like to emphasise 3 or 4 points on Dr. Sengupta's paper. From the experience one finds that one of the major weaknesses in our data collection, or for that matter even control, is inadequate and unreliable communication links. Some day we will have to put in a lot of more effort. I happened to be in one of the State Load Despatching Centres, when there was a massive black-out in the region. For four hours, when I was sitting there, one could see how primitive our method was in trying to rectify the situation, largely as a consequence of communication weakness.

Another point on which I think stress needs to be laid is load forecasting. Dr. Sengupta has referred to the efforts being made to improve things. The Regional Load Survey Group are collecting an enormous amount of data. My own experience unfortunately is that this is one of the weakest areas in spite of the efforts being made to collect a lot of information. I think it is also relevant which method is used for load forecasting. I totally disagree with the end-use method, which we are using, whether at the micro level or unit-to-unit level, because no industry or no consumer is really thinking in terms of more than 1 or 2 or 3 years at any time. Even large manufacturers donot think beyond 3 years, because there are factors like market recession, boom periods, export-import situation, government policy, availability of funds, etc. There are so many factors which really prevent any firm decision on long-term needs of power being properly identified. Apart from that, just as he mentioned about the Electricity Boards exaggerating their need, every manufacturer does exactly the same thing. So what you get ultimately from the enquiry is something which is totally unreliable. Dr. Pachauri and a number of other consultants once did a fairly detailed study for one of the Regional Boards to see what precisely could be done. They found that we were not going to need quite as much copper as they first thought. There was a reduction in electrical requirement for the copper mining industry. It happened to be a certain percentage of the original estimate. But then all states were asked to reduce their requirements by that percentage, quite regardless of the fact that many of them did not have copper mining. Besides there is the question of aggregation involved when it comes to the load forecasting by the Planning Commission.

The other point is about data needed for thermal power station performance. I think in our country and probably elsewhere, brown outs have started occurring. The data system which would help to analyse the causes of these things, thereafter maximise output is something which needs to be strengthened. About managerial aspects, the Rajadhyaksha Committee has done a broad survey of our power industry, and made one point about over staffing. Now it is possible to have indexes about how many people you ought to have for 'x' amount of work and so on. This information people are compiling on an *ad hoc* basis. It is not available on a reliable basis.

MUBAYI

Two or three brief points. One is the inadequacy of the methods for load forecasting that has been mentioned here both by Mr. Shah and Dr. Sengupta. This end-use method is perhaps valid at the most for one or two years, because nobody knows that beyond what he is really going to do. Typically for planning of the electricity sector, one needs at least 10 or 15 years type of load forecast simply because it takes 5 to 7 years and in the case of hydro stations even more, to construct the generating facilities to meet that load. So, the need for long range planning. I think in all fairness to the Electricity Authority of India it should be recognized that in other countries around the world, I know particularly the situation in the US, that load forecasting there by a variety of econometric and other methods has also suffered from similar types of problems. For example, in the middle 70s, a number of national studies were undertaken by various expert bodies, and almost all of their forecasts were greatly in excess of what was the actual load. As a result there is now a staggering amount of over capacity in the US system right now. I think a judicious mix of econometric methods with a suitably modified input output method, as has been adopted by the Planning Commission—but the way they are doing it perhaps needs to be improved—could be taken up as load forecasting technique.

The second point is that I notice that Dr. Sengupta mentioned generation planning as one of the important functions in the medium and long-term but not transmission planning. Just T and D loss were mentioned, but no transmission planning, which I think is very important especially for India where we are going to have a substantial upgrading from 220 kV to 400 kV and perhaps by 1995 to 1000 kV. EHV-DC, is just being initially talked about. For investment analysis this is perhaps almost as important or may be only slightly less important than generation planning.

SENGUPTA

In my paper I was not particularly critical about the Indian methods. I said that in three places they were undertaken and the gaps are narrowing down. This is not typically an Indian problem. The Central Electricity Generating Board (CEGB), in the UK found itself in an embarrassing position of having a very large surplus installed capacity in 1970 because the industries did not come up as expected.

Possibly it was an oversight to leave out transmission planning in the presentation. It is there in the paper. This is precisely the role of Regional Electricity Boards. They almost act as mediators, trying to coax Tamil Nadu to part with some more power or energy; asking Kerala, 'well, we have a deficit in Karnataka State, will you provide some power?' The tie line exchanges are very carefully monitored every half hour. This is a very important use of medium term data.

RAMESH

Data systems have to be organised differently for different levels and types of requirements. Dr Sengupta has ably dealt with these varying requirements and I will only highlight some aspects which I consider important.

In some cases, the type of tariff levied by the state utilities for certain cateogories of consumers makes it difficult to organise an efficient data system. For example, many state Electricity Boards in India supply electricity to private tube-wells and pumpsets on a horsepower basis, that is, a fixed rate per hp, per year is levied as the tariff. There is no metering of energy supplied, and the total electricity bill remains the same irrespective of the quantum of energy consumed. This really implies that, at the margin, the consumer gets electricity *free*, that is, the additional energy consumed is supplied at a zero price.

This is economically a very drastic proposition. But owing mainly to administrative problems, problems of collection and political pressures from the farm lobby which obviously benefits from such a tariff, this sort of unmetered supply is increasing. Recently, Andhra Pradesh too has introduced this, perhaps because of the coming elections. This, I am afraid, is one of the bad points of our democracy. *Better* metering arrangements were mentioned in Dr Sengupta's paper. But actually here some metering arrangements are needed, because there are none wherever a flat rate of tariff is in existence.

Another type of data is the station level data which is required mainly for management purposes. In the case of thermal stations, this has details such as coal consumption, coal quality, oil consumption, auxiliaries consumption, and so on. This is of great importance in monitoring thermal station performance. For instance, auxiliary consumption very often is 10 per cent—in some cases, 12 or even 13 per cent. If there are meters even in these areas, in some cases it has been found, that you can reduce it from 12 to 10 or even 9 per cent. The improvement in the country as a whole can easily be imagined.

Financial and commercial data were also briefly mentioned. These are specially important for monitoring of performance of state utilities. It has been found that electricity sales data are available only by broad sectors. You will be surprised to hear that in India we do not have detailed data for the rural sector for domestic consumption. We have data for pumpsets and tubewells consumption, because all tubewells and pumpsets are in rural areas. There is no problem there. But where domestic consumption is concerned, we have estimated that about five per cent of the total domestic consumption is in rural areas. But we do not know exactly how much this is, because there is no separate classification in the State Electricity Boards. Instances like this can be multiplied. But what has to be examined is whether commercial data must cover all these categories required for policy-making both at the state and the national level. We have to reassess the whole situation and determine what type of data is required and at what level and then go into the details of the reporting system.

Before concluding, I would like to make a few general observations. The level and extent of data collection and reporting are most important for management purposes. Here there has been some improvement, but sometimes there is too much data being reported at some levels. As a result, efficient management is impeded, particularly at higher levels of the C.E.A. Perhaps, many of the details which are now being asked for may not really be required. But this requires reassessment at each level of what data are required. Many of us actually involved in the power sector may think that cartain types of data are required but if somebody from outside questions these assumptions, perhaps we may change our point of view. This point has been rightly highlighted here.

About computerization, I want to say briefly that perhaps one of the recommendations of this workshop could be a concrete suggestion about the extent of computerisation: where micro computers could be used and some broad assessment of the costs. I do agree that this will be a small portion of what we are spending in the entire power sector.

MUNASINGHE

I want to join the chorus of congratulations to Dr. Sengupta for the really very illuminating and interesting speech he gave covering a large amount of ground in a short time. I cannot but agree with practically most of the things he said. What I would like to do is to supplement some of the areas which quite naturally could not be covered because of the time allotted to this vast subject. Each of the speakers before me emphasized that there is a vast amount of data associated with electric power systems that can be collected. It can be cumbersome. At least one of the effective criteria should be how data can be used to improve decision making or its utility to improve the overall efficiency of production and use of electricity. I think if you focus the data gathering and analysis on a particular task associated with decision making, then that is a great help in identifying the items that have priority over other things that perhaps can be ignored or relegated to a later date.

There are two broad areas in decision making that are associated with the electric power sector. One is, of course, supply management, which deals with basically production of electricity and there is an equally important area of demand management which aims at influencing the patterns of electricity used among consumers. I think Dr. Sengupta has covered very comprehensively the technical areas of data, particularly relating to the supply management side. Traditionally, I must say that in the power sector there has been an emphasis on the technical data and this is quite easy to understand, because the demand side data, the non-technical, the socio-economic data, are very difficult to find, and tend to be much less reliable. This is the area I want to concentrate on today, put on my economist's hat so to speak, and also to stress that because of the interaction between supply and demand, the demand side data, the economic data, the non-technical data, are as important in decision making as the supply side data. Therefore, one should perhaps put a little more emphasis, a little more effort, into improving them.

Figure 2.7 shows the sequence of decision making in the power sector that the data system has to support. We have to start with the whole area of load forecasting. Lot has been said about the difficulties in this. There are a number of non-technical data items that go into this process. For example, we have to have ideas of price, both electricity price and price of competing fuels; we have to have data on income levels and economic activity; we have to have some information on population and other demographic data. Built into this is also a host of behavioural assumptions such as price response, or price elasticity, to use the technical term, income elasticity and other parameters which can critically affect the forecast. Then the next stage in decision making is basically the area of designing of the least cost system.

We have the area of system planning and operation and this area has been covered quite extensively by Dr. Sengupta. But let us be quite clear, that in addition to the technical data,

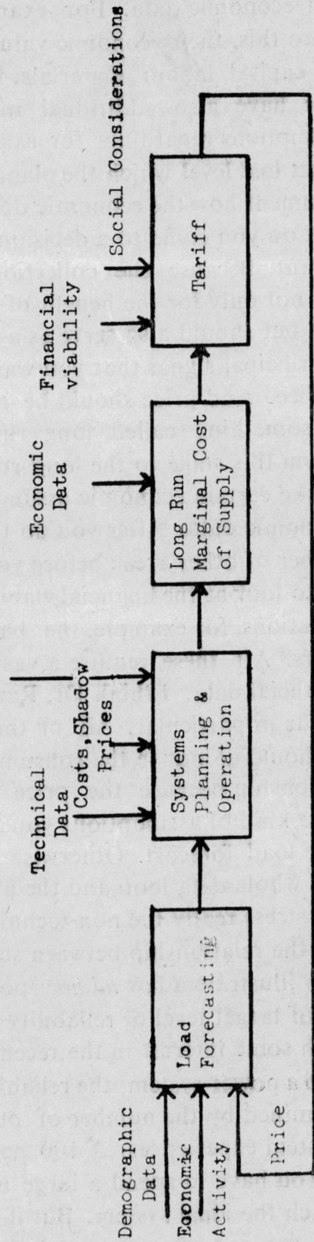

FIG. 2.7 : *Sequence of decision making in the electric power sector*

you need a lot of economic data. For example, the cost of inputs that go into this, their economic values, sometimes the shadow prices of capital, labour, materials, fuels, that go into this process. You have also additional information that is required for assumptions regarding, for example, target reliability level, target loss level which the planners aim for and I will show in a moment how the economic dimension affects the decision. Moving on you come to a decision on pricing, which is equally important. Because the collection of data in the power sector is not only for the benefit of those who operate the power system but should also serve as a signal to users of electricity. The principal signal that you want to feed to users of electricity is price. And price should be related, in economic theory, to something called long run marginal cost of supply. To go from this stage to the long run marginal cost, you need to make certain economic assumptions, you need a further set of economic data. After you do this, you still have to look at a number of other areas before you come to a final tariff. You have to look at the financial viability of your utility, at social considerations, for example, the basic needs of low income consumers. All these require a vast amount of socio-economic and financial data. I think Dr. Ramesh has mentioned the financial side in particular. All of these gives you the price signal that should go out to the consumer. Now you can see the inter-relationship because the price signal should be consistent with the kind of assumptions you made at the beginning regarding the load forecast. Otherwise there is no self-consistency in the whole data loop and the analysis.

I am trying to stress really the non-technical items of data that are required, the relationship between supply and demand and so on. Just to illustrate a few *ad hoc* points, let us take the specific item of target level of reliability—a theme which I have explored with some interest in the recent past.

When you plan a power-system, the reliability or the quality of supply, is determined by the number of outages and other things that the system experiences. A 100 per cent reliability, is unachievable. You have to spend a large amount of money and you never reach the ideal system. But if you have a low level of reliability, you can spend relatively less on this system. Figure 2.8 shows that the cost of supplying a given amount of electricity rises as you increase the reliability. What target should we aim for? Traditionally, this has been established on a rule of thumb basis, so many transmission and distribution trips, outages per year, and so on. This is based mainly on technical information. You need another curve which tells you the consequence of low reliability, which is the cost of power failures to society, the interruption in industrial production, inconvenience to households, and so on. This curve is labelled outage cost in Figure 2.8 and shows the cost of interruptions to society. Now the optimum level of reliability which you should aim for is the minimum of the sum of these two curves, because you incur increasing system cost in order to buy lower cost of outages. Unless you have the economic and behavioural data associated with outages, you cannot make this decision purely on a technical basis. This type of decision also requires substantial amount of additional work in the areas of social and economic data, particularly dealing with human behaviour, a difficult area, but nevertheless we should not give it up because it is required for the decision.

Now the same thing goes for losses. One can draw entirely the same set of parametric curves as in Figure 2.8. Instead of reliability here, you have losses. As you know, when you decrease the level of losses, you have to invest more in the system, more into operation, maintenance, and so on. Figure 2.9 shows the technical system's cost associated with decrease in the level of loss in the system. Also shown is a curve which tells you the value of the loss that you save. Every kW and kWh that you save in the system has a certain value, based on a long run marginal cost of supply. You need additional economic data to estimate the value of your loss. Two of these curves will give you the total cost, which will help you again to identify the optimum level of loss. So if you simply monitor the loss by itself, and say : today the loss is 8.23 per cent of generation, it does not help you very much. The question you should ask is well, it is 8.23 per cent, but what should I aim for ? Somebody will say, well, 5 per cent seems to be a reasonable figure. But it is much better if you do this kind of analysis and have a rational economic base for doing that. So my plea is, in addition to all the excellent suggestions made on the technical side, we also spend some effort on the non-technical side to

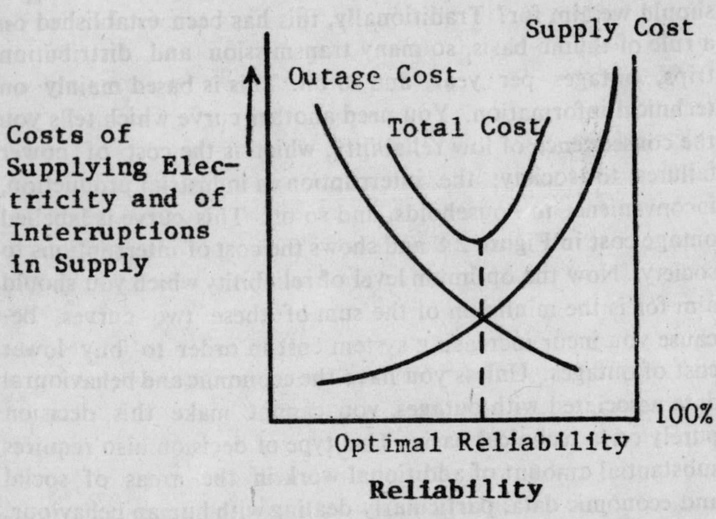

FIG. 2.8 : *The optimum level of reliability of an electric power system*

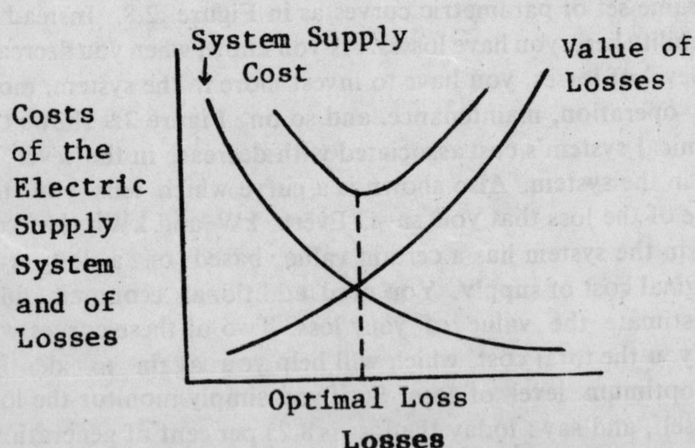

FIG. 2.9: *The optimum level of losses in an electric power system*

develop the kind of information system that will improve the decision making at every stage in the power sector.

YINGZHONG

I would like to speak, as a sort of introduction, about the electricity situation in China with emphasis on the nuclear electricity situation.

China is a very big country. I was interested to note in Dr. Sengupta's report a map of India which shows various regional grids in this country. We have nine individual regional grids in China including the North-East, North China, North-West China, the isolated Sinkiang, South China, Middle China and South-West China and Taiwan. Installed capacity in China is now 60,000 MW and electricity generated per year is about 282×10^9 kWh. By the year 2000, it is planned to increase these figures three or four times. So it is a big job for the electricity industry to expand that much.

In China we have plenty of coal resources, about 600 billion tonnes measured, and also plenty of hydro resources, which are exploitable. So generally speaking, we can expect that we can rely on coal resources and hydro resources to meet the demands of this big nation. But unfortunately our geographic situation is very uneven. Sixty per cent of the coal resources are in the North China region and 70 per cent of such hydro resources are in the South-West region. We anticipate that by the year 2000 some regions will be met with serious energy shortage, actually the energy shortage is already there. We plan to develop nuclear power stations and now our National Commission of Planning is making some plans for power stations. The most recent report from our Planning Commission is that we are now going to have 300 MW in Shanghai and also 900 MW in Hongkong region. So 900 MW in East China near Shanghai and another 900 in the North-East China Provinces, this is the first plan of our nuclear projects up to the year 2000. It means that in these regions about 15 per cent of the electricity will be supplied by nuclear sources. So my opinion is that for a vast country like China, the geographic situation is dominant. We are now talking about connecting regional grids into a national grid. Of course, it will be a huge task for R & D work for such high voltage transmission. First in the South-West and the North-West regions hydraulic power is very abundant. So if we can connect these grids together we can supply the chief hydro electricity for the consuming centres in the eastern part of China. But unfortunately this distance is 1000 to 3000 km. Coal sources and hydro sources will not meet the three- or four- fold increase in demand of all the regions. So we are going to develop nuclear power stations to supply most of the demand. Of course, our Taiwan Province has got six nuclear power stations. This is an overall view of our electricity situation.

SHAH

Well, I think we can now start off on participation by members generally. May I invite comments or questions? Could you kindly indicate whom you would like to have response from?

MENA

Would it be more desirable to require those in charge of collection and analysis of these data to also transform these data and make them available in an appropriate form for the specific purpose?

SENGUPTA

If I have understood your question correctly, my comments are as follows. Referring to Fig. 2.3 for the utilities which have been indicated rather at the bottom, the present procedure is as follows. The Regional Power Survey, which I have indicated in the block, have a set of questionnaires which I could not present in the paper here because of lack of space. At present, they send this set of questionnaires to various utilities, industries, and so on. From the responses, the Regional Power Survey particularly look for what are the projected demand and the 'peaking'—when the maximum demand is expected to appear. These arc the two points which they specifically concentrate on.

There is another point which I have also made in the paper. It is just not an indiscriminate statement of their demand that the Regional Power Survey takes note of. They also clear it and check it with DGTD, that is, the Directorate General of Technical Development, which actually gives licences to them. Check from there that when a particular industry claims that this is going to be their expansion, and so on, whether that expansion has been sanctioned. I do not know whether the task of handling these data could be left to the various utilities and what kind of job they would make of that. I personally feel that if the standardization is made at a central level, like the Regional Power Survey, they would know specifically what kind of data they want and for what purpose. So, if the various utilities just follow the format, that might make data acquisition more systematic. Of course, it is always a good idea for any utility which handles bulk power to project its demand and particularly the conservation.

As an aside, in Karnataka State, Indian Aluminium consumes 1/5 to 1/6 of the total electricity generated in the state, and there I know, they undertake quite an exercise themselves, so far as future projection and conservation measures are concerned, because they consume such a large amount of energy. But whether such duties could be left with a vast number of utilities of one MW+, is doubtful, because as I have mentioned that these utilities, which consume less than one MW, go by the trends of how they have developed. But anything larger than one MW has to fill in the questionnaire and send it to the Regional Power Survey. Concluding, I could make this remark, if a particular utility or industry is a very large power consumer, they ought to do this exercise themselves. But so far as the smaller units are concerned, this is something which should be centrally controlled by the Regional Power Survey. They do not have a computer as far as I know, and the computer naturally will be invaluable to handle and collate data.

SMITH

I would like to address my remarks to Dr. Munasinghe who presented a very excellent framework of social-economic information. Actually many of the data problems and management problems in this area stem from the fact that electricity cannot, except in a small way, be stored. There is one aspect of reliability question, quite rightly emphasised. This reliability question is going to become even more important as less reliable sources of energy are going to be tapped for electric power, particularly renewable sources of energy some of which are stochastic in nature or at least subject to variations because of weather. We will have to understand the cost, like reliability, before we can decide how much storage to put for a power plant, for example. Again, in a place where there is a lot of power available, because of nearby cheap hydro-electric sources, the rate structure, taxing structure, and planning could encourage the electric industry to go to that area. You can also provide areas in the country which have reliable power. Indeed in parts of the US, for example, Silicon Valley Area in California, where there is a large amount of semi-conductor computer work, electricity is extremely reliable—certainly the most reliable place is the US. It is not accidental. Semi-conductor manufacturers are very sensitive to loss of power as also are other batch processors like pharmaceuticals. The interaction between the utility and this lobby results in extremely reliable power. In a country with central planning, like India, you could designate an area as a highly reliable one and, therefore, not have to spread the large cost of high reliability through the country. There are a lot of interesting possibilities in the reliability questions.

Secondly, there is the reserve aspect, in addition to the fact that electricity can't be stored. In addition to forecasting load, you need to forecast load duration, that is, relative amount of daily peaks and seasonal peaks. This can be quite important, for example, the Province of Taiwan, as you probably know, has gone in a very large way for nuclear power. They have the most aggressive nuclear power programme in world based on their national economy. They find themselves in a very awkward situation. Many other countries too find themselves embarrassed like Taiwan with by an over capacity of base load. They have in fact to dump heat during off-peak time from the nuclear power plant. They have five times too little pumped hydro in order to actually follow their load demand. This is partly due to too much nuclear power but also partly due to unfortunate neglect of load management.

Finally, I think it is interesting to understand the economic drive for accurate forecasts. In the US, and in many other countries, the present approach is to build power plants that require only short lead times. Because one cannot predict, say, what will happen 10 years from now, you build power plants that indeed can have a higher cost of electricity associated with them. It is embarrassing and expensive to have either extra or too little capacity. There is a significant value to accurate forecasting and one of those values is the ability to buy larger units which take longer to construct and it will be interesting in an argument to build a data system to have some handle on economic drive for accurate forecasting.

MUNASINGHE

I cannot but agree with everything Dr. Smith has said. I want to stress one particular aspect based on reliability dimension. I said price is the signal that is transmitted from the producer of electricity to the users. The strength and the power of that in demand management should not be underestimated and our data systems, should aim at providing correct price signal to consumers. Because if you have an interruptible load that is at the lowest level of reliability, you can provide power at a low price. Other areas like Silicon Valley in the US where you want high reliability, you provide the service and you charge them a higher price. These are all tools useful for demand management.

There is another dimension to demand management which is brought about by revolution in solid state devices. For example we know sophisticated forms of load control are possible through the transmission of information between the user and the producer of electricity. You can have little micro-processors which are being used, particularly for large electricity users, to switch devices on and off from a remote point. This basically gives the utility, the ability, to effectively shed load on a selective basis in a number of places. Again an advantage, not so much of data gathering, but of monitoring constant data on line, and instant transmission of data on a computerized basis. I think this is part of the wave of the future—I hate to use a cliche. Traditionally utilities have been constrained to assume a certain load and then supply electricity to meet that load. Now we are getting a slightly different emphasis with the improvement in signalling and solid state switching and other devices where you can tailor to some extent the load to follow the supply or some mix thereof.

SHAH

The World Energy Conference in its Munich meet a couple of years ago also spent a lot of time discussing pricing signals to influence usage patterns, as Dr. Munasinghe was pointing out. It is the one thing to which consumer responds to most readily. They are useful to promote conservation and effective substitution of fuel. The use of pricing mechanism was something that was very strongly recognized. These load control systems to which he referred, are also coming in. In India too they have started installing them. It is as much on the point of keeping peak demand under control as to influence what sort of load you want to encourage and keep things straight.

NIKHIL DESAI

I have three questions. The first, is there data on the distribution of consumers of power ? Second is about theft. . . . Everybody is stealing a little. Some people are stealing very little and some people a lot and that too for a particular purpose. The third question was, is it possible to give any information on safety and environmental records on plants in this country, particularly nuclear power plants?

SENGUPTA

I think we can take the questions one by one. The REC does collect fairly detailed information about the state of electrification. I have given in the Appendix some typical questionnaires which you have to answer if you go to RECs. They collect volumes of data regarding details of what kind of distribution lines have been drawn, what rating of transformers have been used, etc. These data are available. Data availability is one, data retrievability is something totally different, however. You have to spend really a month to know what has been achieved, but I can assure you that data are collected.

NIKHIL DESAI

By distribution, I meant not distribution of electricity but the statistical distribution of. . . .

SHAH

I think you are referring to connected load block, like how many consumers, what kind of connected load and so on. Even that data is being collected at utility level. Unfortunately it is not categorized or grouped in a manner one would like to see. When it comes to tariff formulation, it is difficult to know what will be the impact of increase in such and such a block. Usually it requires special effort to retrieve that, as Dr. Sengupta has pointed out; it is a most difficult thing.

SENGUPTA

Actually in Table 2.2 of my paper you will find some information. But I cannot really vouch for the authenticity of the other questionnaires which are filled in. The second point about theft, please don't ask me; I just cannot answer. I do not know whether the people who are responsible for collecting the information, know either.

SHAH

It was thought to be in the region of about one to $1\frac{1}{2}$ per cent of the total loss, until last year in most cases. There are in some areas pockets where it could be much more. To some extent this depended upon the nature of the people. Some areas are populated by not very law abiding citizens. There the situation becomes different. This $1\frac{1}{2}$ per cent level which people generally recognize, I think, has gone up to three to four per cent in the last year. Again there is no means of confirming any such figure. The reason why one feels that way is that the losses which have increased rather quickly over the last year cannot only be the result of system parameters. There must have been an increase in theft. The reason for this again is that there have been tariff increases, and also shortages. If you cannot get power legally, you can try and get it illegally. You cannot afford to have the equipment kept idle. So these factors have increased the losses, may be, to around three to four per cent by now. One other factor which induces additional uncertainty is the increase in unmetered supply through political pressure. More and more districts even states, are insisting that agricultural supply is to be given without metering. The losses ultimately are nothing else but the difference between total energy sent out and total energy sold. So if you do not meter, you can put any number. I think in the figures which he gave, I could spot one or two states where they are making use of this unmetered supply to show lower losses, by merely showing that so much more has been sold to agriculture. That is one of the reasons we have uncertainty in the loss figure.

SENGUPTA

The third question was about. . . . . .

NIKHIL DESAI

Safety and environment.

SENGUPTA

All that I could find is the questionnaire which the CEA sends out. They ask questions about how many deaths have occurred, how many accidents, etc. The Annual General Review gives details of accidents that took place or any electrocutions, even of animals in some cases.

NIKHIL DESAI

Does not the CEA have any authority over nuclear power plants?

SENGUPTA

I really do not know.

NIKHIL DESAI

By environmental, I meant emissions from coal plants.

SHAH

In some states they have environmental controllers. Unfortunately I find that means to measure precisely the concentrations are not there. It is only when there is a terrible situation, that the smoke is too dark, or everybody around the place is grumbling, only then there is some effort made to bring the situation under control. Although the standards do lay down what ought to be. But unfortunately the precipitators do not work as well as they ought to in most of the thermal stations and that creates a certain amount of problem.

SENGUPTA

In many thermal stations you will even find that you really do not even have the pH value of water that is being used even though it is extremely important to know that. So I think one really has to identify that these are absolutely essential data and have to be kept.

AHUJA

A comment and a question. The comment is that one of the ways consumers respond to 'proper' price signals is by taking to the streets. The question is, what is the thinking in China about long-term waste disposal of nuclear plants, now that so many plants would be coming up.

YINGZHONG

As far as I know, we have several production reactors that have been operating for years. We have reprocessing plants and made good progress in the waste disposal problem. Now we have developed some new methodology for the disposal of nuclear waste. Besides we have plenty of room to throw this waste in the desert region of China. So we feel that the waste problem is not the major problem in the development of our nuclear energy. The major problem is investment problem because it requires several times investment for other sources of the same capacity.

SHAH

I think we would like to hear Mr. Agarwal from the CEA if he has some comments to offer.

AGARWAL

There should be a linkage I feel from the industries to the state electricity boards also. Apart from the operation data on the day-to-day load demand and energy consumption of the industries, their projected development programmes are also noted by the state electricity boards for their own planning for power development. If these are properly analysed and then fed to the regional power survey and the CEA for proper long range planning that would reduce the time period of the analysis at CEA level and the Regional Power Survey Level.

NANGIA

I just have some comments to make about what Dr. Munasinghe said about long run marginal cost. It is not that data do not exist for it to be computed. More often it is the independent researchers (I am not talking about the decision makers who have hardly any time to do such policy oriented work) who are just not able to obtain the data. It is a question of not only having enough data but making it accessible to people who have to do the actual work. I think more often that has been the problem in doing any such work.

MUNASINGHE

A point was raised about people taking to streets. I just want to stress that the reason why I brought the issue of price in this workshop is that price is one of the main means of using information to manage demands. There are other means and tools of managing demand, for example, technical and physical control. If you shed load you are automatically managing demand. That is an unkind way of doing it. Price control is a more effective way, where you try to influence the users into a desirable pattern of electricity use. At the same time you have to have also other flows of information particularly education and propaganda where you explain to people that you are using the price method not because somebody in the CEA or the SEB said it is a good idea at that time, but because the cost of producing electricity is going up, you have to finance major investments, etc. There are additional contents to this information flow besides price, namely educating the public which is also very vital in demand management. Not only the long run marginal cost operates in setting the price, you have also social and other considerations to moderate price increase. But all of this play a role in codifying collecting data and transmitting it to the users.

SHAH

Summing up, in Dr. Sengupta's paper we have a very extensive overview of the current situation with respect to information flow, collection, and usage. I would like to concentrate entirely upon his recommendations especially on the use of the computer. I am aware of the systems which have been started in Electricity Boards where the hierarchy starts with mini or micro computer at the subdivision level, to gradually higher levels, with the main frame in the headquarters group, from where information can flow to the CEA. I believe it will be very useful if the CEA or the Ministry would take this up as an area of study and evolve, as Dr. Ramesh was recommending, some guidelines for people to develop this approach, whereby the information flow would be rendered more accurate, and timely.

Dr. Sengupta has also made a reference to rural electrification where a lot of money is being spent unnecessarily. In the sense that you spend a lot, but ultimately wind up with little consumption. I think it is extremely poor usage of resources that we have. I think that will need to be revised especially with the alternative sources of energy which are coming up. I feel strongly that the CEA or the Ministry of Energy would find means of substituting rural electrification by what I would call rural energification.

Dr. Ramesh referred to the end-usages both at the micro level as well as state level for managerial as well as financial planning and monitoring purposes. I believe in India at least these are the current problem areas. Keeping in mind the constraints of resources, whatever use one can make of information availability towards improving project management and improving thermal station management, I believe will be in the right direction. Perhaps one of the tasks that we could set for information specialists would be to see how this can be most quickly and most usefully tailored to the needs that we have.

It was unfortunate we could not go more into the commercial, administrative, or managerial information systems and data arrangements, perhaps we can do so on another occasion. Dr. Munasinghe brought in the non-technical components, required for various decision making purposes and for optimising reliability and reduction of losses. His closing point about price being used as a major information element is also very valid.

Professor Lu gave us something that most of us were totally unaware of, namely, the electricity situation and the anticipated development of nuclear power in China. I suppose this is as important an element of information exchange to understand and know what is happening in other countries. Similarity of circumstances do exist and it is there that one learns a lot from the solutions developed elsewhere. I am not sure why it has been so late, one should have expected, with enormous background in nuclear energy, China would have had nuclear power in a bigger way by now. Anyway, they are now picking it up and going rapidly ahead with it. I wish we can at least think of having the same kind of pace of development of nuclear energy in our country.

TECHNICAL SESSION 3

COAL DATA SYSTEMS

Chairman : G.S. MARWAH

Theme Paper : S.K. BOSE AND D. SENGUPTA

Discussants : S. SUBBA RAO

VINOD MUBAYI

THREE

COAL DATA SYSTEM IN INDIA

S.K. BOSE AND D. SENGUPTA

In the energy scenario of India, half the energy input is supplied by non-commercial fuels, while the remaining half is the commercial energy inputs, of which coal shares 65 per cent. Coalfields occur in 10 states, mostly in the eastern half of the country. Coal mining in India has a long history starting from 1774 in Ranigunge and its output has progressively increased to the level of 125 million tonnes in 1981-82.

With the industry of coal mining well established, the Indian Mines Act was enacted as early as 1902. The Coal Controller is the Statistical Officer on coal in India. With the advent of nationalisation of the industry in 1971 and 1973, the industry has been organized with Coal India and its subsidiary companies under the Government of India operating in nine states and producing about 90 per cent of all coals in India. The remaining production is covered by Singareni Collieries Co. Ltd., operating in Andhra Pradesh, and the private sector steel companies operating in Bihar. The Coal India holding company, formed under the Ministry of Energy, Department of Coal, has been operating coal producing companies as Eastern Coalfields Limited, Bharat Coking Coal Ltd., Central Coalfields Limited and Western Coalfields Limited, and the Central Mine Planning & Design Institute Limited, Ranchi which is engaged in geological exploration and mine planning design activities.

The Coal Data Systems in India have been operating for many decades and have now got themselves reshaped into the present structure of the nationalised coal industry mentioned above. The exploration of coal is initiated by the Geological Survey of India under the Ministry of Steel and Mines. The entire spectrum of coal data system begins from the resource assessment followed by quality evaluation by the Central Fuel Research Institute and goes into all aspects of production, manpower deployment, transportation and many other details, as has been presented in the items with the concerned organisations named as below :

1. Geo-Data System	Geological Survey of India, Central Fuel Research Institute, Central Mine Planning & Design Institute Limited and Mineral Exploration Corporation.
2. Coal Mines Statistics	Coal India Companies, Singareni Collieries Co. Limited and Deptt. of Coal/Coal Controller.
3. Coal Production Data System	Coal Companies and the Coal Controller.
4. Manpower Data System	Coal companies and the Directorate General of Mines Safety.
5. Mines Safety Data System	Coal companies and DGMS.
6, Welfare Data System	Coal companies and Coal Mines Labour Welfare Organisation.
7. Industrial Relations Data System	Coal companies, Department of Coal and Ministry of Labour.
8. Coal Transportation Data System	Coal companies and Indian Railways and the Coal Controller.
9. Mine Equipment Data System	Coal companies and DGMS.
10. Coal Consumption Data System	Coal Controller and the consumers of coal.
11. Finance Data System	Annual reports of Coal Companies.
12. Conservation Data System	Coal Controller

13. Coal Utilisation Data System	Coal Controller
14. Coal Projects Investments	Department of Coal.
15. Environmental Aspects in Investments	Department of Coal.
16. Perspective Planning Data System	Coal India Ltd., Central Mine Planning & Design Institute Limited and Planning Commission.

This paper has attempted to bring out some broad aspects of data generation in each of the above sixteen systems.

Geo data system

Ninety-Nine per cent of Indian coals were formed in the Gondwana times, 230-250 million years ago. The Tertiary coals of recent age cover the remaining portion and they are located in the north-eastern corner of India. The coal seams originated from deposits of vegetative substances that had drifted and were deposited in old lakes or estuaries, and then covered by rock formations. Today, the coal belts are deep seated within the earth crust but occasionally they out-crop at the surface. Geological exploration locates these coal evidences by surface mapping and regional drilling at wide intervals. These are conducted by the Geological Survey of India, established as early as 1851. As per their last geological reappraisal of coal resources, presented in 1978 for briefing the Working Group on Energy Policy, the coal resource in India has been estimated at 111,877 million tonnes, up to a depth of 1200 metres from the surface and for coal seams of thickness 0.5 metres and above. A summarised statement of the coal resources is shown in Table 3.1.

Following the GSI regional explorations, the Central Mine Planning and Design Institute Limited (CMPDIL) chooses the exploration blocks for detailed drilling, keeping in view the location, the volume of the resource, quality, mineability, consumer demand for coal, transportation logistics and many other aspects that are involved for investment decisions. Detailed drilling is conducted either by CMPDIL or by the

TABLE 3.1 : **Total resource (in million tonnes) of coal in India (for seams 0.5 m and above in thickness and up to a maximum depth of 1200 m with major deposits)**

Field		Total Resources (Million Tonnes)
A. Gondwana coalfields		
West Bengal		
Raniganj	—	26154.42
Bihar		
Jharia	—	19521.64
Bokaro	—	8288.21
Karanpura	—	15609.72
Ramgarh	—	1055.56
Rajmahal	—	3656.73
Madhya Pradesh		
Pench-Kanhan-Tawa Valley	—	1015.09
Patharkera	—	315.67
Sonhat	—	223.43
Hasdo-Arand	—	1742.30
Johilla	—	977.07
Bisrampore	—	462.18
Jhilmili	—	255.58
Chirimiri	—	303.72
Korba	—	794.18
Lakhanpore	—	250.96
Singrauli	—	10263.62
Sohagpore	—	824.46
Orissa		
Talcher	—	3530.98
Ib River Valley	—	2376.31
Maharashtra		
Chanda-Wardha	—	2207.78
Kamptee	—	622.21
Andhra Pradesh		
Godawari	—	8504.60
Total (including other fields)	—	1,10,985.83
B. Tertiary coafields		
Arunachal Pradesh, Assam, Meghalaya, Khasi and Jaintia, Nagaland	}	892.02
GRAND TOTAL		1,11,877.85

Mineral Exploration Corporation (MEC). Table 3.2 brings out the detailed drilling by various agencies totalling 215,542 metres in 1981-82, as has been brought out by CIL's Operational Statistics. In addition, drilling is conducted by coal companies for any local problems, as also by Singareni Collieries Co. Limited.

TABLE 3.2 : **Drilling in coal bearing areas**

(in metres)

Agency	1981-82	1980-81	1979-80	1978-79
CMPDIL	74,332	78,501	73,220	70,718
MECL	1,08,865	92,563	89,389	1,30,412
State Govt. (M.P. & Orissa)	9,979	10,438	7,731	10,181
Private Agencies	22,366	22,603	19,837	18,821
Total	2,15,542	2,04,105	1,90,177	2,30,132

All exploration borehole coal cores are analyzed in the laboratory of Central Fuel Research Institute, which has stations located in nine different coalfields. The data of this physical-chemical survey are published in the annual reports of the Coal Survey Stations under CFRI. These detailed physical-chemical data have been published by CFRI in nine massive volumes covering the main coalfields.

The coal resources are classified according to depth, thickness of seams, quality parameters, for example, with or without dirt bands. Thus, a complete inventory of coal resources from all aspects of geology and chemical composition is prepared for the geo-data systems. Now the Coal Mining Research Station is attempting to study rock mechanics from the borehole scores in some of the coalfields.

For detailed planning of Jharia coalfield, the most important coking coalfield in the country, entire geological data have been recorded in a computer. These are useful for preparing a master plan for reconstruction of the Jharia coalfields mines. Such massive mine planning process not only requires complete geological physical-chemical information but also data on fire areas, surface features that affect mining activity below ground, river and water flows, surface road and rail network and plans of old workings below ground and many other details that are required for mine design work.

Coal mine statistics

Following the geological explorations, mine boundaries are established and production units developed. Coal India is operating 384 mine units with average production of 2,85,400 tonnes per mine per year. Table 3.3 indicates the details of the coal companies with their mine-groups or areas and the mine units, as published in the Operational Statistics of Coal India. To these should be added the mines of SCCL and Steel Plant companies to get a comprehensive picture. The coal mine statistics cover the mines under operation or construction. These basic information from the coal companies are also compiled by the Coal Controller and Director-General of Mines Safety.

TABLE 3.3 : **Area and colliery size as on 1.4.1982**

Company	Number of		Average size (1981-82)	
	Area	Colliery	Area M. tonnes/yr.	Colliery '000 tonnes/yr.
ECL	14	122	1.73	198.6
BCCL	14	97	1.61	237.3
CCL	11	62	2.74	485.5
WCL	11	99	2.87	318.8
NEC	1	4	0.7	176.0
Total CIL	51	384	1.98	285.4

These statistics can be assimilated in terms of geological locations, sizes of units, density of coal resource, land as acquired from various sources, etc. Mines operating or kept in abeyance and many other data can also be dug out from the data submitted by mine units from time to time to Coal Controller and Director General, Mines Safety.

Coal production data system

Daily production figures are generated at the mine level and compiled in the Coal Company Headquarters. The monthly statistical data on production from the open cast and underground mines of different quality grades are prepared out of the same data and reported to Coal India Headquarters, Coal Controller, Director-General of Mines Safety, Ministry of Energy, etc. Similarly, SCCL and Tata Iron and Steel Co. also process their production data for information to the Ministry and to the Planning Commission. Table 3.4 indicates the production data for Coal India. The actual daily data on production of different units quickly grows to be voluminous. These data form the common thread that keeps numerous levels of management and several offices fully linked to one system. Yet there will be many areas of gaps in data, specially in the annual figures. For example, we do not know the depth of production year to year, so also the quality variations over time, the changes in coking properties of coal produced, the changing underground mine environment, the changing rock mechanic properties, the changing surface structures that affect underground mining, the changing coal particle size and preparation problems due to mechanisation, market needs and many such data that one needs for design and policy considerations. Perhaps the annual data systems need a review for more meaningful data collection.

TABLE 3.4 : **Overall company-wise coal production**

(*Figures in million tonnes*)

Company	1981-82	1980-81	1979-80	1978-79	1976-77	1974-75
ECL	24.22	22.70	20.52	22.05	26.46	23.16
BCCL	23.02	21.48	20.08	19.72	20.68	17.74
CCL	30.11	27.47	24.15	23.43	20.73	18.31
WCL	31.56	28.75	26.13	24.23	21.04	19.26
NEC	0.70	0.61	0.56	0.62	0.57	0.52
Total CIL	109.61	101.01	91.44	90.05	89.48	78.99

Manpower data system

Similar to production data system, manpower data are also generated from the producing mines and other activity centres and compiled in the Area Offices and Coal Company Headquarters. Synopsis of the data is sent to the Ministry of Energy. Table 3.5 gives the manpower position in CIL as a whole on yearly basis. It shows that the total manpower in CIL has remained more or less constant for the last six years at around 6 lakhs of workers.

TABLE 3.5 : **Overall manpower position in CIL**

Position as on	Officers	Permanent workmen	Temporary/ Casual	Total
1.4.80	9047	557288	34591	600926
1.4.79	8391	549559	34690	592640
1.4.78	8101	531652	48233	587986
1.4.77	7457	519073	59523	586053
1.4.76	6766	527447	71766	605979
1.4.75	5657	513436	79556	598649

Mine safety data system

Details of accident figures and safety standards are recorded at the mine level including the related enquiry reports. Office of the Director-General of Mines Safety is responsible for looking after the safety in the mines and, as such, keeps the safety records as generated from mine units and those generated during mine inspections. Data is compiled and analysed by the Director-General of Mines Safety. National Safety Council collects the data in a comprehensive form from the Director-General of Mines Safety and publishes the data in their bulletins along with the safety records from other industries. Table 3.6 shows company-wise fatality report for the years 1973-81 and Table 3.7 shows company-wise serious injuries for the same years. Safety data generated are voluminous. Yet it will not be possible to know the exact nature of fatality or details of injury to persons nor whether fatality is due to inadequacy of medical attention or extraneous reasons. Many of these data are recorded but not published perhaps due to their volume. It is worth assigning research scholars to dig out the reports and bring to surface useful information for new directions towards improvement of safety.

TABLE 3.6 : **Overall (company-wise) fatalities**

Company	1981	1980	1979	1978	1977	1976	1975	1974	1973
ECL	24	39	40	31	43	52	56	56	41
BCCL	38	50	29	56	60	101	76	53	51
CCL	26	14	22	23	33	35	26	23	28
WCL	34	34	33	38	46	44	74	59	43
NEC	2	4	19	3	7	5	1	1	2
Total	124	141	143	151	189	237	233	192	165

TABLE 3.7 : **Overall (company-wise) serious injuries**

Company	Calendar years								
	1981	1980	1979	1978	1977	1976	1975	1974	1973
ECL	327	389	180	253	286	264	301	172	267
BCCL	191	210	228	271	297	332	380	456	423
CCL	111	118	74	129	135	90	133	149	136
WCL	327	389	478	410	415	425	672	642	505
NEC	8	18	12	5	5	19	10	23	14
Total	964	1124	972	1068	1138	1130	1496	1442	1345

Welfare data system

The mine unit-wise employment, construction of housing and arrangement for water supply is prepared by different coal companies. These are also recorded by the Coal Mines Labour Walfare Organization, Dhanbad. A typical data on provision of housing at different coal companies is shown in Table 3.8.

Industrial relation data system

The data on strikes, go-slow tactics and other industrial relations are [generated at the mine units and same are assimilated by Coal India and the Ministry of Labour and Employment.

TABLE 3.8 : **Company-wise housing facilities**

Name of company	Pre-nationalisation	As on 1.4.80	1.4.79	1 4.78	1.4.77	1 4.76	1.4.75	1.4.74
ECL	37845	54467	51847	49664	46565	42516	38471	37846
BCCL	37383	49535	47768	45939	43154	39304	38470	37383
CCL	22850	37404	34899	32514	29750	25743	23386	22850
WCL	18405	33781	32558	30549	27547	23547	19547	18405
NEC	1884	2523	2389	2279	2243	1884	1884	1884
Total	118368	177710	169461	160945	149259	132994	121758	118368

While these data give an idea about industrial relation problems that have happened and perhaps useful for just record keeping, they do not indicate the real iudustrial psychology prevailing in a mine or company.

Transportation data system

Two generating sources of coal transportation data are the coal mines and the originating railway centres. Information regarding daily supply situation of wagons are collected at the Railway Headquarters and the compiled data sent to the Department of Railways and the Coal Controller. Coal mines also keep the record of wagons supply situation and loading position and send the information to the Company Headquarters. This information is compiled in Coal India for the Ministry of Energy. Table 3.9 illustrates year-wise movement of coal and coke by various modes.

Mine equipment data system

Coal India mines are operating a large spectrum of machinery and that includes over 1,000 dump trucks and over 130 shovels for open cast mining. The operation of equipment of different types is shown in operation statistics of CIL in Table 3.10. Besides investors, these data are relevant to all those in the management of central workshop, inventory controls, procurement and finances.

TABLE 3.9: **Movement of coal and coke by various mode**

Year	Rail	Road	Conveyor Belt	Ropeway	Own wagon	Total
1975-76	59.81	13.38	1.25	1.87	1.69	78.00
1976-77	61.61	13.23	1.39	2.13	1.94	80.29
1977-78	63.03	16.36	1.40	1.96	1.98	84.72
1978-79	57.31	18.04	2.20	1.53	2.21	81.29
1979-80	54.83	23.33	2.43	1.53	2.11	84.23
1980-81	56.16	26.26	2.71	2.05	1.94	89.12
1981-82	65.43	26.45	2.92	2.24	1.80	98.84

Company-wise overall coal despatches and offtake

(*Figures in million tonnes*)

Sl. No. Company	1981-82	1980-81	1979-80	1978-79	1976-77	1974-75
Despatches						
1. ECL	21.57	20.18	19.528	20.721	24.782	21.602
2. BCCL	21.77	18.85	18.461	18.779	18.714	17.497
3. CCL	28.50	26.19	23.492	21.492	19.744	17.502
4. WCL	30.82	27.15	25.260	22.464	20.145	18.661
5. NEC	0.63	0.60	0.552	0.613	0.573	0.521
Total	103.29	92.97	87.293	84.069	83.958	75.783

TABLE 3.10: **Basic underground equipment population**

Year	Equipment							
	Coal cutting machine				Haulages			
	ECL	BCCL	CCL	WCL	ECL	BCCL	CCL	WCL
1979-80	346	73	152	341	1130	505	176	823
1978-79	524	76	150	340	1219	490	167	759
1977-78	503	44	143	332	1095	419	148	759
1976-77	488	51	108	308	1063	438	171	—
1975-76	417	—	146	302	966	465	160	—
1974-75	413	—	—	—	952	—	—	—

TABLE 3.10 : (*contd.*)

Numbers of locomotive winders

Year	Equipment							
	Coal cutting machine				Haulages			
	ECL	BCCL	CCL	WCL	ECL	BCCL	CCL	WCL
1979-80	39	—	8	50	282	170	12	46
1978-79	39	—	8	50	282	170	12	46
1977-78	80	—	9	50	274	160	12	46
1976-77	49	—	9	50	274	160	8	46
1975-76	40	—	9	—	274	155	8	46
1974-75	—	—	11	—	—	—	—	—

Population of basic open cast equipment

Company	Population as on					
	1-4-76	1-4-77	1-4-78	1-4-79	1-4-80	
Shovels						
ECL	6	9	12	17	30	
BCCL	4	7	10	21	33	
CCL	48	51	88	76	89	
WCL	14	17	24	25	42	
Dumpers						
ECL	26	30	55	63	101	
BCCL	14	18	42	75	115	
CCL	268	352	423	431	524	
WCL	79	91	108	154	242	
Dozer						
ECL	21	22	34	49	56	
BCCL	6	7	11	17	31	
CCL	53	61	102	102	134	
WCL	12	20	23	49	84	
Drills						
ECL	11	14	9	5	8	(Excl. Diesel and Wagons)
BCCL	4	5	8	11	21	
CCL	41	41	47	48	85	
WCL	16	20	26	31	42	
Draglines						
ECL	—	—	—	—	1	
BCCL	—	—	—	—	1	
CCL	4	4	4	6	7	
WCL	6	6	7	8	8	

Coal consumption data system

The office of the Coal Controller receives coal consumption data from units of coal consuming industries through an elaborate machine tabulation system, that produces coal consumption data by different sectors of industries, for example power, steel, railway, cement, textiles, brick-burning, domestic coal and other industries. Sector-wise coal consumption in India is shown in Table 3.11.

TABLE 3.11 : **Sector-wise despatches of coal and coke**

(*Figures in lakh tonnes*)

Class of Consumer	1979-80	1978-79	1977-78	1976-77	1975-76	1974-75
	COAL INDIA LIMITED					
Steel : Washed coal	76.19	81.49	86.25	84.69	75.39	67.25
Coking coal (including Blendable)	29.17	29.56	31.68	24.36	23.22	19.89
DSP and Chasnala						
Washeries	14.65	13.88	15.83	15.32	16.23	13.66
Boiler coal	20.05	17.19	17.46	16.75	19.29	12.23
Power House :						
Raw coal	291.70	247.41	242.88	227.44	221.88	191.48
Middlings	15.34	16.55	17.68	16.41	12.24	10.29
Loco	108.72	107.89	118.94	119.08	129.90	128.23
Cement	29.74	34.65	36.81	34.82	35.00	34.83
Paper	15.34	17.92	18.07	15.79	14.87	12.97
Textile/Rayon	20.52	23.17	24.70	23.28	26.35	22.51
Fertilizer	18.12	16.51	13.36	10.53	9.33	9.53
Heavy Engg.	1.83	1.90	2.21	2.52	—	—
Chemical	5.04	5.42	4.94	4.53	—	—
Coke Oven	3.94	4.31	4.88	7.02	—	—
Export	0.94	2.67	6.58	6.23	4.00	5.28
Soft Coke	23.02	22.95	28.92	31.64	27.75	28.01
BRK (Rail only)	5.26	15.83	34.80	34.20	—	—
Others	162.10	153.62	141.23	128.24	161.81	170.94
Total	841.67	812.91	847.22	802.85	777.26	727.10

Finance data system

The published annual reports of the coal companies furnish all data relating to capital assets, profit and losses, major loans, etc. Table 3.12 shows the wholesale price index of a tonne of coal in rupees with 1961 as the base year. Five-Year Plan investments are also shown in the annual reports of the Department of Coal and the Five-Year Plan document of the Planning Commission.

TABLE 3.12 : **Index numbers of wholesale prices of coal**

(*Base: 1961-62=100*)

Year	Coal
1971	169.1
1972	173.6
1973	189.7
1974	230.8
1975	288.0
1976	333.4
1977	333.4
1978	354.0

Conservation data system

Coal is a wasting asset. The manner of its extraction, protection against unwanted fire, losses due to restrictions from surface structures, rivers, roads, etc., losses in the process of minining due to the technique of mining adopted or due to leaving behind all coal seams of a lower quality that may not have ready market are all various ways in which maximum extraction of coal for actual market sale may not become feasible. In Jharia and Ranigunge coalifields over 100 coal fires have arisen out of spontaneous coal heating and some of them have been raging for decades. With coal seams lying below the other, these fires create restrictions besides being wasteful of resource.

As a part of conservation and also to improve safety in mining of coal seams sand is often stowed or packed in the voids of coal extracted areas. By levying a small cess on every tonne of coal despatched from mines, a conservation fund is created every year. This, through a Conservation Committee provides support to coal mining activities by way of provision of subsidy for stowing activity, protections against fires, arrangements for conservation means in underground fires, etc. This conservation fund is coordinated by the Department of Coal and the Coal Controllers Office. The annual report of the Department of Coal indicates the extent of the expenditure.

Coal research data system

Coal utilisation research is conducted by the Central Fuel Research Institute under the Council of Scientific and Industrial Research and also by the Regional Research Laboratories at Hyderabad, Bhubaneshwar and Jorhat. Coal use research is also partly supported by Coal S & T grant under the Department of Coal and is coordinated by the Central Mine Planning and Design Institute Limited. The annual reports of the Department of Coal, CSIR, CMPDI, CFRI and other laboratories indicate the progress of research work on different coal utilisation areas.

The Central Mining Research Station at Dhanbad and CMPDI at Ranchi undertake research on coal mining practices. The specific problems are raised by the coal companies and then the solutions are sought by the research laboratories. There is constant coordination between the mining establishments and the mining research stations. The annual reports of CMRS and CMPDI underline the progress of the mining researches.

Coal projects investments

Since the nationalisation of coal industry, investments have been increased not only for maintaining the production levels but also for creating major new mines and reconstruction of some of the old ones. In coal India, the project investment proposals are first prepared by designers of CMPDI and same are approved with or without modifications, successively by Coal Company Boards, Coal India Board, Inter-Ministerial Group, Public Investment Board and the Government (Cabinet). Singareni Collieries Co. Limited similarly process their investment proposals for final approval by the state government and Ministry of Energy.

Environmental aspects in investment

The considerations of environmental aspects for mines are in their initial stages. Every project investment proposal brings out the environmental implications and tries to evolve the solutions of such problems in an economic manner. The investor does consider fulfilment of the environmental examinations in this way, before approving new investment proposals.

Perspective planning data system

Government of India's Working Group on Energy Policy had presented a report in 1979 indicating the forecast of coal demand during 1982—2000 A.D. Table 3.13 indicates the summarised forecast which totals up to 427 million tonnes at the end of the century. Based on these forecasts, the coal production levels have been assessed at 400 million tonnes in A.D. 2000 bringing up a production growth estimate of 9.6 per cent during 1978-83, 6.2 per cent during 1982-88, 6.1 per cent during 1987-93 and 5.3 per cent during 1992-20C0 A.D. These forecasts are based on a rough estimate of economic growth at 5 per cent till 2000 A.D. Obviously, the energy growth is dependant upon the economic growth of the country. For any variation in economic growth, especially for such long period of two decades, there will have to be automatic adjustment in the forecasts of energy and coal demands. It is envisaged that the energy policy groups will meet from time to time and evolve forecasts for the future.

The Central Mine Planning and Design Institute Limited under Coal India has prepared a report known as Black Diamond, indicating the forecasts of production increases till 2000 A.D.

Coal India has also taken up a large scheme of computerizing the data system originating from the mine level, grouped into areas and linked to the Company Headquarters, to Coal India, and to the Ministry of Energy. This system which is under construction, will have a large number of communication links through which data will be processed.

TABLE 3.13 : **Forecast of demand for coal for energy and non-energy uses (1982-2000)**

(*in million tonnes*)

Consumer category	1982-83	1987-88	1992-93	2000-01
OPTIMAL LEVEL FORECAST				
Railways	12.4	11.0	7.8	3.4
Steel	34.0	49.0	72.0	115.3
Household	5.2	7.2	10.0	16.8
Other industries	40.9	56.0	74.8	123.0
Colliery consumption and other uses	4.3	4.8	5.8	7.0
Total Direct use in Energy Sector	96.8	128.0	170.4	265.5
Power Generation*	50.6	72.3	100.1	158.7
Total coal demands for energy use	147.4	200.3	270.5	424.2
Non-energy use	3.0	3.0	3.0	3.0
Total requirement of coal	150.4	203.3	273.5	427.2

*For power generation the use of middlings has been assumed to be as follows :

1982-83 — 6.8 million tonnes.
1987-88 — 9.8 million tonnes.
1992-93 — 14.4 million tonnes.
2000-01 — 23.0 million tonnes.

Finally, the coal industry in India is old and established and it produces a large volume of data. Yet, there are areas where data bases can be expanded for better appreciation of the health of the industry. There is also scope for research studies in the records.

DISCUSSIONS

SUBBA RAO

I have a few comments to make on this and a couple of questions. Generally, as discussed yesterday and today, this paper also shows that there is no dearth of information, on various aspects of energy industry, production, or its capacity. But there have been some significant gaps. Mr. Bose has listed 16 different systems or sub-systems, which provide different aspects on the data available on coal. Since data are being collected, processed and reported at various levels by different organizations, I would like to know to what extent has there been a coordination of the efforts, standardization of the reporting of the data, and compatibility of the data reported by different sources. For researchers, it is a confounding experience to find data from two different sources to be at variance and they would like to find out what sort of efforts have been made, or are to be made, to ensure standardization or consistency of data from the different sources. Of course, the Coal Controller produces voluminous coal statistics. They bring out information on coal statistics, production despatches by collieries, by mines, by regions, and so on. If you take some other publication, you will also find the same information. But there will be some amount of discrepancy. This is one of the central questions that if there is going to be a centralized data system for the coal sector, how is this going to be organized, what mechanism will it have that the data of these various different organizations are collecting will be compatible and standardized.

There appears to be a gap in the demand side, just as for the electricity data system. A good amount of data is being collected and is available on the supply side. Most of it relates to the production of coal. What happens to demand, what do we do with demand estimation, how do we estimate demand for coal. One answer is since we consume whatever is produced, there is no need for demand estimation. But there is a need to know about the demand side for coal. I do not see any one of the systems providing an adequate base for estimating the demand data for coal in the country. Possibly Coal India, the premier marketing organization for coal is doing this. But this has not been brought out. Perhaps Mr. Bose could clarify this aspect.

The second gap which I think exists is in the materials handling and the management aspects in the coal industry. It is a very large and pressing problem from I hear from the people in the country. What kind of data systems are available or envisaged to handle this materials problem in the coal industry?

Thirdly, I am sure a lot of data is also processed to some extent. It would be very interesting to clarify to what extent the data is being processed, and for what purposes, by the various organizations. If you have to have a total integrated data system for the coal sector as a whole, one should start again with what are the objectives, the reasons for which we need this data, who needs it, when they are needed and periodicity. These things have to be properly settled. Then, if you use computers, may be data collection and dissemination and retrieval would be much more meaningful.

A clarification is needed on the last part of the section on Coal Production Data Systems, which says: we do not know the depth of production year to year, so also the quality variations, or the changes in coking property of coal produced, the change in underground mine environments, the change in rock mechanic properties, the change in surface structure that affect underground mining, the change in coal particle size and preparation problems due to mechanization, market needs and many such data that one needs for design and policy considerations. Perhaps the annual data systems need a review for a more meaningful data collection." The first portion of the first sentence raises questions about forecasting of various aspects; how can we address this problem? Once again, two main gaps I find in the data system are the coal demand system, and the materials handling system.

BOSE

You have asked if the various data collecting organizations are coordinating amongst each other or following some standardised or consistent data presentation system. First of all, the organizations compile the data for different purposes. They all do it mainly for appreciating the health of the industry in their respective spheres. Now, I won't say that it is standardized throughout, but by and large the data is fed into similar sort of terminology groups. Let me clarify with one or two instances. We have already standardised safety as fatality or injury rate. They are to be presented calendar year-wise, and month-wise, and company-wise, and also in terms of fatality of per 300,000 man-shift. That base is there. It is also presented in terms of per million tonnes of production; so all the bases relating to that item are standardized. But somebody may submit his form in one manner or other in his actual reporting. Otherwise, it is fairly standardised.

Coming to the production area, there are two items which are generated every day. One is a production per day from different areas that shows the progressive growth of the operations day-to-day in the month. There is also a second area of rail transportation. About 85 per cent of coal is handled by the railways. Coal production of 125 million tonnes itself is a big quantum per year, accounting for almost 2/3rds of the entire mineral industry in India. Similarly for the railways coal is a very major item for transportation and they go together. If rail transportation is good, production also picks up. That is why the data for this railway wagon supply per day is needed. They have a standardized figure, one wagon load means 22 tonnes. In reality, wagons may carry 55 tonnes or sometimes 24 tonnes. But the standard one wagon load means 22 tonnes. This is accepted as the norm and adopted by the entire industry and also by the railways and the government. Everybody knows it. They keep a watch as to how the wagon supplies are changing and in which zone they are decreasing. For example, if a particular yard does not carry every day wagons committed to it by the Indian Railways, at least somebody is monitoring it. This is also standardised in a major way. The reporting to their own organizations, to the Ministry of Railways, and by the coal producers to their Ministry of Energy through the various channels, changes a little bit. But the base is clear, that is 22 tonnes a wagon. Every day, it is to be reported and day means the calendar day.

Regarding compatibility, let us take production data, which come daily to different places including government organizations like the Coal Controller, coal company headquarters, and the ministry. These daily figures are provisional figures and even when the figures are added at the month, the sum is still a provisional figure. The reason is that coal, as it comes out of a mine, is never weighed physically, unlike many other minerals or many other products. The measure is in terms of so many mine cars or tubs coming out of the mine. They have average multiplication factors for them. For conveyor mining, coal is assessed on the basis of progress of the longer face or drivage of the galleries. So there is an element of assessment in the production. But at the end of the month, production is computed from the railway despatches which are weighed. Every wagon is weighed; they in turn measure the actual coal that is going out. So from the knowledge of actual coal that is going out and the measurement of stock left behind is derived the final figure. That is why there is some element of variation possible and it is natural in the process of handling the production figures.

It is true that a lot of data are available. But it is also true that if you want to get some particular data, you have to dig out the records and many of the records are not there. Now one question was, how demand is assessed. Basically, it is assessed every year and specially in five-year periods. This demand for sectors like power, is assessed by contacting each of the power generating electricity boards. Similarly the railways have an annual figure. They are always very close to their assessment. Because they know how many locomotives are working every year and how many are going out of circuit. Coal consumption is steadily going down in the railways. The locomotive coal assessment is done on an yearly basis. Similarly for the metallurgical plants; as and when new plants are coming up, their requirements are added up. Based on the schedules of major stoppages of furnaces, and on how much coal is required there is demand assessment. In certain areas like cement plants, textiles, domestic consumption or brick kilns, demand is not computed on unit basis. It is computed on progressive percentage basis. By that way some demand assessment is made out. Incidentally, all our previous experience has shown that where it is assessed on an overall basis like percentage of this industry or that, that demand is more correct. Where it is assessed on individual consultation with the power industries concerned and with the steel industry concerned, the vagaries of actuals and demand vary widely.

We have data on materials management but I think I have not brought it out. It is purely an internal company matter. The new system of our computerized data system has got certain chosen areas where more emphasis is given for collection of data. One area is materials management. Another area is finance and accounts management. Then there is the daily sales marketing management. These are special areas where the data will flow in a bigger way when the system comes on line. Some of these things are being worked upon, but I have not brought them out in my paper.

I am not going into the question of periodicity in detail it is the responsibility of the man who gets the data to know what he wants. In most cases people do not know what they want. It just comes and it goes on. That is the reality. But there is a possibility of introspection into why a data element is needed, whether it is needed or not. In each area, there is always scope for rethinking. About year to year changes which I mentioned there is work going on so that we may know the pulse of the industry in greater depth.

MARWAH

Thank you, Mr. Bose. I would like to supplement this response by one comment. About six to seven years ago, may be a little longer than that, the issue of making this data more compatible and reducing the number of organizations which were collecting and publishing data, was raised and discussed jointly between the various organizations concerned. I was at that point of time with the Directorate-General of Mines Safety and involved with this exercise. Basically at that time there was a lot of overlapping of work between the Coal Controller's Office, the Director General of Mines Safety and the Indian Bureau of Mines. For the last five to six years, the area of overlapping has been considerably reduced.

MUBYAI

We have heard over the last two days that a lot of voluminous data is required in different areas of the energy sector and there are many users of this data each with different needs. I would like to confine myself to a few brief remarks on the data needs in this sector which are important for energy planning on a national and regional level and not go into those aspects which have been very comprehensively covered by Mr. Bose.

Let us envisage a network of coal from resource extraction to the final consumer. We start with the extraction of the production of the basic raw material. The next stage would be, perhaps, benefication of coal, coal washing or one could summarize it under a short of pre-conversion activity. Then we have transportation of the coal by different modes, may be railways, ropeways, roads, etc. Then we should include storage, at one or two points in the system, because the coal may be moved to centralized depots or coal stocks. The next stage is the conversion stage where the flows of coal go into various industries like power generation, or metallurgy, coking coal for the steel mines, or coal going to different users for thermal use. Then we have the final end-use industries with all their utilising devices like boilers and furnaces and so forth where the coal is utilized. Now, for each part of this network, one can envisage different data sets which from the planners' perspective gives a bird's eye view of this flow and losses of this important commodity as well as all of the steps that are needed in order to meet the anticipated or the likely demand in the future. One can also envisage economic and environmental data sets, which cover not just the vicinity of the mining areas, but also the broader aspects of coal combustion. Now in his brief Mr. Bose described the technical data system of resources extraction, the geology, the thickness of seams, etc., but he did not indicate which system of classification of reserves (US or Soviet) is being followed here. Also perhaps, for lack of time he did not burden us with those details of how much is proved, indicated and so on.

Another comment I have is that he did not describe much of an economic data system on the production costs of mining coal in India under different conditions, whether open cast, underground, at various depths, etc. My question is asked, partially out of ignorance of Indian coal systems: whether it is possible to construct a supply curve in the economic sense, for the different mine fields for evaluating incremental or marginal cost of production for the different mines. This is an important figure for energy planners who are trying to develop new or extra resources to meet future energy demands from a variety of different energy sources. Secondly, on the transportation data, my feeling is, there is good data available on the economics of different transportation systems to evaluate, let us say, paise per tonne-mile cost for different forms of transport and for different regions of the country. Who is in charge of keeping such data and for updating it in a consistent and coherent frame-work and through time?

Then from some fragmentary information I have, there appears to be some lacunae in coal consumption data in different sectors, I think we have some experts here who can comment on whether there are in fact significant differences in the liftings of coal and the recorded consumption in different sectors of the industry and the economy. Who is in charge of evaluating and commenting on this lacunae, so that they can be eliminated to the extent possible?

On the environmental parameters, Mr. Bose's paper describes a brief aspect of the environmental aspect in the mining areas and the environmental implications of solving problems at the production end. But as the coal consumption in the country reaches within a relatively brief space of time of 15 or 20 years to something like three or four times its current figure, I think the time has come, as shown by the experience of currently developed countries which are also important coal consumers, that we will have to begin to pay very detailed attention to the environmental aspects of coal use and coal combustion in different areas of the country. The emissions and effects on air and water quality, the extent of sulphur compounds of different types, meteorological aspects of transport of these emission products into different areas will have to be looked into.

I also notice that there is some need for a data bank on new coal technologies which may come about in a big way if we decide in the next 20 or 30 years to go in for gasification, liquification, or some of the new technologies. Assuming that these developments become logistically, technically and economically viable, perhaps it would be a good practice to institute a data bank from the technical side of the performance of such technologies worldwide on the costs and projected costs and the manpower requirements.

Finally, regarding forecasting coal demand I find that this is a weak area where we planners are forced to tread. So it may be an exercise which is inherently difficult. The approach which Mr. Bose describes seems to be the end-use approach. Is any attempt being made to use econometrics to forecast coal demand? The problem is that for purposes of energy planning one does need to have a consistent methodology among the different sources which are, to a certain extent, inter-changeable. For meeting thermal needs in different industries, one can have a variety of different sources. For generating electricity too one has technology choices although they may be very constrained and limited as in the case of India. But in any case for forecasting the demands for these different sources, one does need to develop a consistent methodology and I wonder to what extent the methodology being followed by the different energy sector organizations, now centralized under the Ministry of Energy, are consistent with each other or are planned to be consistent with each other. We heard of oil yesterday, of electricity this morning, of coal now; it does appear that there is some need for some consistency in this approach.

MARWAH

Thank yon, Dr. Mubayi. Before I request Mr. Bose to make his response, Dr. Bowonder wanted a clarification.

BOWONDER

I would like to seek a small carification regarding Table 3.13 which respresents the total coal requirements. But Tables 3.9 3.11 refer to only Coal India production. What about the Neyveli Lignite Corporation and the Singareni Collieries? Are both of them included in this Table 3.13?

BOSE

Yes, that is correct.

BOWONDER

But in Tables 3.11 and 3.9, are they not included?

BOSE

No, they are not.

BOWONDER

Does the Coal Controller include this also in the energy production as well as the Singareni Colleries production in the total forecast ?

BOSE

The Coal Controller's figures are all-India figures, not just Coal India figures.

BOWONDER

So the figures given in Tables 3.9 and 3.11 are not of the Coal Controller.

BOSE

No, they are not. The Coal Controller does not do demand assessment. The demand assessment figure as in Fig. 3.13 is an assessment by the Working Group on Energy Policy. That is the total figure, the production expectation or aim.

MARWAH

May I now request Mr. Bose to give his clarifications.

BOSE

Thank you Dr. Mubayi for your very valued comments. Coming to the points which you mentioned, classification of the resources is not very different from the US system or the Russian system. Proved resource is proved by the continuity of the actual knowledge, say, within a radius of about 200 metres from a bore hole or from the working mine. Estimated is around 1000 metres away from the point of knowledge, and the rest is inferred.

Economic data for different mining systems are with the different mining companies; they have their own cost evaluations. I do not think we have a very clear picture of the economic comparison of the methods of mining adopted at different places showing the comparative evaluation as computed to some base—somewhere the manpower may be in excess. That type of evaluation is to be done on a specific basis, as a research item. But the figures are available at the mine units, yet not reported as such to headquarters in any connection.

Perhaps Mr. Banerjee will answer whether it is possible to construct some supply curve for evaluating marginal cost of production in general, today we are on a steep production curve increase, i.e., every area where production can be increased is automatically tapped. The production increase in the existing mines automatically gets priority to the extent it can.

Data on economics of different methods of transportation are not reported as such; one has to dig out actual records before understanding how much each means of transport costs. In our country, the railway transport system is the cheapest, prevalent, widespread, flexible, and most dependable. It is economical even for very short distances, beyond five miles. Yet, the cost on transportation in reality is different from any figure that may be reported on paper. For example, cost of transportation on road by trucks is a theoretical estimate. Something is reported but the rate fluctuates very sharply with the vagaries of the seasons, with the market demand and other loads that the trucks transport. When you want to take an economic decision based on actual data, I think much more assessment has to be done to arrive at a conclusion. These are not reported higher up, but available in different mine unit areas.

I agree there are some lacunae in the data on end-use of coal. The Coal Controller is supposed to oversee the demand assessment and production of coking coal every month. They have monthly meetings with producers and consumers to settle the quantum that is assigned for the next month. At the same time, any lacuna is adjusted as soon as it is detected, but very often it is detected after long span of variation takes place. But I feel, that something going wrong outside of the system makes a much bigger impact on the variation than what day-to-day coordination can correct. During the late sixties, oil predominance hit us. Similarly, it is the oil price which pushed up demand of coal after 1970-74. The present better availability of oil is perhaps allowing ourselves a smoother curve on the production and coal demand rise. But these are basic issues of economic nature or of overall energy nature which affect the variation of coal demand in a much bigger way than normal coordination between the two parties.

I agree with Dr. Mubayi that environmental aspects are important. I have done a study myself on these, reported at an international congress organized by the East-West Centre in Moscow. The study shows that we are having environmental problems from increased road transportation in a much bigger way than the normal things that you worry about in Western Systems. The road transport has increased five to six times in a decade in some of the coalfields, mainly due to the inadequacy in rail improvements. We have a natural systems of making soft coke which creates a lot of smoke around but this has not increased in 20 years. But the road has added a lot to the environment. On the environmental aspect of coal use, like in power generation, studies of particle emission, of oxides of sulphur and nitrogen, etc., have to be done specifically on chosen coal fields. We have chosen some coalfields and have done some studies. I would say that we have not made very great headway, whether in the control of pollution of air, water, or of land. Studies done at the Central Mining Research Station at Jharia Coalfields for a period of seven years from 1974 to 1981 have brought out that the environment is deteriorating. Of course, the impact of environmental deterioration can physically be seen, instead of seeing the study report. But this is an area which needs to be strengthened.

We have gasification plants in Orissa and Andhra Pradesh. The corporations that are handling these plants are compiling the data but these are not reported in general. The figures that I have seen, are not bad; they are quite good. They are quite clean plants on the whole.

At the moment the situation is such that increased coal demand can be physically seen in different areas. Even if we try to know the total energy demand increases in the respective areas, my personal opinion is that econometric analysis for coal demand assessment may not be so helpful. I am talking of our situation here.

Because electricity is in short supply at most of the places, any more electricity that is generated goes into the system. More the electricity, more is the demand for coal. Sixty per-cent of electricity generated today is by use of coal and the percentage is constantly increasing. Ten years back, it used to be 40 per cent only. This is an area where low grade coal is getting fully married to the electricity generation system. In Coal India, we have a specialist econometrician who keeps on assessing this in a total way. But it is a very personal view, perhaps the coal demand itself should be studied all the time. Perhaps Mr. Banerjee may like to supplement.

MARWAH

While he supplements the comments by Mr. Bose, I would like Mr. Banerjee to emphasize particularly the environmental aspect. He is the current Chairman of the Mine Environment Committee of the Department of Environment.

BANERJEE

Thank you Mr. Chairman. Professor Subba Rao mentioned about the compatibility and accuracy of the data we get. This needs a lot more attention. To give an example here, in Table 3.10, you will see an item called 'haulages', in the list of equipments which has been given to indicate the degree of mechanization. In Eastern Coalfields Ltd., the number has increased from 952 in 1974-75 to 1130 in 1979-80. That shows mechanization is increasing. But what I cannot understand is that in 1978-79 it was 1219. Certainly 100 haulages had not been withdrawn from the mines. I will reinforce what Mr. Subba Rao said on this. The accuracy of data becomes questionable in some cases.

Mr. Bose has mentioned about the statistics on recoveries and conservations. Yes, the coal companies give the figures and the Coal Controller compiles them. But who checks the accuracy of the figures ? There should be some centrol checking of the recovery statements given by coal companies.

Regarding employment, I find that the total employment in the coal industry, in coal India particularly has been steady. Mechanization has increased; production has increased in the same years, by almost 40 per cent. But we have no figure of output per man-shift. The production increase is 40 per cent, the employment is constant, mechanization has taken place, the minimum increase in output per man-shift or productivity, should be 40 per cent. If we had that figure we could have had a cross-check.

I want to ask one more question only. This production planning to meet sector-wise demand talks of coal demand. In a workshop like this, where the focus is on energy, while we are talking of increases in coal production, of 8 to 10 per cent, we should ask what is the growth rate going to be in terms of gross calorific value. Because it is only likely, given the emphasis on open cast mining and consequently also on inferior coal, that the rate of growth in gross calorific value may not be the same as the rate of growth in coal production in tonnage terms. So I would suggest that there should be more information on this, on the product mix. Different sectors cannot all be using the same quality coal. Now I will correct Mr. Bose in one place. The analysis of all the bore hole data generated by MEC and CMPDI is no longer tested by CFRI. It was decided that CFRI should do a little more utilization survey than the routine analysis of bore hole coal analysis, etc. This work has been taken up by coal companies themselves, by private companies and the CMPDI in combination. But very few samples now go to the CFRI.

One of the questions asked of Mr. Bose was on the long run marginal cost. We are doing that exercise ourselves for purpose of finding out the rational coal pricing base. We do realize that long range marginal cost is not the only method of fixing the present day cost of selling price of coal particularly when you know that, unlike other industries, coal mining has one peculiarity of geo-mining where working conditions are changing all the time. The future mines will perhaps be deeper mines, whether opencast or underground. The technology will also be very much different. This is what causes the problem of finding out the best way of fixing a price. We can only project what kind of price will be 20 years hence or 15 years hence. But fixing any coal price now on the basis of long range marginal cost is very difficult. Therefore, do we have any alternative ?

Regarding the environmental aspects, there is a committee of the Government of India which examines each project report before it is cleared by the Public Investment Board or the Cabinet. Examination is not as detailed as in the USA or in some other countries. I know, in the USA, between an investment decision and the actual opening of the mine after public hearings on environment, the time gap may be as long as seven years. We do not want to go into all that right now. We are trying to clear it within two to three months. In that time, we really cannot do much of an impact study. A beginning has been made and the government is quite firm in not clearing any project, unless it has been looked into by the Environmental Committee. We also do not want to become a bottleneck by becoming too futuristic and we try to find out some quick solutions to some of the obvious problems. The coal companies themselves are aware now. The project reports now coming to us already have a lot of provisions, lot of anticipated environmental problems and their solutions. But I agree with you that we are nowhere near the standard reached in other countries. But do not forget that our coal price also is the lowest in the world.

MARWAH

So we come to the conclusion of the session with perhaps a thought that there are several gaps existing in our data base which is available publicly and more data is perhaps needed relating to economics of mining, benefication, transport, etc. I would only like to add one thought that in the context of the micro-aspect of energy planning, the energy cost of transport of coal, by railways, by road, is also an important aspect and needs to be examined.

TECHNICAL SESSION 4

INDUSTRIAL ENERGY DATA SYSTEMS

Chairman : K.S. RAJAN

Theme Paper : D. BHATTACHARYA

Discussants : DAVID JHIRAD
DILIP R. AHUJA
S. RAMANI

FOUR

ENERGY ANALYSES OF THREE ENERGY INTENSIVE INDUSTRIES IN INDIA

D. BHATTACHARYA

INTRODUCTION

A RESEARCH grant on future sectoral energy consumption was awarded by the Ford Foundation to the National Council of Applied Economic Research, New Delhi, in early 1980. For the industries sector, a few energy intensive industries like integrated iron and steel, cement, ferro-alloys, PVC, polyethylene and chlor-alkali were chosen to forecast the demand for energy of Indian industries at the end of this century, on the assumption that a few energy intensive industries such as those mentioned above would explain a large proportion of future energy demand.

This study was based on a methodology known as energy analysis. The topic that I am going to discuss is broadly based on this study already done by the NCAER, but confined to data needs related to the use of energy analysis as a tool for energy planning. More specifically, (a) the method of energy analysis ; (b) data requirements and problem in data collection ; (c) the findings related to the Indian integrated iron and steel, cement and ferro-alloys industries and (d) the relevance of energy analysis would be taken up in that order.

METHODOLOGY AND DATA BASE

In any process, with application of energy, a set of given quantities of inputs changes its form to approach the final form of output(s). As a result, the energy content per unit of materials keeps changing. Energy analysis aims at quantifying the energy content(s) of the final output(s) known variously as energy cost, net energy content, energy intensity or specific energy consumption. Since the underlying principle[1] of energy analysis is that the total energy inflow to a given process through inputs should equal the total energy outflow from the process through output(s), the energy intensity of an output becomes :

$$e=(\Sigma x_i e_i)/y$$

where, e is energy intensity,

x_i is quantity of ith input,

e_i is energy content of the ith input, and

y is quantity of the output.

Every input, x_i, excepting those derived from the earth, in turn is an output of another process, with its own energy intensity. Thus, energy input to a process producing a product y has two components, primary and secondary. The primary energy inputs to a process come from those derived from the earth such as coal, hydro-electricity, petroleum crude, natural gas, etc. The secondary ones come from inputs such as thermal electricity, derived fuels like gasoline, diesel oil, coke, furnace off-gases, feedstocks like naphtha, synthetic gases, derived utilities like steam, compressed air, pumped water, oxygen, and even such energy embodied items as sintered ores, ferro-alloys, steel scrap, etc. Each secondary input can further be analysed at its own primary and secondary input levels. This process can continue to any level backward.

Energy analysis requires that system boundaries and the level to which analysis is to be carried out are defined prior to analysis. Since every production process in a country is intricately connected with virtually all other production processes in the country and even the rest of the world, it is necessary to define a boundary around the process to make it a system.

A system boundary at level 1 would contain only the final process stage[2] and energy analysis would yield the quantum of energy required to carry out this single step leading to the production of the final product. The boundary could be further extended to level 2 to include the processing steps just prior to final processing. At level 3, for example, energy requirements for producing capital equipment are included in the energy analysis. The choice of system boundary and level of analysis depends on the type of problem, data availability and extent to which the analysis is desired to be carried out.

Given a system boundary and a level of analysis, energy analysis requires adequate data as regards quantity of final output produced and corresponding consumption of inputs, both primary and secondary. To ensure the dependability of findings of analysis, sufficient number of observations either in time series or cross-sectional form are also required.

In India, different manufacturing industries collect and maintain their own operating and manufacturing data using practices and traditions established in the pre-energy crisis period. Moreover in many industrial units historical operating data, if maintained, are considered as much confidential as cost data, and are not even divulged to institutions willing to analyse these data, the result of which might ultimately prove beneficial to the industry itself.

Collection, maintenance and analysis of historical operating data to a large extent depend on the attitude of management and tend to vary according to size and vintage of the unit, type of produce, type of management (public or private sector), extent of modernisation, sophistication in instrumentation, etc. For example, the four integrated iron and steel plants maintain very detailed operating and production data. Monthly operating and production data of the constituent subunits of individual iron and steel plants are printed and published.[3] Availability of operating and production data for Indian cement units is comparatively much less. One governmental[4] and one private publication[5], publish unit-wise monthly production figures for clinker and cement, despatches of cement, capacity and capacity utilisation. However, these do not publish any regular information as to consumption of even principal energy items like coal and electricity. For the ferro-alloys industry in India, there are hardly any regular published data, excepting for annual production of different types of ferro-alloys produced by units which feature on the Bombay Stock Market. Similar is the story for virtually all other industries, and there are, in fact, very few industries in India which collect and maintain detailed operating and production data regularly and are willing at the same time to share these with others wanting to analyse the data.

The energy analyst in India, intending to estimate energy intensity of various products following the methodology described before, is thus confronted with the problem of paucity of published data, also known as secondary data. The only alternative left to him is to collect sufficient and detailed primary data from several units manufacturing the same range of products. Therefore, it becomes necessary to design and canvass suitable questionnaires among manufacturing units and pursue the matter till sufficient responses are received. When the number of units is large, a sample survey instead of census is required to be carried out. It is needless to mention that the sample chosen ought to be a true representative of the population on the basis of certain statistical attributes.

Specifically in the present study, Indian integrated iron and steel industry presented no data collection problem, because of availability of yearly published data on production of process intemediaries and final product and consumption of various inputs by the constituent main units such as coke oven, sintering plant, blast-furnace, steel melting shop, blooming or billet mill. Such published data were collected for the year 1970 through 1979 for all the four public sector integrated iron and steel units. For cement units, the NCAER questionnaire contained questions on yearly production of clinker and cement and corresponding consumption of various forms of energy and energy carrying utilities. The observation period spanned 1975 through 1979. The questionnaire was mailed to all the 53 operating units. Finally responses from 32 units were received. Unlike integrated iron and steel units in Indian public sector, the cement industry in India is highly hetrogeneous because of adaptation of two distinct methods of manufacturing with wide distributions in capacity and age. Therefore, the question whether 32 responses were a true representative sample of a population of 53 units, in terms of process, age and capacity was required to be answered with suitable statistical tests. A somewhat similar questionnaire asking only the latest (1979-80) ferro-alloys production data and consumption data on related inputs was sent to all the 12 operating units which manufacture tonnage ferro-alloys by electrothermic process. It should be mentioned here that unlike integrated iron and steel units, where government controls a large proportion of production, and pricing and distribution of iron and steel entirely, unlike cement industry where now about 1/3 of production is in the public sector and pricing and distribution are totally under government control, public control of ferro-alloys industry is minimal. In this context, it was considered prudent to keep the number of questions at a minimum to elicit maximum response. The review period was only one year. This questionnaire contained a question on normative and actual consumption of various inputs required to produce a unit quantity of a particular type of ferro-alloy. This is because many ferro-alloy units manufacture more than one type of ferro-alloy. Nine out of 12 units responded to the questionnaire.

ESTIMATION OF ENERGY INTENSITIES OF SOME PRODUCTS

Based on the methodology and data obtained as described above, estimated energy intensities of several products, final and intermediate, of Indian integrated iron and steel, cement and ferro-alloy industries are given in Table 4.1 along with energy intensity figures, whenever available, for the same or similar products produced in other countries.

There are a few points worth mentioning now :

— Before estimating the energy intensities of main products or intermediates, it was necessary to estimate the energy intensities of some of the inputs, especially fuels like coal, utilities like steam, compressed air, oxygen, pumped water and energy-embodied-items such as ferrous scrap and ferro-alloys in case of steel. Energy intensity figures of utilities are in fact estimated from the same yearly operational statistics[3] of respective iron and steel units.

TABLE 4.1 : **Energy intensities of some selected products—Indian[6] and foreign [7], [8], [9], [11], [12]**

Products	Energy Intensities of Indian Products	Energy Intensities of Foreign Products	Remarks
1. *Integrated Iron & Steel Units*			Indian figures are the average of figures of four public sector units
(a) Coke	8.164	7.731 (UK)	
(b) Sinter	1.238	0.645 (US)	U.S. figures are evaluated based on calorific value (and not energy intensity) of coke.
(c) Pig Iron	7.746	3.894 (US)	
(d) Raw Steel	9.598 (open Hearth)	9.488 (Converter)	
(e) Bloom	10.240	—	
(f) Slab	10.953	—	
II. *Cement Units*			
(a) Dry Process Cement	1.683	1.919 (US) (Long Kiln); 1.673 (US) (Preheater)	Indian figure is the average of figures from 10 dry process units
(b) Wet Process Cement	2.147	2.133 (US)	Indian figure is the average of figures from 22 wet process units.
III. *Ferro-alloys Units*			
(a) Ferro-manganese (High Carbon)	19.829	—	Indian figure is the average of figures from six units in the same year.

(b) Ferro-manganese (Superior grade)	47.642	—	Indian figure is that of one unit only
(c) Ferro-silicon	64.316	29.69 (Norway)	Indian figure is average of figures from 3 units in the same year
(d) Ferro-chrome (High Carbon)	26.603	—	Indian figure is the average of figures from 3 units in the same year
(e) Ferro-chrome (Low Carbon)	69.970	—	Indian figure is the average of figures from 2 units in the same year.

— To estimate the energy intensity or net energy requirement of a product from a process having either energy and/or materials recycling facilities or producing other co-products, it is necessary to credit the gross energy requirement with energy or energy equivalent of materials recycled. Typical examples of these can be found in iron and steel study[6]— coke ovens and blast furnaces produce gases as co-product of main materials; open hearth furnaces of steel melting shops have facilities to recover sensible heat from furnace waste gases.

CONCLUSIONS

The above exercise could lead us to one question: why energy analysis? At least three reasons can be cited :

— To ascertain whether with a given technology, the right amount of energy is spent. If we campare Indian energy intensity figures with those of some foreign products, the disparity in ferro-silicon energy intensities for example, stick out, even assuming that electro-smelting furnaces in Norway have the advantage of economy of scale. The conclusion is, therefore, the ferro-alloy industry in India needs an energy check-up.

— With a given volume of production of a commodity, one can estimate the total energy requirement of the industry, once the energy intensity of the commodity is known.

— Comparison of the existing technology of production with any recently developed technology becomes possible. In cement making, for example superiority of dry process over wet process is obvious. In the field of iron making, direct reduction of iron ore with non-coking coal to produce a highly metalised product, which could be charged in the steel smelting furnace instead of ferrous scrap is beginning to gain importance. One of the complaints against direct reduction process is its high energy consumption, for example, 3.9 G-cal/tonne of product, in case of coal based, rotary kiln process without waste heat utilisation.[10] If the above mentioned figure is the process energy requirement[2], that is, sum of specific energy supplied by all the utilities like fuel gas, steam, electricity, compressed air, pumped water, fuel needed for transport of materials, and so on, it does not appear to be much higher than that of pig iron produced in conventional integrated iron and steel plants, whose average value during the last decade was 3.03 G-cal/tonne.

REFERENCES

1. *Energy Analysis*, Thomas John A.G. (Editor), IPC Science and Technology Press Ltd., U.K., 1977.
2. IFIAS Workshop Report, Energy Analysis and Economics, *Resources and Energy*, Vol. I, 1978, pp. 151-204.
3. *Operating Statistics*, Printed and published yearly by all the integrated iron and steel plants in India.
4. *Cement Production and Despatches*, published yearly by the Cement Contiollor, Ministry of Industry, Government of India.
5. *Cement*, Quarterly Journal of the Cement Manufacturers' Association, India.
6. *Energy Studies—Industries: Integrated Iron and Steel, Cement and Ferro-alloys*, National Council of Applied Economic Research, New Delhi, 1982 (Unpublished).
7. Eriksen, *Energy Recovery in Ferro-Alloys, Industry*, Elkem, Seminar on Smelting, Rio de Janeiro, April 1979.
8. Chapman, Energy Costs : A Review of Methods, *Energy Analysis*, Ibid.
9. Chapman, Leach and Slesser, Energy Costs of Fuels, *Energy Analysis*, Ibid.
10. Matsey, Steel Smelters of the 1980s face a New Era of High Demand for DRI, *33 Metal Producing*, April 1980.
11. Elliot, Uses of Energy in the Production of Steel, *The Steel Industry and Energy Crisis*, Szekely (Editor), Marcel Dekker Inc., New York, 1975.
12. Chiogioji, *Industrial Energy Conservation*, Marcel Dekker Inc., New York, 1979.

DISCUSSIONS

S.S. SRIVASTAVA

I am working in the Central Statistical Organization and I thought that in spite of time lag I should mention the series of records we have been bringing out. We present energy inputs in these papers, state by state and you can have data for a period of almost 20 years, that is, from about 1955 onwards to 1973-74. There are data subsequent to that also which are available in the series up to 1978-79 which could be used for estimation purposes. May be you are using some different kinds of variables. Could you elaborate as to what exactly are the energy variables?

BHATTACHARYA

Perhaps you are referring to the Annual Survey of Industries (ASI). We have seen them and we make use of them. However, the aggregation in the ASI publications is not acceptable as far as energy analysis is concerned. For the ferro-alloy industry I found that, for example, in Maharashtra and in AP where the maximum number of units are located, the ASI publications have not given unit-wise consumption and production figures but instead data have been compiled for the total ferro-alloys production and consumption of corresponding inputs for the entire state.

SRIVASTAVA

Inter-industry input comparison should be possible. They are going to four-digit levels of classification and clubbing is only at that level.

BHATTACHARYA

In spite of all these, we do not get the data we want. Going into the specific case for ferro-alloys again, take the example of one particular unit, FACOR in Andhra Pradesh. They manufacture a wide range of ferro-alloys; that particular industry's data, taken from the Annual Survey of Industries, would give the consumption of various energy inputs, and total production of all types of ferro-alloys. I do not know how to apportion the input data for different outputs.

SRIVASTAVA

I do not agree with you. Actually they give you a breakdown, not the total of energy inputs.

BHATTACHARYA

OK, but when it comes to the output, there is no breakdown and clubbing is there for all types of ferro-alloys produced by that unit. That is of no use to me. But if it went down to such disaggregated levels as ferro-silicon, ferro-manganese, that report would be of some use. A second thing, and equally important, there is always a time lag in the collection of data and publication, when it is available to users. Any significant technological change that has taken place since is not reflected. I am sorry, I should have mentioned the Annual Survey of Industries. But I forgot to mention it in my theme paper.

SHAH

I have three questions to ask. First is about energy elasticity of usage which I believe is ratio of energy consumed by a unit of GDP. This has been found over the last 15 to 20 years in India to be around 1.8. But the same figure in advanced and developed countries appears to be around unity. The question is, what are the factors leading to a difference of the order of 80 per cent? What is the time frame by which we can hope to achieve such a unity figure ? and how?

My second question pertains to data on energy use for a unit of production. I found in various sources a number of differing figures. One that indicated the possibility of certain amount of fabrication of figures was from the Bureau of Public Costs and Prices. Which is a dependable source of what may be called norms of unit of consumption of unit energy per unit of production, given the tendency to over state requirements?

My third question is in relation to the fact that we are overconsuming; you have mentioned a number of instances where our consumption levels are well over international levels. Is it possible to bring any regulation to bear upon this, like the laws in a number of countries like Japan and South Korea? South Korea has adopted Japanese laws, and made them more stringent, like having an energy inspector, and an energy manager. Every organization is supposed to report actual consumption, etc. If some such mechanism is desirable, what form should it take in order to reduce the energy intensity and energy use or to promote conservation?

BHATTACHARYA

Regarding the first question of energy elasticity, energy consumption versus GDP, would Dr. Desai, my colleague at NCAER come to my rescue?

ASHOK DESAI

The energy elasticity tends to be slightly below 1 for industrial countries. Darmstadter has done a cross-sectional study for different types of countries. For developed countries, it tends to be slightly below 1 because the energy requirements for a unit of output fall with technological improvement. It tends to be greater than 1 for developing countries.

SHAH

What does automation do to it?

ASHOK DESAI

Automation does precisely that; it reduces the energy requirement per unit of output. That is what brings down the elasticity below one. In developing countries it is found to be greater than one, partly due to our substitution of commercial by non-commercial energy. So one would expect that as the energy system of developing countries get more commercialized, elasticity would come down towards 1.

MUBAYI

We have been engaged in a project over the last year on international comparison of income elasticity, precisely of the type that has been described. These exercises have been done in addition to the old work done by Dr Darmstadter which I think suffers from many problems. The World Bank and other prestigious organizations have been trying to do the same work. One of the problems in this inter-country comparisons, which is far bigger than the energy consumption figures themselves, is in comparison of GDP. Because these are based on BTUS and so many things per GDP generated. If you adopt international comparison of purchasing power and use this for intercountry GDP comparison, you get a fairly different set of results. And that is what we have been finding. We have not reported it because we are in the process of constructing tables, and so on.

In doing all these comparisons, we have found that they have been a little overstated. The direction of your commonsense argument is that, as the country industrializes and develops, eventually the productivity, including energy productivity, improves over a period of time. This ratio or this figure will tend to drop. In fact in most developed countries it has already dropped below 1, especially after the recent oil price. It is now down to between 0.7 and 0.8. But it is not as high as originally thought of. The Indian figure is a kind of anomaly to me and I would appreciate if someone would seriously check if the number 1.8, which is mentioned in the Working Group Report on Energy Policy, is accurate.

Because in different countries for the same GDP level we are finding a somewhat lower number for this figure.

SHAH

Industrialization is supposed to reduce this to 1. The Working Group was making the point that when the figure comes down, the need for electricity also comes down. But industrialization certainly increases the need.

ASHOK DESAI

I am afraid I have checked the Indian figure and it is really high. It is anomalously high. I tried to partition it. A large proportion of the variance in energy intensity is due to structural changes mainly in the industry. The energy intensity in transport has actually been reducing. In industry it has been rising due to structural changes.

There is also the other factor, that agriculture has been getting commercialized.

NIKHIL DESAI

One thing, in response to Dr. Mubayi, even if you use Krayvis dollar, the energy intensity would appear to be different. But I had done this in some cases. If you take the growth rate, there is no variation in it. If you take GDP growth rate, whether you use the conventional dollar measure or not, growth rates do not often change at all.

MUBAYI

That is not true. At any point of time the official market rate of exchange at which you are converting is fluctuating in some unknown way, which is perhaps unrelated to the way the domestic price index is fluctuating. So, how you value a GDP at a given point of time makes a big difference.

NIKHIL DESAI

Perfectly agreed.

SENGUPTA

Just want to make a comment on the energy intensity which Dr. Bhattacharya was referring to, particularly with regard to disaggregation. We have undertaken a study on energy conservation in industries and the greatest difficulty we have been facing in some of the modern industries is that they do not bother to keep track of the energy inputs into different parts of the activity or product and all that they are interested in, if they are using electrical energy, is in one energy meter which indicates to them that 15,000 units are used per day; and into which products they go is very difficult to disaggregate. This is where one really has to make a guess, unless one does a proper time and motion study, goes whole hog and makes an analysis. We found very great difficulty in a number of industries in Karnataka State.

RAJAN

I would like to add to this. How far this industrial energy data collection is accurate in India is a very problematic question. In the Directorate General of Technical Development (DGTD) at one stage we thought of getting this information from the industries along with their production statistics, not only for energy conservation but also for material conservation and planning. The DGTD was doing that portion of the job also. And I must regretfully say that in spite of our efforts for the last three years, as Dr. Bhattacharya said, the industries have not been responding sufficiently seriously to give correct data, correct analysis, of the energy consumed by them. It is a rather big charge I am making. But it is a fact. Whereas public sector units—which have nothing to lose, and nothing to gain, the energy returns from all of them—including the Railways—are extremely correct and complete enough so that you can do the analyses. In fact, we had put it in as one of the infrastructure inputs that the industries must report once in six months along with the returns. Fifty per cent of them do not report. After one year, when we found that nobody was reporting this, we tried to analyse why they were not. There is the question of shortage. They did not want to tell the truth in many cases. Some of them are over-producing. Also they do not want to report the full production figures vis-a-vis inputs which they have consumed. The third point was that if they did tell the truth—with the system of cuts and other things envisaged—they could land up getting less than what they were getting earlier. The result was that everybody was inflating their actual consumption. So there is both use and abuse.

Again coal, gas or fuel oil consumption figures without full information of the quality are probably meaningless, if you want to know the efficiency of use. And we are so far away from the question of the consumption and quality in the industries that I do feel that the time has come when they *have* to do it. The last speaker also mentioned that he did try. In fact legislation has been passed that an industry can be prosecuted if it does not give these returns correctly. But it is a long process in every case. Nothing has been done about it. But legislation is there today. It has been passed. The industry opposed it. But it has been made into law that the government has a right to demand the production statistics as well as the consumption of raw material or the energy or whatever inputs industries are making for that purpose.

BHATTACHARYA

In answer to Mr. Shah's second question of authenticity as far as energy analysis is concerned, I think perhaps only the NCAER so far has done work on energy analysis of a number of industries. At least we have not got any published energy intensity figure from any other institute in India. I know, the R & D unit of the Steel Authority of India had done a similar energy analysis exercise for one year for the four public sector steel plants. They arrived at certain figures for energy intensities of finished steel, and process intermediates. We at the NCAER have improved upon some of their assumptions. We have not yet published our figures. I do not know if anybody else like NPC has done any work in this regard.

SHAH

It is just merely a question of publishing figures of consumption per unit of production—not analysis of that consumption. Even those norms like how much the textile mills consume per metre of cloth vary with the type of fabrics. The aluminium industry is one big example where is varies all over from 12,000 to 20,000 kWh/tonne. This kind of thing, especially when it is an input or output—one can go off the mark if we do not have very reliable norms. Norms are published by a number of people. I already know three Government of India organizations doing it. I am sure, others must be too. I do not know if the DGTD has its own reference point.

BHATTACHARYA

You mean, those technical norms, coefficient. . . .

SHAH

Well, end result, not analysis of consumption per unit of production for a given industry, like cement, paper, sugar, etc.

RAJAN

Secondly, at present I think it is not really usable unless you compare a unit with that unit itself, is it not? But a cross comparison with one unit becomes very difficult because technology in many of them differ. Secondly inputs received like coal particularly are very different. We have been able to lay down at least certain things, their own consumption, their own best performance when things were very tight and they were doing much better. You can compare their own performance and you can fix some norms.

SHAH

This was an approach strongly advocated for the purpose of load management. Restrictions have to be imposed and by and large they are *ad hoc* restrictions. This is not really very helpful. If one says instead that you are producing so much, you will get only that much of electricity, it will be a whole lot better.

BHATTACHARYA

Now, I can give another example to the question raised by Mr. Shah. In connection with another ongoing study in NCAER, we had to estimate the original equipment demand for conveyor belting. For this we adopted end-use technique and our approach was to cover all the significant consuming industries for conveyer belting. We had these consumption norms that we estimated from the data, that is, the norms of technical coefficients at the various disaggregate levels—conveyor belting, in terms of widths, number of plys, type of construction, and so on. The end-use now would answer questions like, what is the metre of consumption of conveyor beltings per annual tonne of their production. Particularly for the power plants, we have come to figures with narrow deviations. I do not recall exactly the figure. The figures for the newer units, like its NTPC, do not vary much from those of the older power units. This is an example of technical norms. That is the case with so many other industries also. Anyway, this is not connected with the present study.

NYATI

I just want to respond to the point Mr. Shah raised. I am from the National Productivity Council which has been doing a number of energy conservation studies. It is extremely difficult to evolve norms of energy consumption at aggregate level. It is easy to meter electricity, it is easy to meter oil, but coal is altogether a different matter. Most of the industries go on the basis of the Railway Receipt (RR). And everyone knows, there is a big difference between what is the figure in the RR and what one gets. On top of that, you have the variation in terms of quality. And the problem is that when you take all these energies put together, the figures would certainly vary. So it is difficult to arrive at a norm. Number two, as the Chairman has very rightly mentioned, the process technology is varying. It is really difficult until and unless you classify further down each industry group or each industry sector, with respect to process technology. Only then it is possible to arrive at norms.

There could be an industry where the total energy concept is in operation, whereas in a similar industry producing the same product may not have that. Now this would make all the difference. In fact, in our country the industrial sector generates through co-generation only eight to nine per cent of the power that it consumes. Whereas in the case of industrialized or developed nations like the UK the figure is 20 per cent, in Germany 25 per cent, and in Sweden 30 per cent. Now, the fact is that so long as these types of variation in the process technology and other energy conservation measures or technologies exist, it is difficult to arrive at norms or even compare them with international figures. It will not be meaningful until and unless we disaggregate them through process technology.

RAJAN

But almost all industries do it. Their efficiency depends upon it. But they are not made available to the authorities concerned. Not all industries are inefficient enough not to do it. They have been analysing it to improve their own efficiency. It has not been put into a concrete form or made available to organizations which could make use of it. Somehow or other it is kept a closely guarded secret because of various considerations.

KHURANA

Just a comment on this elasticity. I noticed that the energy elasticity in the developed countries in the last three years was slightly on the increase. So we asked an international energy expert. The answer he gave us was that this is because of the fall in proportion of oil in the energy mix. Because oil is more efficient than the substitute, particularly in power generation and industry. So it is surmised that the difference between 1.00 and 1.8 is perhaps explained by the very low proportion of highly efficient oil in the energy mix of a developing country like India.

RAJAN

May we go on to the discussants?

DAVID JHIRAD

I just wanted to make some remarks regarding the kind of data we need in order to apply these analyses. Starting at the very fundamental engineering level, we find that there has been a big movement over the last ten years to do an analysis of industrial processes from the point of view of the second law of thermodynamics, an energy analysis, and when available, a work analysis. I know the National Productivity Council (NPC) here is doing some of that. If one is going to really identify the potential for energy conservation in the industrial sector, one really has to look at each industrial process from the point of view of available work, which is a much better yardstick for gauging the efficiency of the process than simply using BTUs of energy. One has to look at the various steps in the process in terms of how available work is being degraded at all these steps. But this is a big operation—and it has not been done exhaustively even in advanced countries. But pieces of the puzzle are in, in the sense that there is now very exhaustive temperature spectrum that is available at four significant digits for various industrial processes. That gives one some handle on how effectively the energy is being utilized.

So one plea I would make is that all engineers should be trained to examine these processes and to record data using criteria other than BTUs-in and BTUs-out. Because that will not give one a correct handle on how to use potential improvements and efficiency. For example, the major losses in industry are due to three causes. One is the combustion of high grade fuels to produce low temperature heat; heating boilers on premium fuels to make low temperature heat is extremely wasteful of available work. The second one is loss of high grade waste heat to the environment and the third is the particular kind of manufacturing process that is used. If one looks at how one might save available work, using these major sources of inefficiency, technologies are quite proven and there is nothing very exotic about them. They are matched to how these inefficiencies could be improved, for example, heat recovery, heat pumps, co-generation, knowing the thermal and electrical load, and so on and so forth. Summing up all these, one comes up with the very staggering figure for what is possible. For example, in the US, it is quite clear, that something like 25 per cent of the total industrial energy used, could be saved over a period of five years, using proven technology with capital investment less than or equal to that of generating new sources of supply. This kind of figure would not have come out of a first law analysis; one really has to look at the second law analysis and inefficiency in order to get the potential.

Of course, there is another economic argument which is how do you gauge the economies of the system both from the firm level, from the enterprise level, and from the national level. My first point is really on the engineering level. These criteria are also very important if one is trying to look at renewables in the industrial sector because one really needs to know the temperature spectrum of the heat and the kind of electrical requirements in some detail in order to decide which renewable technology can be used.

The second point relates to decision making at the firm level and I think the kind of data required involves some economic and financial data, as Dr. Munasinghe was saying earlier. Even in cases where the technology was shown to be cost effective there are some case studies in the US, where the installation of a bottoming cycle unit in a cement plant did not meet the internal rate of return criteria for that company because the rate of return criteria for energy conservation was set much higher than the rate set for other investments in the company. So if one is anxious to implement in the industrial sector some of these very exciting energy saving approaches, one should really know what are the criteria that are being used by the managers of the firm for investment in energy conservation technology.

At the national level, the matter becomes very shrouded because information is hard to get. For example one study that we have access to at Brookhaven is the unpublished Canadian study. What it tries to do is to look at 40 energy conservation projects in the industrial sectors broken down into various categories, like waste heat recovery, co-generation, process controllers, and so forth, and look at the capital cost per unit of energy delivered. The unit of energy is defined qualitatively and quantitatively and then the question asked is how does it compare with delivering that unit of energy if one would drill for an extra barrel of oil or an extra barrel of gas? What are the costs? What is the cost comparison? And one can essentially draw a graph which basically shows the cost of energy conservation investment versus the amount of energy saved. As you see, (for) more and more energy (saved), the capital cost is expected to go up, and at some point which is around 30 per cent saving in the Canadian example and similarly in the US example, the capital cost of investment becomes comparable to domestic oil and gas capital investment. I think it will be very useful for government decision-makers here to know what these figures are—capital cost for barrel of oil delivered for oil equivalent, for oil, gas, coal, electricity, and so forth, so that when we are making decisions to save energy with the national criterion in mind, we are using figures where we can really compare the energy conservation investment against something else. So then this brings up the fact that investment in the energy sector and even in the industrial sector do require information about all other sectors, and specifically the capital cost to produce that barrel of oil equivalent from all the various sources of energy.

The third point, this is getting away from the orientation of the workshop, is that to make sensible decisions at the national level, one requires access to data from all the various sectors, oil, coal, gas. Going back to the firm level, as I said, we need essential economic and financial data, about the criteria that the firm uses to make these decisions and there is the other category which is technology data that I find is hard to get. A reliable data base on what these units are actually costing in practice is needed. I think that data based on case studies of heat recovery systems and co-generation in the industrial sector is a very important part of an industrial data base. Then, of course, there have to be various tools—as Mohan has mentioned—some of them to reconcile firm level and national level behaviour. And this includes pricing and non-pricing tools pricing regulation, and also direct financial incentive. An intelligent government policy to implement some of these measures could be devised because one would have the criteria designed at both levels. Legislation based on some analysis should encourage technologies that are going to have national significance, and benefit, and for a national cause.

So, just to quickly summarise, the information requirements are not really as simple as they seem but they do fall into categories—from the engineering category, the financial and internal rate of return at firm level to the capital investment possible at national level. Then one hopes techniques can be devised using pricing, financial incentives and regulations, to reconcile firm level to investment level decision. We have to have that kind of data base to have a sensible national policy.

RAJAN

We will now have a discussion on Dr. Jhirad's remarks.

C.L. GUPTA

In our country there is a large scope both for industrial and rural systems to have renewables. But I think it is useless to install solar systems in a place where the tanks and pipes are not insulated. The second point is about the way one sells renewables in this country and partly so in the US as well. Instead of calculating the cost per unit of energy of the renewables in a hybrid system—most of the solar and renewable systems in industry will be hybrid due to reliability requirements, it should be per unit heat. If one says, I can give you a solar collector which gives 80°C, it makes nonsense of both the first and second laws. The question is how much solar fraction you can give, at what cost, what energy cost and at what financial cost. It is not the temperature because it does not matter with oil, if you heat from 50 to 70 or from 80 to 100 and people who are in industry should take notice of this before they accept a system.

AHUJA

I have been asked to speak about environmental effects and the kind of data needed to ensure environmental protection. Since this is a very broad topic, I will try to restrict my remarks to the kind of information needed to know the extent of enviromental degradation caused by industrial processes. There are two basic ways by which industry can contribute to this enviromental protection; one is by consuming less energy, thereby reducing the harm associated with the centralized energy producing stations and the other is by reducing the amount of noxious effluents that they themselves release. Different kinds of data and policy are needed for the two ways and I shall take them up separately.

As we have just heard there is enormous potential for energy saving in the industrial sector. For some companies the energy costs are 10 to 15 per cent of sales. There are three major ways by which energy can be saved in industry and in order of increasing expensiveness, they are: by improved housekeeping, the other is by recovery of waste heat and the third is by redesign of industrial processes to use more efficient ones.

Improved housekeeping does not sound glamorous but we generally underestimate how much energy can be saved by simple things like furnace maintenance, fixing steam leaks, shutting off motor and lights when not needed, etc. There is plenty of anecdotal evidence from the US of companies having reduced their energy consumption between 20 to 60 per cent over a three or five year period just by these kinds of things. But this requires constant attention. Otherwise the benefits soon slip away.

The second major area is recovery of waste heat. This was referred to by Dr. Jhirad and Mr. Nyati. This involves retrofitting like installation of heat exchangers, construction of co-generation facilities which use the steam used in industrial processes to drive a turbine and to produce electricity as well. Up to half the energy used in industry in the US is used to produce steam and two-thirds of that goes up the chimney as waste heat. For at least the major industrial users of energy, there should be an elaborate system to keep track of the amount of steam and electricity used in various manufacturing steps so that the monthly data on cost of energy per pound of product is available. The most successful company in this regard has been Dow Chemicals which has been doing this since 1954. One of the necessary conditions for achieving energy conservation in industry is that the senior management must be convinced first of the desirability of doing this. Also the data is best collected by the industry themselves. Many privately held companies would consider this as proprietary information and would not wish to divulge how much energy they are using per pound of output. I differ with some of the comments that have been made earlier this afternoon. Instead of making it illegal for them to withhold this information, we have to make it in the economic self-interest of the company to collect this data to save energy. The savings should accrue to the company. And our interest is in energy conservation rather than data collection and for that we have to devise ways to make it profitable for them, to do this. We have to devise policies so that national interests and private interests match.

The other kind of monitoring, for the restriction of effluents is best done by government, because most industries are not enlightened to do so themselves. This is the metering of effluents from smoke stacks, exhaust liquid pipes, etc. I would like to take up a minute to discuss the rationale for environmental protection—why do it ? In the final analysis, I think, the justification for environmental protection are unfortunately anthropocentric, if we don't do anything, then our actions will come to harm us, if not now, in future generations, in terms of excess disease and adverse health effects. Most economists are now convinced that it is not proper to use a high discount rate—and some even use zero or negative discount rate—for adverse health effects in the future. Thus far we have treated air and water as free commodities, as infinite receptacles for our wastes. It is true that cost of environmental protection will increase inflation somewhat. But I think that is how it should be. Fair allocation of costs demands that those who benefit from an industrial activity should also bear the burden. And right now, the burden of polluting activity falls disproportionately on those who live down wind and down stream from a plant and not on those who benefit from the activity, namely the consumer of the product and the shareholders. So the prices should go up so that the profits of the shareholders go down and the consumers pay the right price for the product.

Now what kind of data are needed ? We need to measure the concentrations of pollutants in the air and in our water streams. These are called ambient concentrations. Then we need to set ambient standards so as to protect human, animal, and plant health and ensure that the actual concentration do not exceed these. However, to achieve ambient standards, we have to know how much stuff is coming out of the pipes. What is the source inventory, who is putting out how much ? Once we know all the sources for a given region then we can set effluent standards and limit the amounts that can come out of individual pipes.

Let me emphasise here the extreme variability of environmental data and some of the difficulties in metering effluents. Both spatially and temporally, data are very variable. The measurement made in the centre of the room will differ from the one made in the corner. If you take a watersample from the bank of the river, it will differ from the sample taken from the middle of the river and that will differ from the one taken at a depth of 5 metres. Readings averaged over one minute will be different from readings averaged over one hour or one day. All this necessitates that all recording be standardized if any comparable or meaningful data are to be generated. There are similar difficulties with metering technology. The effluents will vary from day to day and will vary with production levels. It is not an easy task to climb a 100 metre track and sample hot gases. The samples of the exhaust liquids sometimes take five to 100 days to analyse in the laboratory. Moreover, who should do this ? I think this kind of data is best collected by government or one of its agencies because I do not believe that it is in the economic self-interest of the industry to undertake these measurements. This is true of both public and private sector industries. Their record of environmental protection is uniformly or equally bad whether you take Russia, or the US, or public sector companies here.

Another point is that once environmental controls are instituted, one is left with sludge or concentrated liquids. Instead of having it dispersed all over the place, you have it in one place. Previously you had air pollution or water pollution problems. Now you have a disposal problem. This is still an improvement. You have all these toxic chemicals and active sludge and you have to dispose them of somehow and for this government needs to collect information of properties of suitable sites. This again I believe is something that government should do and provide. There are many other complications in setting environmental standards. Other things being equal, the fear is that uniform national ambient standards will attract industry to those places where air and water are relatively clean. Thereby diffusing the industry and making everything equally dirty. There are distributive aspects. There are questions whether occupational standards should be different from ambient standards. But all these are later questions. First we have to make an effort and know how much of the noxious substances are out there.

RAJAN

We will go on to Professor Ramani.

RAMANI

I shall try to complement Dr Bhattacharya's paper by a few additional remarks. Dr Bhattacharya referred in his table concerning ferro-silicon where energy intensity is 64.316 for the Indian product as against 29.69 for the Norwegian. My own experience in Hindustan Zinc Smelter in the Electrolysis Plant is very similar and the reason is not far to seek. It has been partly answered by some of the participants. We did not have the basic meter and computing devices on line and the process frequently goes out of control, as one of the participants rightly pointed out. Automation no doubt reduces the energy requirements but a feasibility study has to be done to prove that such automation is economically viable. It is not just a question of reducing energy consumption. It has to be justified as economically viable, even though it may be financially feasible.

Recently, I did some modelling on products where I evolved a measure of the total productivity in terms of four components, viz. the labour, the material, the equipment productivities and all other factors taken. Subsequently I realized that energy productivity is becoming important, and I modified the Cobb-Douglas production function to introduce energy productivity also. I analysed two industries in the city of Madras, and my conclusion with the experience of these two industries is that if an industry has got further resources to spend for increasing productivity, energy productivity seems to be the most sensitive. So, in other words, your first priority should be to allocate resources for energy productivity. If we can generalize for a few more industries on this, that should be an important pointer for this workshop.

I will come to some of the key areas requiring decision support systems. For example, to answer the questions of how energy cost can be controlled, what should be the functions of energy management at industry level, and at governmental level, we have to evolve a strategy like we have in materials management of making an ABC analysis and prioritising industries based on the data available in terms of measures for energy conservation. In fact, in a small study I found that the energy cost as a percentage of the total cost of production is increasing and also as a percentage of the cost of total raw material used was increasing. Over the last ten years, the energy costs are going up in relation to other costs and in all industries, excepting perhaps in the fertilizer industry. Energy costs again need constant monitoring in industries and energy data system will have to support such analysis.

Control of energy costs has to be through management awareness, management commitment, and suitable reporting systems. We must strive to reduce energy consumption per unit of output with monitoring, leakage plugging, better scheduling, minor capital investment, etc. In a few industries that I studied about 5 to 15 per cent reduction in energy consumption was possible. But with an energy management plan, high capital investment, etc., 15 to 30 per cent is possible. I am talking about textile and sugar industries. All this goes to show that if a good industrial data system is available, it is going to considerably enhance our decision-making.

I would like to take a minute to discuss the role of the industrial data system itself. Ultimately the main purpose is to take effective decisions. Anatomy of decision-making at industry level is that basically we have data which is processed into information. This information has to be fed into some type of model. It may be an operational, a purely statistical, or a verbal model. This model takes in the information uses certain criteria, evaluates alternatives and enables us to choose the best alternative. In short, decision making, in management parlance, is choice among alternatives. But however good your model may be, unless your information supports the input to your model, your decision making can be ineffective. And it is here that decision data support systems have to play a vital role. The mere existence of data alone does not guarantee effectiveness of decision making. We have to have a suitable model and that has to be in parallel with decision data support system.

Concerning government's role requiring data system for decision making, I have one or two comments to make. I think government can encourage energy conservation at industry level by providing incentives and rewards based on data systems and information generation. An energy conservation fund can be created; in India, perhaps the Industrial Development Bank and the Industrial Credit and Investment Corporation can provide tax incentives and depreciation for energy auditing, monitoring, conservation, etc. Perhaps DGTD, DST can collaborate with other agencies and using the industrial data system can evolve an energy conservation Act. Of course, I am now told that Acts do not mean much. The Act can provide for higher degree of energy conservation equipment and perhaps tax concessions for energy conservation equipment. The industrial data system could support industrial energy management alerts which can be published by industries, by government, or in combination with several other agencies. Energy standards could be set to identify and certify energy efficient equipment—in developing countries this is very important—and introduce life cycle costing. Dissemination of energy conservation information and monitoring energy productivity on a continuing basis, at least for A-class industries should be undertaken. And, of course, we have to resort to optimal short-range, as well as long-range investment. I can see corporate planning taking deeper routes in Indian industries and corporate planning strategies of industries are increasingly including energy planning in their studies. There is, of course, a need for a total system approach in this context and the objectives of industry must mesh with national objectives and priorities. It is very important that our data systems support such a synchronisation of objectives to the extent possible.

Data systems for technology forecasting which are required by industries, particularly the corporate level, require a considerable amount of improvement. My study recently on a hydrogen energy use projection, using Delphi technique, and a cross impact analysis, done for the Government of India, Department of Science and Technology, using projection for the year, 2020 had to suffer from a lack of data for hydrogen requirements both currently as well as in the future. Already many industries seem to be too late in taking decision for hydrogen as a carrier of energy. The recent Third World Hydrogen Energy Conference conducted in Tokyo last year, more than amply revealed this. The state of art has to be well documented before we resort to modelling of the supply and demand.

Recently I spent a few weeks in the International Institute for Applied Economic Research Analysis where the thinking seems to be 'beware of solar energy claims'. Industry seems to be confused regarding the data which is available for taking a decision regarding solar energy. There is a notion that it takes nearly 80 per cent of the life of the solar collector to recover the energy that went into fabrication of the solar collector itself. So one has to look at this problem from the total perspective. There are many textile industries and others who are thinking of, or have already gone in for solar energy. I think one has to do a total life cycle costing here to synchronise the individual objectives of industries with national objectives. If conservation of scarce resources is a national priority and objective, I think industries will have to modify accordingly their policies. Of course it is very difficult to evolve norms for energy consumption but I am glad to say that one of the participants used the word 'difficult' rather than 'impossible'. Recently, we had the Tandon Committee's Report in India, where norms for materials management have been given. Even though it is a very difficult job, still they came out with some norms. There are many exceptions to norms, of course. But I feel the Tandon Committee Report has introduced a certain amount of discipline among industries, in terms of inventories. Similarly, if you evolve norms, even though they may not be very accurate, it will discipline industry to some extent.

K.S. RAJAN

Any questions?

MUNASINGHE

My intervention is more by nature of a comment than a question. I strongly support the proposition put forward by Drs Jhirad and Ramani that the economics of the process is an important criterion for decision making. I think there is a distinction between the engineering and the technical efficiency of the process and the data (including non-technical data) required to make that calculation from the so-called economic efficiency analysis of the process. Certainly, when it comes to the technical efficiency the second law based analysis is superior to one based on the first law. Because the second law recognizes the quality of heat and inputs into the process as opposed to the first law, which only recognizes BTUs, whatever the origin. But I think one has to go beyond this to the second law efficiency and also look at the total economics of the process. The simple minded diagram in Fig. 4.1 illustrates the point.

This is a plot of efficiency versus total costs, in a process where you have defined the boundary and the energy flow and so on. There is a theoretical limit for the process in terms of second law efficiency. You cannot go beyond it because of physical reasons. The cost of achieving a particular efficiency, rises very sharply as you cross a limit because you are dealing with the cost concept which is a function of energy as well as other inputs such as capital, labour, and material.

The efficiency range that you want to aim for is the one that minimises the total cost, and which recognizes that the other inputs also have economic opportunity costs. So if you are below this range of efficiency, (in A) then it pays you, in economic terms, to improve the thermo-dynamic efficiency of the process. But if you are above this range, (in B) it is not economically efficient to improve the efficiency at that cost, at a high cost of all the other inputs. The argument is really of economic efficiency and minimising the cost of all the inputs where we value the inputs in terms of the so-called economic opportunity costs of shadow prices. This is an important criterion in the decision making process. And simply pursuing technical efficiency for its own sake, may sometimes lead you in the wrong direction. Thank you.

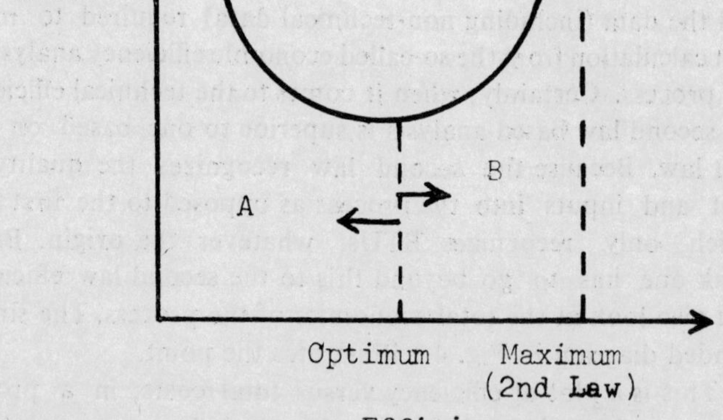

FIG. 4.1

K.S. RAJAN

Gentlemen, I think we have had an useful afternoon discussing the industrial data collection system. There is no doubt in all our minds that collection of energy data is essential and the self discipline by the industry is something which probably will get worked out in this country very soon. There is some talk of incentives being given. I think the last budget itself carried some minor incentives for energy and environmental pollution control etc. I think there is a lot of activity, a lot of thinking in the government that we should give tax rebates for investments on energy conservation as well as for pollution control. I think those will take shape very soon. There is a lot of awareness. How soon it will materialize is very difficult to say. But to get this awareness concretized, probably a system of excellent collection of data is needed. At present we are still groping in the dark in many areas.

[illegible]

Gentlemen, I think we have had an excellent discussion on the fact that the industrial data collection system ... there is no doubt in all our minds that collection of correct data is essential and the act of disclosure by the industry is something which probably will get worked out in this country very soon. There is some talk of incentives being given. I think the last budget itself carried some minor incentives for energy and environmental pollution control etc. I think there is a lot of awareness, a lot of thinking in the government, that we should give tax rebates for investments on energy conservation as well as for pollution control. I think these will take shape very soon. There is a lot of awareness. How soon it will materialize is very difficult to say. But to put this awareness concretized, probably a system of excellent collection of data is needed. At present we are still groping in the dark in many areas.

TECHNICAL SESSION-5

INTERNATIONAL ENERGY DATA SYSTEMS

Chairman : PROF. MAURICE LEVY

Country Papers : CHARUAY BOONYUBOL (*Thailand*)
SUCHAI LIMSUMALEE (*Thailand*)
LU YINGZHONG (*China*)
HUGO ALTOMONTE (*Argentina*)
JOAO LIZARDO R.H. de ARAUIO (*Brazil*)
K.C. SEAL (*India*)
D'ERRICO *et al* (*Dominican Republic*)

Discussant : DAVID JHIRAD

FIVE

THAILAND'S NATIONAL ENERGY DATA SYSTEMS

CHARUAY BOONYUBOL

BACKGROUND INFORMATION

THAILAND is an importer of energy, currently importing about 75 per cent of its energy needs amounting to 78.6 Million Barrels of Oil Equivalent (MMBOE) in 1979. Oil and oil products accounted for 97.7 per cent of this total with hydro electricity (from Laos) and coal providing the balance. Although still considerable, Thailand's dependence on imported energy has fallen steadily in recent years, from 83.8 per cent in 1970 and 81.3 per cent in 1975 to the current 75 per cent.

In volume terms, oil and oil product imports rose 62.2 per cent from 33.2 MMBOE in 1970 to 53.8 MMBOE in 1975, and 42.7 per cent to 76.8 MMBOE in the four years to 1979. Because of the sharp increase in oil prices, the oil import bill has soared from only US $226 million in 1973 to as much as US $2.9 billion in 1980. It is anticipated that the oil import bill will reach US $4 billion mark in 1985. Oil's share of total imports has, consequently, jumped from 12.3 per cent in 1973 to 34.9 per cent in 1980, while 45.3 per cent of export earnings (vs. 14.9 per cent in 1973) now have to be used to finance the enlarged oil bill. Similarly, oil imports amounted to only 2.1 per cent of GDP in 1973 but 8.9 per cent in 1980.

Because of the surge in global oil prices in 1973, the trade and current accounts have deteriorated from a surplus position in 1973 to deficits of US $2.4 billion in 1980. Hence, it is readily seen that the most fundamental problem facing Thailand is the rising price of imported crude oil and oil products. Thailand has been spending a great portion of export earnings on importing this type of energy which has resulted in more and more deficits in trade balance. However, it is envisaged that the country's dependence on costly imported energy is expected to fall in the 1980s because of the increased exploitation of domestic energy resources, especially natural gas, production of which has been on stream since October 1981.

ENERGY POLICY

The National Energy Administration (NEA) set up in 1953 within the Prime Minister's Office is responsible for formulating energy programmes and policies and for regulating and supervising energy activities. The NEA's operations come under the Ministry of Science, Technology and Energy. It is headed by a Secretary-General and is guided in carrying out its functions by the National Energy Policy Committee, which comprises government ministers and heads of various government organizations. The NEA cooperates with various energy-related government departments and agencies : Electricity Generating Authority of Thailand (EGAT), Petroleum Authority of Thailand (PTT), Department of Mineral Resources (DMR), the Office of Atomic Energy for Peace, and the Ministry of Defence's Energy Department. The PTT, headed by Prime Minister Prem Tinsulanond, was set up under the Petroleum Authority of Thailand Act in December 1978, with the responsibility for overseeing development and operations in the petroleum and natural gas sectors, including refining and distribution.

A detailed national energy programme has not as yet been introduced. However, according to the 5th Social and Development Plan (1981-86), the following may be classified as Energy Policy and Targets as spelled out in the Plan.

1. Accelerate the survey and the development of various kinds of indigenous energy resources properly and efficiently.
2. Enforce energy conservation measures, especially in the transportation and the industrial sectors.
3. Adjust energy pricing to be more realistic and provide higher degree of fairness among producer, distributor and consumer.
4. Provide and reserve enough energy for normal and emergency demand consistent with the economic growth and the security of the country.
5. Accelerate the supply of other kind of energy, now proved to be commercially viable, thus reducing amount of imported crude oil and petroleum products.
6. Promote and cooperate with other countries to pool energy reserve and to develop energy sources in order to be able to sell, to buy, to transfer and to exchange among interested countries to increase the economy and the stability of energy supply.
7. Promote both private and public sectors to invest in energy business, which will benefit the country, both within the country and outside country.
8. Accelerate the energy supply and utilization in the rural area with suitable technology and pattern.
9. Establish the operation plan in provision, production and consumption of energy consistent with the environmental conservation plan of the country.

Target

1. The increase in energy demand in the period of the 5th plan is to be kept at the average rate of 4.8 per cent per year.
2. Reduce the consumption of imported crude oil and petroleum products from 74 per cent of the total energy demand in 1980 to 46 per cent in 1986 by using more alternative energy sources and conservation measures.
3. Increase the production of natural gas from the gulf from 250 MMSCFD in 1982 to 525 MMSCFD in 1986.
4. Construct natural gas separating plant in order to increase the benefit from the utilization of natural gas

as much as possible, with a capacity not less than 350 MMSCFD by 1986.

5. Increase the utilization of lignite and hydro power for electric generation, from 210 MW in 1980 to 885 MW in 1986 for lignite and from 1,269 MW in 1980 to 2,013 MW in 1986 for hydro power.
6. Increase the supply and utilization of Non-conventional Energy equivalent to 220-290 million litres of crude oil by 1986.
7. Reduce energy consumption in transportation and industrial sectors through increased efficiency and conservation.
8. Develop international energy projects where necessary and appropriate.
9. Increase the total refinery capacity to about 280,000 barrel per day by 1986.
10. Increase crude oil and petroleum products stockpile for emergency use to not less than 60 days by 1986.
11. Expand rural electrification to new villages of about 31,523 in 1986.
12. Revise energy pricing to be more realistic.
13. Intensify public campaign to educate people on energy problems.
14. Complete Thailand Energy Master Plan by the beginning of 1982.
15. Decrease pollution generated from energy productions and utilization to within acceptable levels.

THAILAND'S ENERGY DEMAND

From 1961 to 1972, energy consumption (including non-commercial) grew 15.3 per cent a year, significantly faster than the economy's real average annual growth of 7.6 per cent. In the 1973-1979 period, annual energy consumption growth slowed to 9.5 per cent, while the economic growth rate increased slightly to 8.2 per cent; in 1980 energy consumption fell by 5.2 per cent; the economy grew 6.3 per cent. Although the sharp drop in energy demand in 1980 was due partly to the slower economic growth, it was largely due to the 80 per cent increase in domestic oil prices in 1979-80.

Tables 5.1 and 5.2 show the cousumption of energy by various sectors in the country from 1974 to 1979.

TABLE 5.1 : **Petroleum products consumption by sector**

	1974		1977		1979	
	MMBOE	% of Total	MMBOE	% of Total	MMBOE	% of Total
Agriculture	6.7	12.9	7.1	10.6	8.0	10.4
Construction	0.8	1.5	0.7	1.0	0.8	1.0
Manufacturing	8.4	16.3	11.8	17.6	13.3	17.3
Utilities	8.0	15.5	12.6	18.8	15.3	19.9
Transportation/ Communications	22.2	43.0	28.8	43.0	32.7	42.5
Commerce/Services and other activities	5.6	10.8	5.9	8.8	6.8	8.8
Total	51.6	100.0	67.0	100.0	77.0	100.0

Source : Nationl Energy Administration.

TABLE 5.2 : **Petroleum products consumption by sector 1979 (% share)**

	Fuel oil	Diesel oil	Gasoline	Kerosene	Jet fuel	LPG
Agriculture	0.01	28.85	0.15	—	—	—
Industry	40.73	9.43	1.56	3.93	—	12.0
Construction	0.12	2.67	0.16	0.39	—	0.1
Electricity/water	56.16	4.37	0.37	—	—	—
Transportation/ communications	0.76	47.65	92.76	6.97	100.0	0.1
Commerce/services and other activities	2.22	7.02	5.0	88.70	—	87.9
Total	100.0	100.0	100.0	100.0	100.0	100.0

Source : National Energy Administration.

Transportation is the largest consumer of energy in Thailand, accounting for 42.5 per cent (32.7 MMBOE) of total petroleum products in 1979 (vs. 43 per cent in 1974). This sector consumed 47.6 per cent of all diesel oil produced in 1979 and 92.8 per cent of gasoline. Increased transportation has been encouraged by a well-developed road network, the construction of which was accelerated during the Vietnam War. The registered number of motor vehicles increased three-fold between 1966 and 1976. A substantial network of inland water transportation has also increased demand for gasoline and diesel fuel. Since the government's conservation efforts in the 1980s can be expected to concentrate on encouraging the use of more fuel-efficient transportation and discouraging gasoline and diesel consumption for private use, transportation of petroleum product consumption can be expected to fall steadily over the next 10 years. Moreover, the government is considering offering incentives to public vehicle owners to use LPG, which will be processed from local natural gas, as an alternative to gasoline or diesel fuel. The government is also encouraging the use of alcogas (an alcohol/gasoline mix).

Industry consumed 13.3 MMBOE or 17.3 per cent of total petroleum product output in 1979 (vs. 16.3 per cent in 1974). This sector accounted for 40.7 per cent of all fuel oil consumed, 12 per cent of LPG and 9.4 per cent of diesel oil. It also consumes about 95 per cent of all coal produced. Energy demand in the industrial sector is likely to grow more rapidly than in other sectors in the 1980s since industry will continue to expand by about 10-12 per cent a year, stimulated by government measures to reduce the economy's dependence on primary production and exports. Moreover, industries that will become increasingly important in the 1980s—chemicals, cement, clothing, electronics, electrical products, and steel—are more energy-intensive than industries that Thailand has emphasized in the past—processed food, textiles. Industry's share of petroleum product consumption can, therefore, be expected to increase significantly.

Agriculture accounted for 10.4 per cent (8 MMBOE) of total petroleum products (principally diesel oil) in 1979, down from 12.9 per cent in 1974. The main reason for this sector's declining share of energy demand is its decreasing relative economic importance. For example, the sector contributed only 25.8 per cent to GDP in 1979 vs. more than 30 per cent in 1974. Energy demand in absolute terms, however, will increase steadily during the 1980s for several reasons: (1) increased farm mechanization; (2) expansion of irrigation networks; (3) increased use of petrochemical fertilizers; and (4) greater rural electrification.

About 9 per cent (6.8 MMBOE) of petroleum products were consumed in the *commercial/services/residential* sectors in 1979, down from 10.8 per cent in 1974. This group's declining share of petroleum products is attributable mainly to the shift by some households from increasingly expensive commercial fuels to cheaper non-commercial energy forms, a trend that can be expected to continue in the 1980s. In contrast, energy demand from tertiary industries is likely to steadily grow, mainly because growing urbanization will boost distribution and service industries, and in turn energy use.

The share of *electricity and water* of total petroleum products demanded was 19.9 per cent (vs. 15.5 per cent in 1974) in 1979 (15.3 MMBOE). The main fuels used in the utilities sector are diesel and fuel oil. In addition to petroleum products, which account for 63.4 per cent of power generation, other energy forms used in electricity generation are hydro (28.1 per cent in 1979) and brown coal or lignite (8.5 per cent in 1979) (see Table 5.3). The industrial sector is the largest consumer of electricity, accounting for 75 per cent of the total generation in 1979. Specific industries that are heavy electricity users are cement, steel, vegetable oil refining, and automobile tyre manufacturing. Sectors other than industry that are significant power consumers are commerce and services (together accounting for 21 per cent of power generated). Agriculture uses less than 1 per cent of all electricity generated.

Electricity's energy demand is likely to grow by about 12-15 per cent a year in the 1980s (vs. about 12.3 per cent annual average growth since 1973) because of (1) the steady industrial growth anticipated over the next five to ten years, (2) the greater emphasis that will be placed on more energy-intensive industries, and (3) the planned expansion of the rural electrification programme (by 1986, 50943 villages are scheduled to be electrified vs. only 17604 at the end of 1979).

TABLE 5.3 : **Structure of installed generating capacity**

Installed generating capacity	The last year of the First National Economic and Social Development Plan 1966	The last year of the Second National Economic and Social Development Plan 1971	The last year of the Third National Economic and Social Development Plan 1976	1980
Hydro (%)	32.03	36.49	35.79	36.83
Oil-Fired (%)	43.26	39.17	52.45	51.55
Diesel (%)	24.71	11.67	5.27	3.35
Gas Turbine (%)	—	12.67	6.49	8.27
Total (%)	100.00	100.00	100.00	100.00

Source : Electricity Generating Authority of Thailand (EGAT).

DEVELOPMENT PLAN OF ENERGY SOURCES

Since Thailand currently imports approximately 75 per cent of its energy needs, it is particularly vulnerable to any interruption in overseas supplies. Hence, the government is exerting its utmost effort to minimise the country's dependence on imported Middle East oil through supply of domestic energy sources which accounted for 23 per cent of total energy needs in 1979. Given the extent of these resources and the likely prospects of developing them, the most realistic estimate is for this share to rise to 35.1 per cent by 1985 and 42.6 per cent by 1990.

The chief energy resources being developed in the country are described in the following sections.

Natural gas

Hydrocarbon exploration in Thailand has been reasonably successful. Union oil first discovered natural gas deposits in the Gulf of Thailand in January 1973. Texas Pacific made its first discovery in the Gulf in 1977. Both companies have since made further discoveries. As of March 1981, 72 wells had been drilled, 41 of which contained hydrocarbons. This represents a success ratio of 1 : 1.8. Known natural gas resources are now estimated at about 11.5 trillion cubic feet (1.8 billion barrels of oil equivalent). At a daily production rate of about 500 million cubic feet, the reserves would last 60 years. At the present planned production rate, reserves would last until the year 2017 (see Table 5.4). The PTT estimates that natural gas potential resources could amount to as much as 20 trillion cubic feet (3.2 billion barrels of oil equivalent).

Exploring for further oil and gas reserves is continuing in the Gulf of Thailand. Interest in searching for oil and gas onshore has also picked up. Esso is currently drilling in the northeastern Korat Plateau. Shell started drilling onshore in June 1981 in the Sukhothai Province, in the Central Plain. Both Esso and Shell rate the chances of finding commercial quantities of oil at one in 10. Shell recently discovered gas and oil in its experimental well, and commercial exploitation at a rate of 5000 bpd is expected to commence soon in early 1983. Esso also, has had a showing of gas in the well it is currently drilling.

The PTT, which is responsible for exploiting and developing the gas, has identified possible options for its use :

1. As an alternative fuel to oil in power generation, If 1,000 million cubic feet a day are available by 1990, fuel applications are expected to take 60 per cent of the availability. Natural gas will also be used in the private sector for power generation. The cement industry will use 40-50 million cubic feet per day starting in mid-1982.
2. A gas separation plant to produce LPG. This is expected to use about 150 million cubic feet/day by 1990.
3. A petrochemical complex. This would use 50 million cubic feet/day.
4. A fertilizer plant based on ammonia and soda ash. This would use a further 100 million cubic feet/day.

TABLE 5.4 : **Installed electric generating capacity as on January 1981 (EGAT only)**

Plant Type	No. of units	Capacity (MW) Installed	Capacity (MW) Ultimate	Average energy capability (GWh/yr)
A. *Hydroelectric Plant*				
Bhumibol	6	420	686	1,550
Sirikit	3	375	500	965
Ubolratana	3	25	25	65
Sirindhorn	2	24	36	73
Chulabhorn	2	40	40	115
Kang Krachan	1	19	19	70
Nam Pung	2	6	6	15
Srinagarind	3	360	720	1,179
Sub-Total	22	1,269	2,032	4,032
B. *Thermal Power Plant*				
North Bangkok (oil)	3	237.5		1,660
South Bangkok (oil)	5	1,300.0		9,100
Mae Moh (lignite)	2	150.0		985
Krabi (lignite)	3	60.0		300
Surat Thani (oil)	1	30.0		210
Sub-Total	14	1,777.5		12,255
C. *Gas Turbine*				
Ao Phai	1	15		13
Nakhon Ratchasima	1	15		13
Udon Thani	1	15		13
Hat Yai	3	45		40
Surat Thani	5	75		67
Bang Pakong (1)	3	180		946
Sub-Total	14	345		1,092
D. *Diesel Power Plant*				
Chiang Mai	3	3		3
Mae Moh	9	9		8
Phuket	4	10.6		9
Nakhon Si Thammarat	2	2		2
Bang Lang	5	5		4
Khao Laem	5	5		4
Sub-Total	28	34.6		30
Total	78	3,426.1		17,409

Source : Electricity Generating Authority of Thailand (EGAT).

5. The manufacture of methanol as a substitute for diesel fuel, the major petroleum product used in Thailand.

The PTT's first two options will have priority. The order in which the other options will be developed will depend on the initiatives that come from the private sector. Bangkok sources say, however, that unless the well-head price negotiations are concluded soon, there will not be sufficient gas for any more than the first two options.

Most of the natural gas produced in the first five years—200 million cubic feet/day of the 250 million produced in the first three years and 325 million in the fourth and fifth—will be sold by the PTT to EGAT to replace fuel oil in generating electricity at the Bang Pakong and South Bangkok power plants. Electricity Generating Authority of Thailand (EGAT) has installed four generating units that will use gas as a fuel, each capable of generating some 60 MW of electricity, at the Bang Pakong plant. It has also modified three oil-fired generators to use natural gas as fuel at the South Bangkok plant. Electricity Generating Authority of Thailand (EGAT) is expected to generate about 840 MW of electricity, using natural gas, each year. Siam Cement, the country's largest company, will use 40-60 million cubic feet/day to generate power for its cement plants in Saraburi Province, north of Bangkok. A spur pipeline has been built, connecting the plants with the main pipeline.

A US $320 million gas separation plant, at the port of Sattahip, will process natural gas into LPG and other gases. Production is expected to be about 450,000 MT a year when the plant comes onstream in 1983. The plant will produce such products as propane and butane, which can be used for household use (about one million litres/day), and ethane and methane (about 90,000 litres/day), which form the principal feedstock in fertilizer production. A US $450 million fertilizer complex has been proposed with the capacity to produce 496,000 tons of ammonia, 396,000 tons of urea, 660,000 tons of sulfuric acid, and 231,000 tons of phosphoric acid a year. LPG is also being considered as a substitute for gasoline for passenger cars and farm tractors. A second LPG plant is planned as natural gas production increases, involving an additional investment of about US$220 million. A special wharf and storage facilities will be constructed as part of the gas separation project.

Several private sector industrial projects have been proposed to use the gas, either as a feedstock or as a fuel. The PTT Director has stated that industries that intend to use gas as a feedstock rather than a fuel would be given priority in allocation. Gas demand for such industrial plants on the drawing board is approximately 145-190 million cubic feet per day. The proposed projects include the following :

1. Siam Ferro Industry wishes to invest US $77 million in a 400,000 ton capacity sponge iron plant, which will substitute part of the scrap iron imported. The IMF has agreed to act as technical adviser and financial coordinator. A feasibility study is currently being prepared. Once the project receives Board of Investment approval, it will take three years to complete.
2. Asahi Glass has announced plans to build a gas-fired sheet glass factory.
3. Thai glass industries is already building one of the largest glass container factories in Asia, close to the pipeline.
4. A US $2 billion integrated steel project to manufacture steel bars and sheets has also been proposed. The plant would require as much as 400 MW of electricity generating capacity. Because of its size, further study of this project will have to be made, which would take at least seven years to construct.
5. Several proposals have been made to manufacture both high density and low density polyethylene, polypropylene, and other resins.

Oil shale

Low grade oil shale deposits amounting to about 8,000 million tons, capable of yielding about 6,000 million barrels of oil shale, have been located in Mae Sot District of Tak Province. The deposits have not yet been exploited because of lack of financial support. Some private sector firms have recently indicated interest in working with the government in developing them. They include two foreign firms, Lurgi of Germany and C. Itoh of Japan. Other firms have been reluctant to accept the government's invitation to join the project because of uncertainty concerning its commercial viability.

Hydro

This energy source supplied 5.7 per cent (5.8 MMBOE) of total energy needs in 1979 (6.6 per cent if non-commercial energy is excluded). Imported electricity from Laos raised this share to 7 per cent. Hydro's contribution is expected to rise to 7.3 per cent (10.0 MMBOE) by 1985 and 11.9 per cent (23 MMBOE) by 1990, based on the number of dams under construction or planned in the 1980s. However, allowing for the fact that not all dams may be completed on schedule and/or operate at anticipated capacity (because of low rainfall), a more likely estimate of hydro's contribution is 8.9 per cent and 18.7 MMBOE respectively.

Hydroelectric capacity is currently 900 MW, provided by several multipurpose dams. The largest is the Bhumibol Dam with a capacity of 420 MW. The 900 MW of current hydro capacity represents only 4.2 per cent of Thailand's 10,000 MW hydroelectric potential.

Hydro capacity is being expanded through the construction of more dams: Srinagarind (720 MW capacity), Pattani (72 MW), Lower Quae Yai (38 MW), and Khao Laem (300 MW). The Srinagarind dam, 155 miles from Bangkok, will meet about 30 per cent of the country's power needs when completed. The World Bank provided US $80 million for the Khao Laem Project, which is located 265 km northwest of Bangkok. EGAT plans to build a further five dams by 1992: Lang Suan (135 MW), Chiew Larn (250 MW), Upper Quae Yai (580 MW), miscellaneous hydro (300 MW), and Quae Yai pumped storage (1,000 MW). Chiew Larn will be built in Surat Thani Province and will be the third and largest hydro-power project in Southern Thailand. Due to be commissioned in 1986, the dam will have a generating capacity of 240 MW. This will complement the Pattani dam's 72 MW hydropower and a further 180 MW from the Lang Suan Project, due to come onstream in 1984. If all the dams are completed on schedule, installed hydro capacity will amount to 4,294 MW or about 45 per cent of hydro potential in 1992. This will represent 35.5 per cent of the country's electricity generating capacity of 12,086 MW.

A major problem with hydro power is that utilization falls short of capacity during periods of drought. Although Thailand's existing hydroelectric plants can theoretically meet 26 per cent of the country's electricity needs, the drought in 1980 reduced this to only 11 per cent. Moreover, the large-scale dams have not contributed as much as was originally targeted since in times of heavy rains, water runs off denuded forest lands. Reservoir supplies have, therefore, to be tapped in times of drought.

Lignite

About 3 per cent (3 MMBOE) of energy needs were met by lignite (brown coal) resources in 1979 (3.4 per cent if non-commercial energy is excluded). This share is expected to rise to 4.4 per cent (6 MMBOE) in 1985 and 10.3 per cent (20 MMBOE) in 1990. However, since all the lignite projects may not be completed on schedule, 5.1 MMBOE and 15.6 MMBOE would seem more realistic estimates of lignite's output in 1985 and 1990 respectively.

Lignite reserves are substantial. Recoverable deposits are estimated at 380 million tons and probable reserves at 400 million tons. Production from the three operating mines more than doubled to 1,356,000 tons in 1979. New lignite mines, financed with World Bank loans, are to be opened up in 1983. The NEA has identified 11 lignite development projects—nine for power generation and two for industrial use. Currently, almost 10 per cent of electricity in Thailand is generated from lignite-fired power stations but the government hopes this will rise to 30 per cent in 10 years.

Three lignite power plants are now in operation at Mae Moh in Lampang Province in Northern Thailand, with a total capacity of 225 MW. Work is continuing on a fourth and fifth lignite-powered plant, each of which will have a capacity of 150 MW. The fourth lignite power plant in Mae Moh scheduled for completion in 1984, will require an investment of US $370 million Funds for the project have been provided by Saudi Arabia, ADB, Switzerland, OPEC, and a commercial bank. Another lignite-fired power plant is located in Krabi province in southern Thailand. The plant has three units each with a capacity of 20 MW. EGAT plans to construct a further four lignite power plants of 300 MW capacity each. When the nine plants are in operation, they will consume about one million tons of ore each month. By 1990, therefore, lignite will be used to generate power of almost 1,800 MW, substituting for about 20 MMBOE a year.

In addition to using lignite in power generation, plans are also being made to develop lignite deposits for industrial uses. The government, together with a group of private firms, has set up a company, the Thai Lignite Company, to develop deposits in Lamphun and Lampang Provinces. The private sector companies involved in the project are mainly local cement firms, the largest potential consumers. Proven reserves in Lamphun are estimated at 20-25 million tons and Lampang at 35-40 million tons. Lignite demand for the cement companies is expected to be about 792,000 tons a year while other industries such as tobacco curing plants would require 400,000 tons a year. This estimated demand would make exploitation of the reserves commercially viable, and would allow the supply to last 60 years. Siam City plans to spend US $15-20 million in the next two years to convert its two cement plants from oil to coal use. Siam Cement also plans to convert some of its facilities while Jalaprathan Cement is already converting its plant.

Local coal may not be sufficient to meet the accelerating demand that will result from the various conversions from oil to coal use. Coal may, therefore, have to be imported. Australia has indicated interest in supplying the fuel. Australian experts have prepared a study on the construction of a coal-handling port southeast of Bangkok and the creation of a national coal purchasing authority to manage imports of up to 12 million tons by the 1990s.

DOMESTIC ENERGY CONSUMPTION

Thailand has, in 1980, a total population of 46.961 million with 14.275 million living in 2.260 million households in the urban area and 32.686 million living in 5.540 million households in the rural area. The annual population growth rate is relatively high, at 2.5 per cent.

The National Energy Administration (NEA), conducted a rural and urban energy survey in 1980 and its findings are tabulated in Table 5.5. The total energy consumed in the rural sector was estimated at 46.808×10^{12} kcal, of which 70.92 per cent were used for cooking, 8.09 per cent for space heating, 7.30 per cent for household industries, 6.35 per cent for lighting, 5.97 per cent for insect repelling, 0.91 per cent for ironing and 0.46 per cent for water pumping. In comparison, in the urban sector, 3.611×10^{12} kcal were consumed, and 42.20 per cent were used for cooking, 24.73 per cent for lighting, 15.43 per cent for air-conditioning, 6.92 per cent for refrigeration, 6.23 per cent for ironing and 4.29 per cent for television viewing.

The distribution of sources of energy used in rural areas is: 41.60 per cent was from charcoal, 44.57 per cent firewood, 5.68 per cent kerosene, 5.08 per cent paddy husk, 1.46 per cent agricultural residues, 0.75 per cent electricity, 0.36 per cent diesel, 0.25 per cent LPG, 0.14 per cent calcium carbide and 0.11 per cent gasoline.

TABLE 5.5 : Estimated domestic energy consumption in rural and urban households, 1980, by energy form and use

(in 10^{12} kcal and percentage)

Covered Sources	Rural[1]			Urban (Bangkok)[2]		
	Total	%Use	%Source	Total	%Use	%Source
1	2	3	4	5	6	7
By Use	*46.808*	*100.00*	—	*3.611*	*100.00*	—
Cooking	33.196	70.92	—	1.531	42.40	—
Lighting	2.974	6.35	—	0.893	24.73	—
Space heating and cooling	3.786	8.09	—	0.557	15.43	—
Household industry	3.414	7.30	—	—	—	—
Ironing	0.428	0.91	—	0.225	6.23	—
Water pumping	0.216	0.46	—	—	—	—
Insect repellent	2.794	5.97	—	—	—	—
Other : Television and Radio	0.090	—	—	0.155	4.27	—
Refrigeration	—	—	—	0.250	6.90	—

TABLE 5.5—*contd.*

1	2	3	4	5	6	7
By Sources :						
Charcoal	*19.469*	*100.00*	*41.60*	*0.477*	*100.00*	*13.21*
Cooking	18.936	97.26	—	0.477	100.00	—
Ironing	0.393	2.02	—	—	—	—
Household industry	0.140	0.72	—	—	—	—
Fuel Wood	*20.863*	*100.00*	*44.57*	—	—	—
Cooking	13.826	66.27	—	—	—	—
Space heating	3.616	17.33	—	—	—	—
Household industry	1.234	5.92	—	—	—	—
Insect repellent	2.187	10.48	—	—	—	—
Kerosene	*2.657*	*100.00*	*5.68*	—	—	—
Lighting	2.647	99.62	—	—	—	—
Cooking	0.010	0.38	—	—	—	—
Electricity	*0.441*	*100.00*	*0.75*	*2.271*	*100.00*	*62.89*
Lighting and other	0.164	37.19	—	0.893	39.32	—
Cooking	0.059	13.38	—	0.191	8.41	—
Cooling	—	—	—	0.557	24.52	—
Ironing	0.035	7.94	—	0.225	9.91	—
Television	0.090	20.41	—	—	—	—
Refrigeration	—	—	—	0.250	11.01	—
Water pumping	0.093	21.09	—	—	—	—
Paddy husk	*2.379*	*100.00*	*5.08*	—	—	—
Cooking	0.246	10.34	—	—	—	—
Household industry	2.040	85.75	—	—	—	—
Space heating	—	—	—	—	—	—
Insect repellent	0.093	3.91	—	—	—	—
Other agricultural residue	*0.684*	*100.00*	*1.46*	—	—	—
Space heating	0.170	24.85	—	—	—	—
Insect repellent	0.514	75.15	—	—	—	—
L.P.G.	*0.119*	*100.00*	*0.25*	*0.863*	*100.00*	*23.00*
Cooking	0.119	100.00	—	0.863	100.00	—
Benzene	*0.051*	*100.00*	*0.11*	—	—	—
Water pumping	0.051	100.00	—	—	—	—
Diesel	*0.170*	*100.00*	*0.36*	—	—	—
Water pumping	0.072	42.35	—	—	—	—
Lighting	0.093	57.65	—	—	—	—
Calcium Carbide	*0.065*	*100.00*	*0.14*	—	—	—
Lighting	0.065	100.00	—	—	—	—

Sources :

1. NEA : Report on the Rural Household Survey (1980).
2. NEA : Report on the Bangkok Urban Household Survey (1980).

While in Bangkok, the capital, electricity was the most used energy form at 62.89 per cent followed by LPG at 23.90 per cent and charcoal at 19.21 per cent.

The survey concluded that the main end-use of energy in the domestic sector is cooking using charcoal and firewood. This has already forced the country to begin importing charcoal from her neighbouring countries, as Thailand has now less than 25 per cent of its total land areas left as dwindling evergreen forest reserves.

ENERGY PLANNING AND MANAGING BODIES

The National Energy Authority of Thailand, now under the Ministry of Science, Technology, and Energy, was set up on January 5, 1953, and empowered to deal with all problems relating to energy including planning and managing the country's energy needs and supplies, as well as developing indigenous energy sources. There are other government organisations that collaborate with the NEA and chiefly among these are:

1. Natural Fuel Department, Ministry of Industry, whose primary task involves granting concessions for Petroleum and Gas Exploration throughout the kingdom.
2. Industrial Oil Department, Ministry of Industry, which looks after all oil refining activities in the country.
3. Petroleum Authority of Thailand (PTT), an organization recently set up to take care of the nation's oil and gas imports and domestic distribution network.
4. The Electricity Generating Authority of Thailand, and
5. The Provincial Electricity Authority.

The national energy data systems are compiled, updated and edited in various forms of statistical reports by the NEA annually. Realizing the importance of the energy problem facing Thailand, the United States Agency for International Development is currently assisting the NEA, by financing and providing expertise, in its 5-year Renewable Nonconventional Energy Project, at the end of which, baseline survey reports on various forms of non-conventional energy will be available for future planning of energy issues by the government.

REFERENCES

Prapath Premmani, *Energy in Thailand*, National Energy Administration, June 2, 1982.

Sompongse Chantavorapap, *Domestic Energy Utilisation*, National Energy Administration, 1982.

Kulthorn Silapabanleng, *Transition from Oil to other Energy Sources in Thailand*, Paper Presented at APESC V Workshop, Honolulu, Hawaii, May 17-21, 1982.

N.E.A. Publication, *Baseline Survey Report*, Volume 1: Survey Overview and Guide to Data, 1982.

N.E.A. Publication, *Annexes to Project Paper for Renewable Nonconventional Energy Project*, USAID/Thailand, July 1979.

Bank of America, *Economic Report*, September 7, 1981.

N.E.A. Publication, *Electric Power in Thailand*, 1981.

N.E.A. Publication, *Thailand's Energy Situations*, 1980.

SIX

ENERGY DATA SYSTEMS IN THAILAND

SUCHAI LIMSUMALEE

INTRODUCTION

PRESENTLY, Thailand is dependent on imported oil which accounts for 68 per cent of the total energy. Though Thailand used about 76 million barrels annually or only 1.6 barrels per capita which is much less than consumption in developed countries, all of oil (crude and finished products) used is imported. This is presently a major item (29 per cent of total commodity imports) affecting the deficit in the balance of trade of the country. An aspect of the general energy policy and planning which has been adopted is to encourage exploitation and development of indigenous energy resources (conventional and non-conventional energy) with a view to reducing dependence on imported oil as well as to promote efficiency in the distribution and use of energy.

Accurate inventory and systematic collection, analysis, and presentation of national energy data system are needed as inputs to energy planning and policy formulation. The National Energy Administration, a government department under the Ministry of Science, Technology and Energy, is responsible for the national energy data centre on both conventional and non-conventional forms of energy.

ORGANIZATIONAL APPARATUS FOR ENERGY PLANNING

The National Energy Administration (NEA), established by Royal Decree in 1953, has very broad powers and duties, including:

1. To procure and to establish energy works for the development of the country and for the welfare of the people;
2. To conduct research, experiments, inspections, surveys, and to compile statistics on locations suitable for establishing hydro-electric power works or places from which materials for the production of electric and other energies would be derived;
3. To lay down policy and devise control for all production of energy, such as electric energy;
4. To provide, control, construct, purchase, sell, lease, or close down any energy production works and the transmission or the distribution systems;
5. To lay down regulations and to ensure safety from dangers resulting from the energy works;
6. To establish standards and to prescribe rates for the sale of energy;
7. To aid and promote the utilization of energy for economic developments such as agriculture, handicraft, industry, commerce, and communications;
8. To bring about other benefits jointly coupled with those accruing from energy works, such as irrigation, flood control, and navigation.

The National Energy Information and Documentation Centre, Thailand (NEIDCT)

The NEIDCT is being developed and implemented by NEA over a period of five years (1981-86). Its objective will be to establish a centralized source of information on energy, from which various government departments, public utilities and private bodies can draw for preparing studies and conducting research needed for planning and policy-making activities. The NEIDCT's responsibilities will include:

— Collection, processing, and publishing accurate, complete and up-to-date energy statistics, and related information.
— Designing a system to answer specific questions put forward by private and public sectors on energy statistics and data.

The other major elements in the energy sector are the four major energy supply institutions (EGAT, PTT, PEA, and MEA), the final demand sectors (namely, transportation, agriculture, industry), and a series of committees and sub-committees. Three major committees and about 19 sub-committees are responsible for coordinating the various entities of the energy sector. These committees and sub-committees are discussed below.

National Energy Board Committee

The National Energy Board Committee was established as part of the original NEA Act. This committee is comprised of representatives from the principal public sector agencies in the energy sector. It is chaired by a Deputy Prime Minister, although the original act shows the Prime Minister as chairman.

This committee has 13 sub-committees. These sub-committees appear to have been established in response to the dynamics of the energy sector. Indeed some sub-committees are replaced by other sub-committees from time to time.

National Petroleum Policy Committee

The National Petroleum Policy Committee was established by the Cabinet. It is chaired by the Prime Minister.

This committee has two sub-committees :

(i) Sub-committee on Oil Fuel Price Determination, Compensation, Procurement, and Distribution; and
(ii) Sub-committee on National Refinery Capacity Expansion.

Power Sector Policy Development Committee

The third major committee established by the cabinet to coordinate activities in the power sector is also chaired by a Deputy Prime Minister. It is composed of two sub-committees:

(i) Sub-committee on Tariff Adjustments;
(ii) Sub-committee on Tariff Structure; and in addition, a Load Forecast Working Group is chaired by NESDB under the Sub-commi ttee on Tariff Adjustments.

ENERGY DATA IN PLANNING AND POLICY FORMULATION

The role of energy data in planning and policy formulation is especially significant. Energy data on resources, production, consumption, pricing, taxation, and so on, are essential not only for sound energy planning but also enhance the importance and relevance of overall national, economic and social planning in the country. Not only is the accurate inventory of basic and analytical energy data needed as an input to energy planning and policy formulation, other information such as the relevant technological, financial, economic and social aspects are also required.

As discussed very briefly, there are two categories of data or information needed for energy planning. The first one is information influencing directly and indirectly the energy system, such as the pattern and level of economic growth, income distribution and balance of payments and so on. The second one deals with information on the energy sector itself covering energy consumption, resources data, energy pricing, data on energy technologies, etc.

In Thailand, the collection, processing and presentation of statistical energy data is carried out by the office of the National Energy Administration, which issues several periodic publications. Three separate statistical reports concerning energy information in Thailand are published annually, namely Thailand Energy Situation, Oil and Thailand, and Electric Power in Thailand. These reports are intended for direct use in the formulation of long-term and short-term national energy plans and policy, for project planning, and in the investigation and study of the development prospects of particular projects such as the natural gas development project and in a study on the capacity expansion of refineries in Thailand. There are also several other periodic reports such as an annual statistical hydro-electric publication, used as an input in the planning and development of hydro-power resources.

The Thailand Energy Master Plan (EMP) Study is the optimised plan of energy demand and supply which will cover the short-term and long-term energy plan for development, provision, production, transportation, utilization, reservation and every energy conservation in the country. The EMP study is being carried out within the framework of systems analyses requiring integrated consideration of the various work components or inputs. The basic data on energy, including specific statistics, special investigation of a geological nature, extraction technology, and the economies and cost affecting adoption of such technology for production, and so on, would be required and modulated in terms of both quality and quantity of available statistics for accomplishing the EMP Study.

REVIEW OF CURRENT ENERGY DATA SYSTEMS

The national energy data systems that tend to be classified into three basic components of the energy sector are sources and supply, transformation and distribution, and consumption. In turn, each of these components is divided into elements. The elements for source and supply components are :

1. *Energy sources materials* : Identifying and measuring primary energy reserves. (Crude oil, natural gas, lignite, hydro-potential, oil shale etc.)
2. *Primary energy supply* : Measuring primary energy supplies, in physical units and in terms of energy equivalence, that is
   - (i) Energy production :
     - (a) production from indigenous sources (including the production of traditional renewables and non-conventional renewables).
     - (b) production from foreign sources (such as production of petroleum products from crude oil import).
   - (ii) *Imports of energy products* : At this point, concerned with energy imported for use as fuel only (secondary form) such as petroleum products,

coal and gas, and not with the primary energy such as crude oil which has already been touched upon earlier.

(iii) *Exports of energy products :* Exports are always negative to reflect a decrease of energy supply.

The second basic component of the energy sector, transformation and distribution, represents various elements: crude oil refining, electric power generation, gas manufacturing, and energy distribution. With the possible exception of some traditional supply sources (wood, charcoal, bagasse, husk), renewables like solar and wind, and organic (biogas, biomass, refuse, and agricultural residues), all sources of supply go through transformation and distribution stages before entering into final demand. This illustrates that final energy users are in effect expressing a demand for the total amount of primary energy sources needed to meet their requirements for secondary sources.

The third component of the energy sector is final demand or end-use energy demand. Firstly, total energy requirement is defined as total primary energy available plus net imports, additions to or subtractions from stockpiles, minus bunkers. That is, the total indigenous primary production minus the net movements of products out of the economy for the relevant period. The final demand is composed of transportation, industry, household and tertiary (including government), and rural sectors.

ASSESSMENT OF STRENGTHS AS WELL AS GAPS IN PRESENT SYSTEMS

The following highlights the elements within main categories where national energy data systems tend to be weak.

1. Adequate data on energy consumption by detailed sectors and end-use categories are lacking, particularly for the industrial and household sectors (rural and urban areas). In the household sector it would be useful to have a more accurate description of the types of commercial, non-commercial and renewable fuels which

meet these demands and the potential for substitution. A special survey should be conducted.

2. The presentation of energy statistics in their original units are clear, but when transferred to an equivalent unit or to primary resources, the conversion and transformation factors, and losses should be stated clearly.
3. It should be recognized that non-energy products and also non-statistical information including market intelligence in the energy field should be recorded and given as much attention as the energy statistics systems.
4. Projection of future energy demand should take into account energy substitutions, particularly increased migration from rural to urban areas that will promote a significant substitution of "non-commercial" for commercial sources of energy. Scarcity of fuelwood in some areas will also increase commercial fuel substitution. An improved historical data base on non-commercial fuels, as well as better information on expected trends affecting energy consumption patterns in rural and urban areas are required.
5. There are various technical terms on energy and fuels such as commercial and non-commercial, conventional and non-conventional, traditional and non-traditional, and renewable and non-renewable, which may be confusing. In presentations or statistics that deal with these terms they should be carefully clarified. The definitions should be stated accurately for commonly used and understood terms. The grouping of types of energy into such terms will be a very useful format in specific statistical presentations, policy making, and so on.

POSSIBILITIES FOR DATA SHARING AND EXCHANGE WITH OTHER COUNTRIES, LIKELY BENEFITS

Cooperation in the exchange of energy data, which could be revealed without any conflict of national interest on either a bilateral or multilateral basis, would result in benefits and advantages to all the participating nations. There already exist exchanges of country data between member countries in international organisations such as the World Energy Conference, the Organisation for Economic Co-operation and Development and various organisations in the United Nations such as ESCAP, UNIDO, UNDP, UNSO, and UNEP, which carry out some activities relating directly or indirectly to energy. For individual types of energy, in particular petroleum, a committee comprising member countries of the Association of South-East Asian Nations (ASEAN) has been established and is known as ASEAN Council on Petroleum (ASCOPE). Its objectives are cooperation in all fields of activity of the petroleum industries and promotion of multilateral assistance on matters of common interest in the field of petroleum. ASCOPE has discussed and agreed to the exchange of appropriate data on petroleum and related matters as well as the establishment of a petroleum data bank for its members. It provides, through these arrangements, joint benefits resulting in the fields of exploration, development, production, transportation, and marketing in the petroleum industry.

In a developing country like Thailand, national energy data are also needed for the consideration, development and implementation of international energy projects such as a hydroelectric project on an international river and the interconnected electricity grid between Thailand and her neighbouring countries.

RECOMMENDATIONS AND CONCLUSIONS

The role of energy statistics in plan formulation is especially significant; it can provide better information for the planning process in its many forms. Thus statistics must be developed which can serve multiple purposes. Improvement of the quality of energy data depends on the creation of the appropriate institutional arrangements from the operational level of energy management up to the level of government decision making. To accomplish progress in this direction, setting up a central organization for collection, processing, analysis and presentation of energy statistics is required. This could be an energy library as well, and could exchange data with other countries and international agencies.

Most developing countries concentrate on collecting energy

statistics data particularly on "Energy Materials" (Energy and Fuel). The Non-Energy Materials such as lubrication oil, asphalt petrochemical products are assuming an increasingly important role, and cause problems directly or indirectly to energy planning. The utilisations of non-energy materials are growing both in quantities and price with the growth of energy materials in the national development. Therefore, to have a complete energy statistical system, both energy and non-energy materials should be included.

DISCUSSION

LEVY

Thank you! Are there any questions or comments on Suchai's paper?

MULCHANDANI

Since Thailand is basically an oil importing country, how is it making efforts in foreign exchange and trade balance?

LIMSUMALEE

Oil imports from foreign countries account now for about 68 per cent of the total energy consumed. Formerly they were about 80 to 85 per cent. We have set up the Bureau of Indigenous Resources to replace the import of oil by using our indigenous resources such as lignite, bio-gas, natural gas, and some non-commercial energy. In our 5-year programme this is how we intend to reduce our balance of trade.

SHAH

Yesterday while discussing the electricity information energy system prevalent in India, it was stated that the bodies directly involved like the Electricity Boards need to do a fair amount of work, including computerization. Now, what is the corresponding situation in Thailand?

LIMSUMALEE

I will tell you about the organizational structure in the power sector in our country. We have three agencies responsible in this sector. One, EGAT, Electricity Generation Authority of Thailand. The responsibility of this agency is to generate electricity and supply to the four regions in the country. We have another agency, MEA, that supplies power to the agencies in far east distributions near the metropolitan Bangkok area. The population in this area is about four to five million. The third agency PAF distributes electricity outside the MEA area and also can generate electricity in the isolated and remote areas where the EGAT system does not reach. EGAT use their own computer for production policy and planning. They also have their own board, the EGAT Board, the Committee of EGAT, to manage their activities. So the policy on the power system of Thailand mostly conform to the decisions of the Board of EGAT and also we have another committee and sub-committee. The functions of this sub-committee is mostly concerned about the tariff, electricity tariff, we call it the Electricity Tariff Committee. This committee is controlling the tariff over the whole kingdom.

AHUJA

What is meant by bunkers? And what is the most predominant traditional fuel in Thailand?

LIMSUMALEE

Bunker consumption means the oil or the energy which is consumed by ships or vessels which come to Thailand and take energy they need for their own use. So they take our energy to use in another place, and it is not our domestic consumption. The traditional energies, I think, means fuelwood, and charcoal in the rural area. You see, our charcoal we sometimes call commercial, sometimes non-commercial. When charcoal is used in towns, it is commercial, but when used in rural areas, we call it non-commercial. So we cannot separate this material into commercial and non-commercial.

PRAKASA RAO

You have mentioned that you have a National Energy Information and Documentation Centre (NEIDC). It is charged with the collection, processing, publishing, accurate, complete and up-to-date energy statistics but one of your recommendations is that there should be a central agency for collection, processing, analysis, and presentation of energy statistics. Do you mean to say that this organization which is already existing is unable to cater to the needs, or are you thinking of only a data bank type?

BOONYUBOL

To answer your question, the National Energy Information and Documentation Centre is only at the beginning phase of being established. This is only a small unit under the bigger organization of the NEA and his suggestion at the end is that a central unit of energy information may be, after the first stage, enlarged to an independent organization that can control, collect, and perhaps have some more powers to collect, process, etc.

The period between 1981-86 is our 5th National Plan period. It is hoped by 1986 this unit will function up to a certain limit and after that it should be enlarged, to have more capability in dealing with energy data system.

PRAKASA RAO

Will this central organization which you have mentioned, give the overall picture of energy balance also?

BOONYUBOL

Yes.

NIKHIL DESAI

I was wondering if you could also tell me whether NEA interacts with international organizations which keep energy data and where will be the source to find all such information? What I have in mind is the IEA publication in 1979 in which 1970-71 data on Thailand are available. After 1977, when I was working in Bangkok, I was finding it difficult to obtain data other than those obtainable from the World Bank. I was wondering if NEA maintain a link with the IEA and whether together they continue to update the information just as in other countries.

BOONYUBOL

I am not familiar with the publication that you have just mentioned. . .?

NIKHIL DESAI

International Energy Agency had a Workshop on Energy Data in developing countries in 1978. After that they came out with two publications, one was the workshop proceedings and the other one was on energy balances and basic energy statistics of selected developing countries. I think there was a total of 17 countries, out of which 11 were oil importing countries. Thailand was one of them. That was a source I had used extensively to get data on Thailand's energy. If I remember correctly, even the last annual report of one of NEA publications, had data only going as far back as 1977. I was wondering if the NEA is maintaining contact with the IEA, and both together have some system of working out and publishing the Thailand data along with other countries on a periodic basis.

LIMSUMALEE

We have an annual publication.

NIKHIL DESAI

What is the last year?

LIMSUMALEE

The last year is 1981-82.

NIKHIL DESAI

If not the IEA, I was just wondering if you have contact with other international organizations, for example, do the Thailand data as I find in the UNU book, come from the NEA or some other part of the Thailand government?

LIMSUMALEE

Most of the international agencies have their questionnaires passed on to Thailand; then the Ministry or Agency concerned sends the questionnaire to the NEA which we try to complete. I know that in some questionnaires, they have some technical queries. But we try to update, follow through the previous questionnaire, because if we do not have this tally, the previous questionnaire may go off base and data cannot be consistent.

SEVEN

AN INTRODUCTION TO THE NATIONAL ENERGY DATA SYSTEM IN THE PEOPLE'S REPUBLIC OF CHINA

LU YINGZHONG

TO MEET the needs of planning in a socialist country with a centrally-planned economy, the statistics of all kinds of the economic activities in the People's Republic of China (PRC) have long been considered of prime importance and an extensive network of statistical organisations was established and large amount of data collected. However, rapid progress towards 'Four Modernisation' requires higher quality in both planning and management, hence more accurate and detailed data base should be provided which seems to be beyond the reach of the present statistical system. New items are being added to the existing statistical data, especially in the sector of energy consumption. The National Energy Data System (NEDS) is now being upgraded to meet the ever-growing requirements of both energy planning and research.

PRESENT STATUS OF THE NATIONAL ENERGY DATA SYSTEM IN PEOPLE'S REPUBLIC OF CHINA

The Dual Statistics System

In order to meet the needs of economic planning, the Dual Statistic System under the centralised leadership of the National Statistic Bureau was established to give cross-checked and mutually-complementary data sets which should also be accurate and reliable. Both systems will collect and submit information to the National Statistic Bureau, where such data will be reviewed, assessed and made available to all users.

The Dual Statistic Systems are shown schematically in Fig. 7.1.

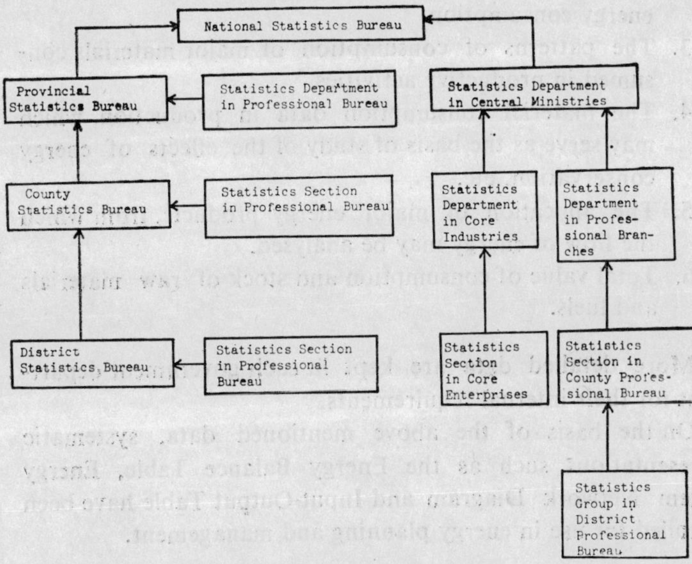

FIG. 7.1 : *The Dual Statistics System in PRC*[1]

Items and data collected in energy statistics

Most Chinese enterprises are government or cooperative owned, thus all the data in productive activities are transferred to the administrative agencies according to some prescribed formats to form a basically consistent and complete data bank of production. On the other hand, the energy statistics in consumer sectors have not been perfected to such an extent, so only the statistics data of government allocated portion of energy products are available. These are insufficient for planning purposes. Statistical requirements on the consumption of main energy resources in each department, provinces, and municipalities have been established since 1980, which include the following items:

1. The consumption and stock of major energy resources which may provide an overall view of the consumption and stock of said resources in all departments and provinces in the PRC.
2. The consumption of major materials in each department, which may provide information according to trades and serve as the basis for further analysis of energy consumption.
3. The patterns of consumption of major materials consumed in productive activities.
4. The material consumption data in production, which may serve as the basis of study of the effects of energy conservation, etc.
5. The allocation of major energy products, from which the flow of energy may be analysed.
6. Total value of consumption and stock of raw materials and fuels.

More detailed data are kept in each government department for their internal requirements.

On the basis of the above mentioned data, systematic representations such as the Energy Balance Table, Energy System Network Diagram and Input-Output Table have been compiled for use in energy planning and management.

THE ENERGY BALANCE TABLE

The Energy Balance Table has been adopted extensively for describing the status of energy supply and consumption at both the national and provincial levels. The objectives of such tabulation are to facilitate:

1. Analysis of the present energy situation.
2. Analysis of the relationships between energy sector and other economic sectors and/or livelihood, hence to provide a basis for energy policy-making, and the scheme for allocation of energy products, as well as to orient the energy-conservation programmes.
3. Analysis of the effects of energy conservation, the

introduction of alternative energy and other important measures in energy policy.

A simplified format of such a table suggested for application in the PRC is shown in Table 7.1.

THE ENERGY SYSTEM NETWORK DIAGRAM

Energy System Network Diagrams at both national and provincial levels have been made tentatively since 1978[3,4]. Such diagrams may take into account the exploitation, collection, refining, transport, allocation, conversion (centralised or decentralised), transmission, distribution, energy consuming equipments and final use of energy. In these diagrams, not only are the rates of flow of energy shown systematically but also the efficiency or loss as well as additional information. On the basis of such a diagram, an optimisation model has been developed which is now going to find wider and wider applications in energy planning. A typical regional network diagram is shown in Figure 7.2.

THE INPUT-OUTPUT TABLE

A National Input-output Products Table of 61 main products for year the 1973 in the PRC, was compiled in 1974, by a group from the Institute of System Analysis of the Chinese Academy of Sciences, the Computing Centre of the State Commission of Planning, and so on. A list of 61 products is given in Table 7.2 and the format of such a matrix is shown as Table 7.3. At present, a more elaborate table of more than 100 main products is being compiled and is expected to provide a powerful tool for planning. Additional data have to be collected for this purpose. At the same time, several provinces and even some large enterprises have compiled or are in the process of compiling such tables. For instance, the Sanxi Province (the largest coal-producing region in China) has compiled both the Product-type and the Value-type Input-Output Table for the year 1979. The progress in this area will have profound effects on the future planning activity in the PRC.

TABLE 7.1 : National energy balance (simplified)[2]

Unit : 10⁴ TCE

	Primary						Secondary								
	1	2	3	4	5	6	7	8	9	10	11	12	13	14	15
	Raw Coal	Washed Coal	Sludge from Washing Plant	Crude Oil	Natural Gas	Hydropower*	Coke	Coking Products	Petroleum Products for Energy Use	Gaseous Fuel	Electricity	Heat from Power Station	Petroleum Products for non-Energy Use	Sum-Total of Items 1-13	Primary Energy Equivalent of Electricity
1. Production of Primary Energy	Constituents of Primary Energy													Total Product	
2. Export (–)															
3. Import (+)															
4. Fuelling foreign ships															
5. Change in stock															
6. Energy consumption														Total consumption	
7. Processing and conversion														Total Amount of Feed for Conversion	

8. Coal-washing 9. Formed coal 10. Electricity generation 11. Petroleum refining 12. Coking 13. Town gas	(−) Input (+) Output		Loss in Conversion
14. Blast Furnace Gas			
15. Self-consumption in Energy Industries			
16. Coal Industry 17. Power Industry 18. Petroleum Industry 19. Coking 20. Town Gas	(−)		
21. Loss in Transportation and Allocation			
22. Deviation in Statistics			
23. Final Consumption		Constituents of consumption	Total consumption
24. Industry 25. Agriculture 26. Construction 27. Transportation			Department Constituents of Energy Consumption

(*contd.*)

TABLE 7.1—*contd.*

	1	2	3	4	5	6	7	8	9	10	11	12	13	14	15
28. Commerce														Department Constituents of Energy Consumption	
29. Livelihood															
30. Urban Livelihood															
31. Rural Livelihood															
32. Others															
33. (Non-energy Applications)															

*For hydropower, each kilowatt-hour is converted into coal equivalent equal to the annual average coal consumption per kw-hr in coal-fired power plant.

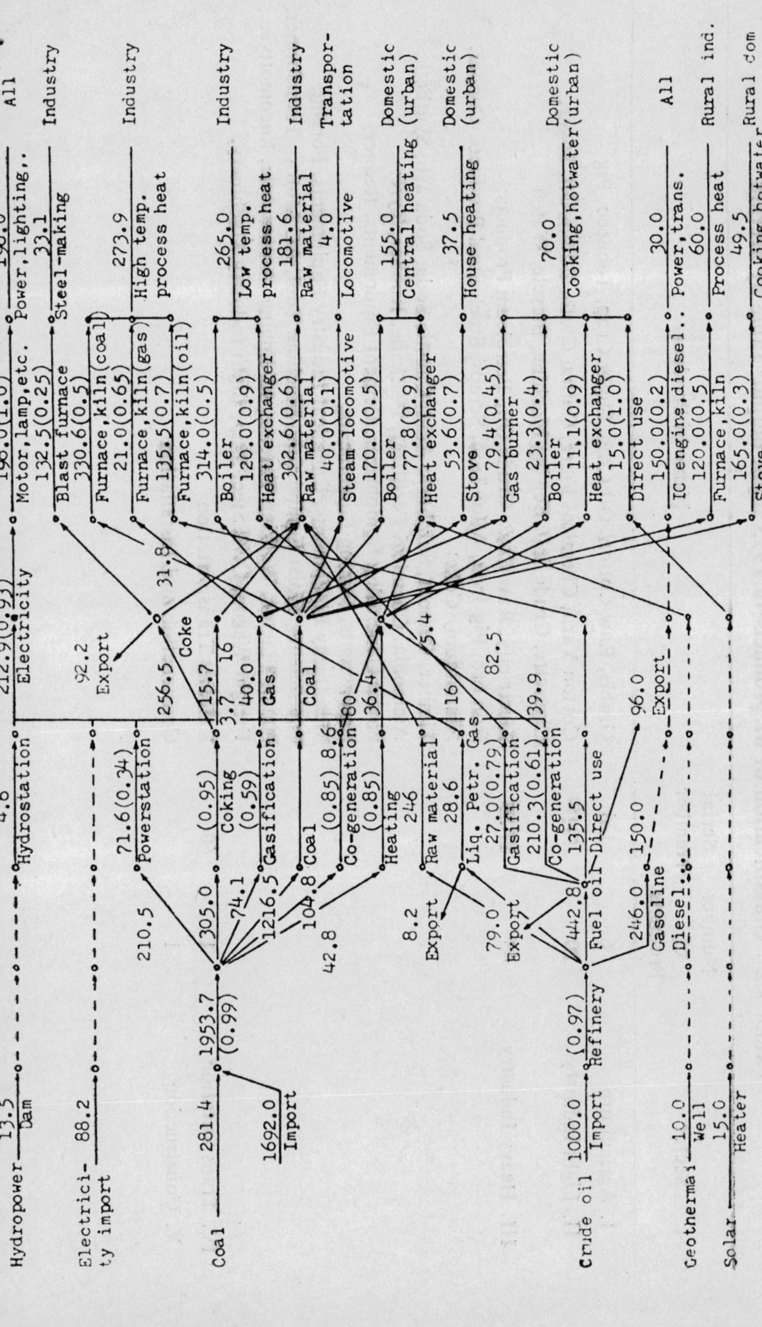

FIG. 7.2 : *Optimisation model*

Note : All figures inside parentheses denote efficiency, and those outside are amounts of energy in 10^4 TCE.

TABLE 7.2 : Classification of 61 products in 1973 input-output[5] table by sectors

Sector	Number of Products	Serial Number	Name of Product
I. Agriculture	6	1-6	Grains, Raw Cotton, Oil Crop, Sugar Crop, Tobacco, Pig
II. Light Industry	10	7-16	Cotton Yarn, Chemical Fibre, Cotton Cloth, Edible Oil, Sugar, Cigarette, Crude Salt, Sewing Machine, Bicycle, Watch
III. Heavy Industry	43	17-59	Electricity, Raw Coal, Crude Oil, Petroleum Products; Iron ore, Pig Iron, Alloy Iron, Scraps, Steel, Steel Products, Copper, Aluminium, Coke; Timber, Cement, Sulphuric Acid, Caustic Soda, Soda ash, Synthetic Ammonia, Chemical Fertilizer, Pesticide, Rubber, Tyre; Power Generating Equipments, Cement Equipments, Refinery Equipments, Mining Equipments, Metallurgy Equipments, Chemical Industry Equipments, Forging Equipments, Hoist Equipments, Machine Tool, AC Motor, Transformer, Pump, Tractor, Walking Tractor, IC Engine, Automobile, Locomotive, Freight Wagon, Civil Steel Ship, Light Industry Machinery;
IV. Transportation	1	60	Railway Transportation
V. Construction	1		Construction and Installation

TABLE 7.3 : Format of the 1973 input-output product table[5]

		Intermediate Products								Final Products								
		1 Grains	2 Raw Cotton	. . .	60 Railway Transportation	61 Construction and Installation	Sub-Total	Other Consumption	Total	Consumption for Livelihood	Increase in Stocks	Increase in National Reserve	Increase in Fixed Assets	Imports	Exports	Losses	Deficit	Total Products
Grains	1																	
Raw Cotton	2																	
	.																	
	.																	
	.																	
Railway Transportation	60																	
Construction and Installation	61																	

CONCLUSION

The National Energy Data System is one of the basis for planning in the Energy Sector in the PRC. Emphasis has been laid on the collection and the subsequent processing of energy data, especially in the area of energy consumption. Compiled data in the form of Energy Balance Tables, Energy System Network Diagrams and Input-Output Tables are finding more applications and are being upgraded. A more self-consistent and complete data bank is under construction, and this will meet the extensive needs in planning in the future.

REFERENCES

1. Cheng Santong, *A Study on the Upgrading of the Energy Statistics in China*, China-U.S. Forum on Energy Study, in Beijing, September 1982.
2. Li Jing Sen, *et al., Data Base for Energy Research*, Lecture Notes for Beijing Training Centre, Inst. of Nuclear Energy Technology, 1982 (in Chinese).
3. Li Jing Sen, *National Energy Flow Diagrams*, 1978, 1979, 1980 (Internally distributed).
4. Lu Yingzhong, *et al.*, An Elementary Model System for Energy Demand Forecasting and System Optimation in China, *Proceedings of the U.S.-China Conference on Energy*, Resources and Environment, 1982 (to be published).
5. Chang Sikang, *et al.*, *Mathematical Methodology and Modelling in Economy*, China Finance and Economy Press, 1982.

DISCUSSION

PROF. LEVY

Thank you very much, Professor Lu. Is there any question?

NIKHIL DESAI

I have some comments and questions. First of all, are there any published sources of Chinese data in English or is there any international organization other than the World Bank to which the Chinese provide energy data?

YINGZHONG

Some references will be found at the back of my paper. But unfortunately most are in Chinese.

LEVY

Are there regular data published?

YINGZHONG

Sure. International Organizations like the World Bank and the United Nations publish every year data on China.

NIKHIL DESAI

No. The World Bank does not publish any (Chinese) data. The United Nations data are fairly limited. One question that I had was about coal production. Now it is very difficult to find the coal production by small mines, which I am told, accounts for half of the coal production. Data on both the quality and the quantity of coal produced are very difficult to get.

Another difficulty I have faced even at the World Bank, where they had some data on Chinese energy, was hydro-electricity production from small hydro plants. The third difficulty I have found was information on transport of coal within China. I was not able to get any such information. The fourth and most important one was in the area of traditional energy resources in China. We had great difficulty just ascertaining numbers. For example, on forestry, it is very difficult to get a clear idea from the NCNA announcements, from Committee reports, or even from the US CIA reports, which are available to some people, on what kind of land the forest is, the afforested area, the new area afforested, the survival rate of seedlings and plants, and distribution of plants. I was wondering if they are inherently poor data, so that even the Chinese themselves do not know whether such data do exist and are obtainable.

YINGZHONG

First our statistics are limited to the commercial energy sources. For non-commercial energy sources, so called traditional sources, we have no exact statistical data but estimates. For you cannot ask some 800 million people to give you data everyday on how many bundles of grass or branches or leaves or straw are burnt, and so on. It is very difficult for us also. Do you mean to ask if the World Bank gave us a questionnaire and we did not respond, or that you haven't any connections with our government for some other data, for instance, mini-hydro or mines. We have such statistics. But, of course, only if they enter the government commercial system. Some small mines, mini hydro plants do not enter the government controlled system.

NIKHIL DESAI

About the traditional sources, I was rather excited by the fuel wood lots programme which China announced last year. That was an ambitious programme. I was wondering, if you really didn't have information on the pattern of traditional energy usages in rural areas, how would you be able to make a sensible plan for this fuel wood lots programme.

YINGZHONG

We were basing it on estimation. You see for some cases, estimation is more reliable than statistics because it is not possible to collect such statistics. We have various communes and village systems which prefer their own plans.

NIKHIL DESAI

Are such estimates available at any place for us? That was my question. Is there any place where we can get such estimates or whatever you call them in the English language particularly.

YINGZHONG

Sure. The World Bank has some connections with our government's statistical bureau.

GHATE

We in this country are interested in your bio-gas plants. We tried to introduce the Chinese type of plant in this country. There have been reports recently that you had maintenance problems with your designs. Could you throw some light on this and also more generally what direction the bio-gas programme is taking? Is it going to be based on small family, or are you going to focus on larger community plants?

YINGZHONG

I see many of you are interested in the rural energy problem in China. I will just give you a brief picture. Our rural region has been short of energy for a long time because our commercial energy only amounts to some 600 m. tonnes coal equivalent. It is far from enough for the urban regions. So there is very little to despatch to the rural regions. Our rural regions rely mainly on fuel wood, straw, mini-hydro plants, and so on, non-commercial energy mainly. So our energy policy for rural areas is to first develop firewood. Also our cultivated land is very limited. There is some non-cultivated land unsuitable for cultivation and we have organized the farmers to plant as much firewood as possible, which is more productive than straw and may be used as fuel.

Secondly, we have to popularize the bio-gas pits. Now we have 6 million bio-gas pits of which I think may be 30 to 40 per cent are not working any more. Not because of its structure. Its structure is okay, because many pits have worked very well. It is due to the lack of technological knowledge of the farmers. Because it is a bio-technology and bio-technology is very delicate. For instance, you cannot use every source of vegetable or branches or leaves to produce bio-gas. You must have a certain ratio of nitrogen to carbon, certain limits of temperature and so forth. You must control them within certain limits. And this is not possible for some of our villagers. To the organized village, we send some technical people to see that the bio-gas pits are working well; we have even reached some distant villages where the pits are working very well. So we hope to increase the number to 20 million by the year 1990. We are also sure to develop many mini hydro projects as well as the solar-stoves and solar water heaters. In all we use every means to increase the energy supply to the rural region and to a certain extent, increasing the commercial energy supply also.

SHAH

Distribution losses in India are about 20 per cent. I think the two countries are somewhat similar in their general layout as far as electricity is concerned. What is the plant load factor in the thermal power stations, which is the ratio of actual output to theoretical maximum output? Can you give me this number please?

YINGZHONG

What do you mean by output?

SHAH

The kWh output. 8640 hours average per year is supposed to be 100 per cent output.

YINGZHONG

Is it availability?

SHAH

Yes, you call it availability. It is slightly different from availability because capacity also comes in. Any way, if there is number. . . .

YINGZHONG

I cannot reply to your question directly. The loss to my knowledge is around 15 to 20 per cent. In the rural region the loss is much higher, may be 30 per cent so so. They use occasionally, not copper wire, but iron wire for transmission. They also use one wire transmission and the back wire used is the ground, there the loss is very high. I will find the exact values and write to you.

SHAH

I shall be grateful.

EIGHT

ARGENTINE ENERGY DATA SYSTEMS

HUGO A. ALTOMONTE

THIS work is an introductory summary concerning the Energy Information System in Argentina. Some aspects should be developed and analysed in detail in the future. A serious study on this subject that needs to be carried out at a secondary stage must be based on the economic analysis of the information system. This may help to establish a contrast between the cost of the information system and the needs of the users, thereby helping to set up referential criteria for the implementation of the improvements in the system.

REFERENCE FRAMEWORK

Organisational apparatus for energy planning in Argentina

The Energy Bureau (EB) is at the present time the agency responsible for carrying out overall energy planning in Argentina. At the same time, there are national and provincial public enterprises, private enterprises and other State agencies which act at different decision levels of energy planning.

A simplified organisation chart showing the structural connections between these different institutions is given in Fig. 8.1. The main hierarchical relationships existing between the entities can be observed in the chart, although to a great extent, all are indirectly interconnected. For example, although the provincial energy enterprises do not depend directly on the

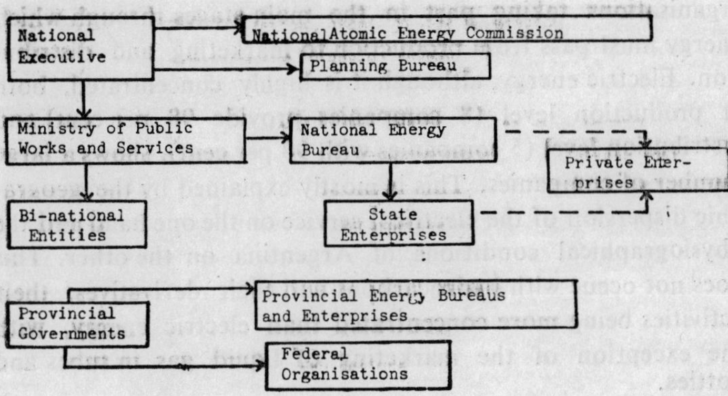

FIG. 8.1 : *Organisation chart of the main entities participating in Energy Planning*

E.B., the latter must approve of, and present to the higher authorities (Ministries of Economy, Public Works and Services) the projects which require outside financing sources, etc.

The decentralisation of the energy system is summarised in Table 8.1 which shows the number of enterprises and

TABLE 8.1 : **Number of enterprises taking part in the production, transformation and marketing of the different Energy Sources**

Sources \ Activity	Production	Transformation	Marketing
Oil	5[a]	7	10[b]
Natural gas	5[a]	1	1[c]
Coal	1	1	1
Liquefied gas	5	—	28
Electric energy	154[d]	—	190[e]
Charcoal	—	—	—
Other sources	—	—	—

[a] A State enterprise, YPF, contracts hydrocarbon production with 32 other companies.
[b] Concerns the marketing of oil derivatives.
[c] Concerns the marketing of network distributed gas.
[d] 8 Enterprises concentrate 98 per cent of the production.
[e] 5 Enterprises concentrate 90 per cent of the distribution.

organisations taking part in the main stages through which energy must pass from production to marketing and distribution. Electric energy, although it is highly concentrated, both at production level (8 companies provide 98 per cent) and distribution level (5 companies with 90 per cent), shows a large number of companies. This is mostly explained by the geographic dispersion of the electrical service on the one hand and the physiographical conditions of Argentina on the other. This does not occur with hydrocarbons and their derivatives; their activities being more concentrated than electric energy, with the exception of the marketing of liquid gas in tubes and bottles.

Argentina has a long history of and experience in energy planning. Sub-sectoral plans and programmes are undertaken according to the different energy policies existing at that moment.

Moreover, at the present time Argentina has a Nuclear Plan, a National Plan for Electric Construction Schemes and an Energy Programme. The first was prepared by the National Atomic Energy Commission and the other two by the Energy Bureau. Participation in the total energy consumption by electricity and fuels for its generation, are defined before establishing the overall policies on which the energy scenarios considered in the Energy Programme are based.

Furthermore, the National Plan for Electric Construction Schemes contains a considerable degree of disaggregation—emphasised more in the analysis of the supply than the demand—and is the object of constant updating. But the same does not apply to the Energy Programme, which is updated less frequently and has wider uncertainty in its parameters.

Finally, as will be seen later on, since a modification to the energy planning scheme has direct consequences on the information system, it is necessary to carry out a characterisation of the information before describing the present state of this system.

Characterisation and classification of the information

Obtaining the necessary, pertinent information for undertaking research work, decision-making and planning, is generally a problem for scientists, businessmen and state officers. A real, immediate need therefore arises to organise information systems which enable the users to obtain the required data within reasonable time.

Characterisation of an information system and its circuits implies description of the phases from generation to utilisation, as well as the agents intervening in each of the stages. In a simplified manner, this could be as follows :

Generation of information	Producers
Compiling, Filtering, Cohesion, Publication	Intermediaries
Direct and Indirect Users	Users

In practice, the information-generating entities are often the users as well, so that in reality there is not a one-directional flow but rather a network of inter-connected flows. Some of them do not reach the user, nor even the second phase of this diagram.

A more realistic diagram would be a matrix in which the producers and the users appear in both the rows and columns.[1] Evidently, all the agents described in Table 8.1 would form part of this matrix, as well as the International Administrations, Professional Organisations, Universities, Research Institutes, and so on.

In addition, it is necessary to define what is meant by information: throughout this paper reference will be made to information in the energy sector as a set of elements necessary for preparing a document refering to the energy sector. This implies that the limits of scientific and technical information, and consequently those of energy sector information, are established by the *use* which is made of information, since it is practically impossible to define it on the basis of the sources which provide it.

Moreover, the information system will be understood to meanthe set of interdependent elements which contribute to the transfer of information from the producer to the user.

As regards the classification of energy information, three large groups exist: primary information, prepared or secondary information or non-existent information.

(i) Primary information is that generated by socio-economic activity as a whole, giving rise to a series of original data or non-processed data. At the same time, this type of information is not usually published and can be sub-divided into :

— Underground, that which is very difficult for the user to obtain and which is finally detected if there is a knowledge of where or to whom to contact for the purpose with access to files, or through 'informal' or 'personal networks'.

— Confidential information, inaccessible for various reasons, ranging from State to Company secrets.

The handling of the series of primary or original data deserves special attention, not only because it is information which must be evaluated before being used but also because of its importance in the construction of Energy Balances, as well as in any other surveys to be undertaken.

The products of the energy industry are heterogeneous; from the physical viewpoint. There is electrical energy and the solid, liquid and gaseous fuels. The first problem arises from this heterogeneity. Statistical information will be given in different units according to the product involved: tons, litres, cubic meters or kWh.

Moreover a single product can be expressed in different units and consequently the measurement unit employed varies with the information source. Thus, for example, in some cases the production of oil can be given in tons, in others in barrels/day etc. Some information sources give the different energy products a common unit, for example, Tonne oil equivalent, Tonne coal equivalent, and so on.

This problem of measurement units would be easy to solve if there were a unique ratio between the different units. This would imply that with the simple application of a coefficient it would be possible to go from one unit to another. In practice, however, this is not the case, since although coefficients exist they allow passing from one unit to another only with a certain margin of error, particularly where putting a common denominator to the different energy types is concerned (T.o.e., T.c.e., etc.). Furthermore, there is not always agreement as regards the conversion coefficients.

Another of the difficulties concerns the gaps in information. Most publications of statistics refer to data which could not be obtained and which, therefore, had to be estimated. Unfortunately, however, very few give the criteria and methods used for making the estimate.

The definition of the variable is another of the problems which besets the user. For example, in some cases coal production is taken to be the quantity extracted from the mine (gross production), while in others it is taken to be the quantity after washing (marketable production). Even the definition of the product is not clear, since this may or may not include lignite.

Something similar occurs in the case of oil since it is not clarified whether the production refers to heavy or light oil, whenever expressed in volume units and in general the average of both is taken.

(ii) Prepared or secondary information is that generated by the organisations dedicated to primary data processing. Apart from the two sub-divisions used in the previous case, there is a third one which is the prepared information (books, magazines, yearbooks) published either periodically or irregularly.

The preparation of primary energy data is vitally important as an initial step for planning work in general and for the construction of National Energy Balances in particular. Construction of energy balances not only permits systematisation of historical information but also creates the necessary bases for the systematisation of future information. This systematisation means that, on the one hand, each of the products to be accounted for must be clearly defined and, on the other hand, it creates a commitment for a fluid interchange between producer, processor and user.

The statistical energy information that the user often requires for preparing the balances can be classified as follows:[2]

(a) *Systematic* : the information compiled periodically on regular lines, which may be either primary or secondary information according to whether or not it is published in the form of yearbooks.

(b) *Non-systematic* : Occasional information, which is not regular either in frequency or in format.

According to the type of information, the method of tackling varies : if it is secondary systematic information, its reliability and homogeneity must be analysed starting from specific statistical verifications. In the event the reliability analysis is not satisfactory, primary systematic information must be resorted to, and the work redone, modifying the preparation criteria in order to submit the data once again to the control tests.

As regards non-systematic information, in some cases an attempt can be made at total or partial systematisation often by consulting persons, institutions or publications.

(iii) *Non-existent information* : Information which is not compiled or published, that is, which does not exist in the form of a statistical system. However, in many cases there are indirect qualitative or quantitative estimates which can be obtained verbally or in writing and in this way a substitute of the non-existent information can be generated. Thereafter, the need to carry out polls on pilot, regional or national scale will be determined.

Later on it will be explained how, in Argentina, it is possible to go gradually from non-existent to systematic information.

THE PRESENT SYSTEM OF ENERGY INFORMATION IN ARGENTINA

Although all types of information are important, in accordance with the definition in the preceding section, this section will focus on the description of the energy information in Argentina. In this sense, two types of organisations will be differentiated which process and publish in different ways the information of the energy sector: the Energy Bureau and the remaining entities and users enterprises.

This differentiation does not prevent the existence of a system, since these organisations act jointly in the production and use of the information, generating a continuous, interdependent flow from the producer to the user.

ENERGY BUREAU

Background and location of the EB Information Sector in the National Statistics System

The origins of the Information Sector of the present General Directorate of Energy Coordination and Information, go back to 1943, when the National Energy Directorate was created, to which it reported as the Energy Antecedents Office. Subsequently, and ranking as a Department, it became part of the ex-Directorate of Energy and Fuels, reporting to a Bureau with the same name. Later it formed part of the ex-Directorate of Technical and Economic Affairs and, later still, with the denomination of Department of Information and Applied Research, it was incorporated into the Sectorial Office of Energy Development. Afterwards it was integrated in the Directorate of Energy and Planning Control and finally in Energy Coordination and Information.

With successive transformations, new and important functions were added in order to meet the requirements of the parent Organisation, and to fulfill the responsibilities inherent in its nature as guiding organ on energy matters, in accordance with the policies laid down by the Executive. Its activities are not confined to official spheres, but also extend into the private fields, industrial as well as research, study, educational, and so on.

In some entities informatics is often regarded as a passive activity, not meriting the same attention as that given to conjunctural problems and the physical and human means for its development are withheld. Its importance becomes evident, however when a series of data is required urgently but cannot be provided since the formation of a series requires, like every activity, a maturing period. Fortunately today a deep change is occurring in almost all the countries of the world, in the sense of increasing efforts to improve the information systems.

As regards the location of the Information Sector, it became part of the National Statistics System in accordance with Article 4 of Law 17.622 which created it, as follows:

"Article 4. The National Statistics System will comprise:

(a) The National Institute of Statistics and Census.

(b) The Central Statistics Organisation, which are :

(i) the statistical services of Ministries and State Bureaux.

(ii) the statistical services of the Commanders-in-chief of the Armed Forces.

(iii) the statistical services of decentralized organisations in the National Administration.

(iv) the statistical services of state enterprises.

(c) Peripheral statistics organisations, which are :

(i) the statistical services of the provincial governments.

(ii) the statistical services of the municipal governments.

(iii) the statistical services of the provincial and municipal autonomous decentralized departments.

(iv) the statistical services of provincial and municipal enterprises.

(v) the statistical services of the inter-provincial entities."

The National Institute of Statistics and Census presently reports to the Ministry of Economy.

Publications

Currently, four periodical publications are produced by the General Directorate of Energy Coordination and Information: The Electrical Energy and Fuels Yearbooks and the Monthly Bulletins of Electrical Energy and Fuels.

In 1957 what could be called the first publication was edited in the form of a yearbook. Previously there had been some publications but not of that kind, nor were they in series form. This Fuels Yearbook summarized the main historical information on the subject, almost all since its origins. These publications have continued without interruption up to the present.

Subsequently and with certain hesitation, since similarly reliable background information was not available but encouraged by the Fuels Yearbook, the first Electrical Energy Yearbook was published with information corresponding to 1957. At the moment, the latest publications are : the 1981 yearbooks for Fuels and the 1979/1980 yearbooks for Electrical Energy. Their contents are given in Tables 8.2 and 8.3.

TABLE 8.2 : **Contents of the fuel bulletin 1980***

1. Oilwells
   Wells completed and metres bored by 31.12.80. Condition of wells, 31.12.80 Y.P.F. contractor companies
2. Oil
   Production
   By oilfield
   By company
   By province and monthly daily average
   By primary and secondary recovery
   Processed
   Broken down by origin
   Broken down by company and distillery
   Broken down by company
3. By-Products
   Preparation
4. Airplane gasoline, common gasoline
   Special gasoline—Solvents
   Turpentine—Kerosene
   Retro fuel—Gas oil
   Diesel oil—Fuel oil
   Lubricants—Grease
   Asphalt—Propane
   Butane—Coke and residual coal
5. Exportation
   Oil and derivatives
6. Sales
   By company and product—Bunker
   By jurisdiction
7. Own consumption
   By product
   By company
8. Natural Gas
   Production by gas field and distribution
9. Distillery Gas
   Production and distribution
10. Coal and similar products
    Production, sales, and consumption and importation
11. Sales prices of fuels
12. FOB oil prices
13. Preparation capacity
14. Storage capacity
15. Duet transport
16. Apparent consumption
17. Service stations, sales, outlets and garages

*The Bulletins may possibly incorporate 15 or 10 day series of each heading indicated.

TABLE 8.3 : Contents of the electrical energy bulletin*

Public Service
Conventional symbols and initials
Clarifications
Installed power, production, energy billed and users in the Federal Capital, districts and/or departments
List of steam power stations
List of gas turbines
List of hydroelectric power stations
List of main diesel power stations
List of nuclear power stations
Classification of national power stations
Classification of provincial power stations
Classification of cooperative power stations
Classification of municipal power stations
Classification of private power stations
General summary of power stations
Breakdown of installed power and production by company
Breakdown of installed power by jurisdiction
Number of power stations and installed groups
Breakdown of production by jurisdiction
Electrical energy billed by company
Distribution of electrical energy delivered to network
Electrical energy billed by jurisdiction and concessionaires
Movement of energy between companies
Transfer of energy by jurisdiction
Energy received from self-production—Importation—Exportation
Number of user by jurisdiction and concessionaires
Consumption of fuels by jurisdiction
Consumption of fuels in t.o.e.
Consumption of fuels by company
Consumption of fuels by type of generation
Consumption of fuels by main thermal power stations

Self-production
Clarifications
VII C Classifications
Self-production of electrical energy by districts and/or departments
Installed power by jurisdiction
Generation in self-production power stations by jurisdiction
Energy generated by branches of industry
Installed power in public service and self-production
Evolution of production per inhabitant.

*In some bulletins 10-day series are published.

These, as well as the other publications, are not an objective of the sector, but rather orderly compilations which reflect part of the activity carried out to meet the requirements of the Energy Bureau and other organisations or connected with energy work, and are made public. Therefore the information published has a high degree of reliability, because it becomes a permanent record, and because of the fear that in its use by the consignees, the slightest error that may appear will be detected.

Experience has confirmed that no finer coherent analysis exists than the handling of data, since no matter how much they are analysed on receiving and computing them, it will never be possible to do so with the detail and thoroughness of the specialist who works on the subject in particular.

As regards the monthly Bulletins, it should be mentioned that the *Fuels Bulletins* was first published in 1964 and *Electrical Energy Bulletin* in 1967. These Bulletins have gradually incorporated new series and others are in preparation, in a continual effort to increase the plan of information.

Fuels Bulletin

This is a monthly bulletin which is normally issued in the second half of each month. Its contents deal with the different stages of oil, natural gas and coal activities. This makes it possible, in a relatively short time, to have available and to make public the most characteristic data, thus making possible an evaluation of the sector's progress.

Electrical Energy Bulletin

Simultaneously and also in monthly form, the Electrical Energy Bulletin is issued on the previous month's generation information and the last billing information which is three months old. These values are accumulated throughout the year and allow the evolution of such a fundamental index as that of electrical energy to be followed.

TABLE 8.4 : Integrated energy balance (OLADE)

				PRIMARY ENERGY								SECONDARY ENERGY												
				Coal	Firewood	Other Biomass	Oil	Nat. Gas	Hydraulic	Uranium	Total Prim. Energy	Coke	Charcoal	Liquefied-Gas	Gasoline	Kerosene Jet - Fuel	Gas oil Diesel oil	Fuel oil	Others oil	Non-energ.	Gas	Electric energy	Total se-cond. energ.	TOTAL
Final Consumption — Energy Sector	Supply	1	Production																					
		2	Import																					
		3	Variation of stocks																					
		4	Total supply																					
		5	Export																					
		6	Not used																					
		7	Internal gross supply																					
	Transf. centers	8	Total transformation																					
		8.1	Coking-Blast furnaces																					Transformation losses
		8.2	Coal plants																					
		8.3	Biogas plants																					
		8.4	Oil distillation																					
		8.5	Gas treatment centers																					
		8.6	Public utility power pl.																					
		8.7	Self prodn.power pl.																					
		9.	Own consumption																					
		10	Losses																					
		11	Fitting																					
Sectorial FINAL Consumption		12	Total final consumption																					
		12.1	Non energy consumption																					
		12.2	Final energetic consumption																					
		12.21	Domestic-Services-Public																					
		12.22	Transportation																					
		12.23	Rural production																					
		12.24	Industry																					
		12.25	Non identify consumption																					
Gross Secondary Energy Production																								

Obtaining the information

Fuels

The corresponding information is obtained directly from the producer companies by means of questionnaires. For experienced organizations with well ordered structures, it is possible to count on a good basic statistical preparation at given stages of the activity. However, it is insufficient as regards the other consumer sectors.

With the exception of the figures regarding the use of fuels used in electric energy generation, which is obtained directly from the utility companies and from some other companies such as railroads, cement, steel, petrochemicals and sugar mills, and so on, it is not presently possible to establish a breakdown of consumptions of the different products, by sectors. In the case of oil and derivatives, the manner of marketing which the different companies use for most of the products has been a factor which to a certain extent has prevented obtaining this information. In fact, this stage of the activity is carried out through agents or distributors who until now were not required to give a classification of the product they sold. It is also logical to assume that the distributor, completely uninvolved in investment and planning problems both in the private and state sector, is not in a position to appraise a task which for him is entirely superfluous and costly.

Electrical energy

The corresponding information for electricity is also mainly obtained by means of a yearly poll of the required data from the public service concessionaires and self-producers. To facilitate understanding, as the respondents are not always familiar with the terms used in this field, an attempt has been made to prepare the form as simply as possible, without departing from the essential technical norms as regards the terminology used. These forms have been modified from time to time and have been broadened as circumstances required, with the constant purpose of increasing their volume of data to achieve objectives established.

Experience gained over the years has shown that one of the basic reasons which impedes and delays processing of the information received is the subsequent consultations and clarifications which must be made in many cases, due to different interpretations of the topics requested and/or utilization of different terminologies by the services organization. As examples, the diversity of tariff schemes could be mentioned, which in some cases do not take into account the final destination of the energy, but rather consumption scale; and different systems for classifying users, and so on.

Regional information

There are presently no problems as regards information relating to electric energy, both for the public service sector and self-production as far as production, consumption, installed capacity, total users, and so on, are concerned, since these are published at departmental level within each province.

As has been mentioned already, the same does not happen as regards fuels, but the collecton of regional information figures in the list of scheduled developments. The fixed location of oilfields and distilleries allows part of the information to be connected to given regions, but this does not cover the wide range necessary for regional studies. Nevertheless at the present time we have begun to progress towards a solution.

Energy balances

This important tool for the first stage of any planning process has been the object of erratic application in Argentina. Energy balances were first published for the years 1960-1961, then there was a thirteen-year period when they were not. In 1973-74 the 1960-1973 series was published and in 1981 they were published using another methodology (OLADE) for the years 1970-74 and 1980.

At the time of writing, the 1970-1980 series of National Energy Balances with the original methodology and which basically corresponds to the scheme in Table 8.4 is in the process of publication.

So that these balances can be compared correctly it is necessary to clarify the criteria which were taken into account for conversion of the different energy sources.

Among the factors causing the greatest distortion, hydroelectricity and new-commercial fuels can be singled out; the latter because of their very dubious evaluation, the former because of the different criteria for choosing the conversion factor. In fact, very many countries, like Argentina, measure each kWh by the amount of fuel which would be necessary to produce it, that is, between 2,200 and 3,000 calories, in accordance with the thermal register and its location in the load diagram. In other words, the so-called substitution criterion prevails, while in other countries this value is reduced to 860 calories which is the thermal equivalent of kWh following the criterion of end use of the energy.

In Argentina, the use of electrical energy as a source of heating is relatively less, that is, electricity is produced for purposes where substitution by another form of energy is presently almost impossible in significant quantities. If energy were used as a source of heat, it would be reasonable to calculate it by using 860 cal/kWh as a conversion factor.

OTHER ENTITIES

In general the energy enterprises possess a level of information sufficiently disaggregated for their own activities and evolution. In addition the information systematically compiled in some cases is published by these companies in their annual reports and balances or special reports and the remainder is prepared by the EB in the publications referred to previously.

As regards natural gas, the information provides acceptable regional and sectorial coverage. In fact the monopolistic structure of its marketing enables the information on the different regional administrations to be centralized and be available with a high degree of reliability. While the sectorial breakdowns of consumption is made for the different user categories—residential, commercial, electric power stations, and official entities—a disaggregation in accordance with the UIIC code of industrial consumptions is not published.

Information on liquefied gas does not have the same level of disaggregation, due to the nature of its marketing and the atomization of the distribution market.

The contents of the Gas del Estado yearly bulletin are described in Table 8.5.

TABLE 8.5 : **Gas del Estado Annual Statistical Bulletin 1979**

Contents

I. *Commercial Statistics*

Acquisitions

1. Natural Gas
2. Liquefied Gas
3. Total Acquisitions

Sales

4. Network Gas
5. Liquefied Gas
6. Total Sales

Users

7. Network Gas
8. Liquefied Gas
9. Total Users

II. *Other Information*

10. Natural Gas
11. Liquefied Gas
12. Branches, Delegations and Agencies
13. Manufacture of bottles appliances and storage tanks.
14. Evolution of Gas del Estado

III. *Social and Labour Statistics*

15. Personnel roll by occupation on 31.12.79
16. Personnel roll by sectors. on 31.12.79
17. Evolution of number of officials.

INFORMATION AND ENERGY PLANNING

Planning scheme

The role of the information system, both energy and non-energy, for policy formulation and energy sector planning, is of fundamental importance. Its usefulness, however, will depend on its scope, disaggregation, quality and geographical coverage.

In addition, the levels of information required are very different according to the planning scheme which one desires to implement. In fact, the Energy Programme put into effect in Argentina consists of two well-defined parts: the projection of the energy demand on the one hand and the general outlines of its supply on the other. The projection of total energy demand is done starting from a model which relates overall consumption to the GNP, so that, only those two large aggregates are required as necessary information: energy and non-energy. Moreover, since for supply only rough outlines are established and there is no analysis of alternatives, the energy and non-energy information required for this second phase is unimportant.

Totally different situation arises when the planning scheme is integral and analytic, when it relates the energy system with the rest of the socio-economic activity as shown in Fig. 8.2.[3] It can be seen from the diagram how the energy system relates to the rest of the socio-economic system, following an iterative scheme until reaching solutions which coincide with the overall development policies and energy policies.

If this scheme is to be applicable, a methodology must exist which can be adapted to it. Reference will be made in this paper in regard to the possibility of applying to Argentina an analytical methodology such as the one proposed by IDEE's "Cooperative Research Programme on Energy in Developing Countries"[4], and to the type of energy and non-energy information required and the changes which should be made to the present information system so that it may be adapted to the requirements of a more analytical planning scheme.

Energy balance

As mentioned previously, the Energy Balance is a fundamental element in the first stage of planning. If this stage follows the formulation given in Fig. 8.2 it is evident that the balance scheme shown in Table 8.4 should be modified.

The modifications should lead to taking into account energy reserves and breakdown of energy by sources and uses, measured in useful energy, as shown in Fig. 8.3. This method of construction allows rapid visualization of the physical flow of the energy system starting from the primary energy Reserve Balance, transformation and production of secondary energy and finally the sectorial breakdown of consumption by sources and uses in net and useful energy.

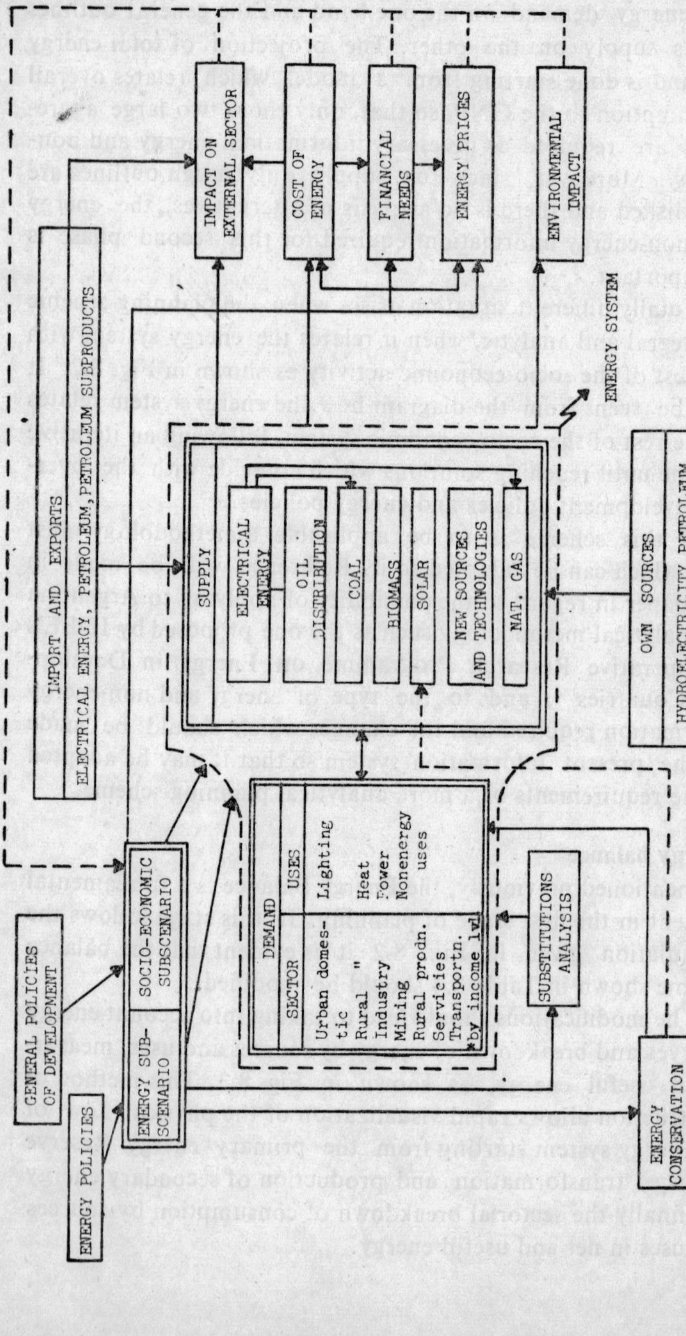

FIG. 8.2 : General description of the relationship between Energy and the rest of Socio-Economic Systems

The present state of the information system compared with the information requirements on Figure 8.3 can be summarized as follows:

BALANCE	PRIMARY SOURCES	SECONDARY SOURCES	USES
RESERVES	(1)		
PRIMARY ENERGY	(2)		
PRIMARY TRANSFORMATION CENTRES		(3)	
SECONDARY TRANSFORMATION CENTRES		(4)	
SECONDARY ENERGY		(5)	
SUPPLY AND CONSUMPTION		(6)	
NET CONSUMPTION		(7)	(8)
USEFUL CONSUMPTION		(9)	(10)

FIG. 8.3 : *Integral energy balance*

Notes :

1. Balance of Primary Energy Reserves
2. Balance of Primary Energy Reserves
3. Balance of Primary Energy Transformation
4. Balance of Secondary Energy Transformation
5. Balance of Secondary Energy Transformation
6. Balance of Supply and of Consumption
7. Balance of Net Sectorial Consumption (by sources)
8. Balance of Net Sectorial Consumption (by uses)
9. Balance of Useful Sectorial Consumption (by sources)
10. Balance of Useful Sectorial Consumption (by uses)

— information published systematically sub-matrixes (2) to (6);

— unpublished primary and prepared information sub-matrixes (1) and (7).

The information required by these sub-matrixes is only partially prepared since, information on reserves is only available for the commercial sources and sectorial consumption information only exists for some sources (natural gas and electrical energy). In turn it must be clarified that information of rural area participation in total consumption of a given source should be provided.

— underground or non-existent information sub-matrixes (8), (9) and (10).

In addition, the sectorial breakdown shown in Figure 8.3 should be accompanied by a "breakdown" (division) in the uses, as follows:

(i) *Heat* : cooking, space heating, water heating, ironing, process heat (high, medium and low temperature);

(ii) *Cold* : ventilation, refrigeration, food preservation, process cold;

(iii) *Power* : electric, diesel engine, auto engine, turbines, steam machinery, mechanical energy (wind or hydraulic), animal;

(iv) *Lighting*

Non-energy information associated with planning

The level of non-energy information required by the planning process taken here as a reference is so vast that the participation of all the organisations integrated in the National Statistics System becomes necessary.

By way of example, a summary of the non-energy information connected with the evaluation of energy requirements will be made here. From the viewpoint of supply or demand both the physical and economic information have a more standard character and the present system responds to a great extent to planning requirements.

The methodology consists of characterization of the determinants of energy demand at module level within each sector. For the domestic sector for example, if the different income levels, the climatic differences of the regions, type of dwelling, equipment structure, etc., are determinant variables of energy consumption, then this sector is divided into homogeneous modules which take into account the differences framed by the determinant variables.

In the list given in Table 8.6(a), (b), (c) and (d), correspond to the information of a general nature, while (e) is the information referring to the sectorial variables.

TABLE 8.6 : **Non-energy information necessary for the energy evaluations**

(a) *Population*
- Total population
- Structure
  by sex,
  by age,
  by area (urban-rural),
  by race or nationality,
  percentage at subsistence level,
- Economically active population.
- Structure of employment and unemployment
- Size and characteristics of families
- Demographic variables—mortality—fertility—migrations.

(b) *Income Distribution*
- Income distribution curves
- Population structures by income levels for each of the identified rural and urban areas
- Population distribution by socio-professional categories.

(c) *Macro-economic Variables*
- Total Gross National Product
- Sectorial structure
- Evolution of most important sub-sectors
- Gross National Investment
- Total and energy imports and exports

TABLE 8.6—*contd.*

As far as possible this information should be available at regional levels and not only national levels. Explicit references should be made to the magnitude of informal activities not included in the national or regional accounts.

(d) *Social Variables*

Must attempt to cover, as far as possible, those variables which are representative of social and/or cultural milieu, not previously mentioned, such as:

(d_1) Education

— Primary School Matriculation
— High School Matriculation
— University Matriculation
— Percentage of drop-outs or completion rates
— Total illiteracy, by sex and specific groups (youth, minorities, etc.)

(d_2) Health

— Total and infant mortality
— Life expectancy at birth
— Medical personnel per inhabitant
— Health facilities per inhabitant

(d_3) Food

— Calories per inhabitant
— Proteins per inhabitant
— Degree of self-sufficiency in food

(e) *Sectorial Variables*

(e_1) Domestic

Basically this concerns the populational variables already indicated.

(e_2) Farming

Evolution of physical production of the main products of each region's or zone's activities (agriculture, cattle, fisheries, forestry, etc.) so as to cover a high percentage of the added value or the value of sectorial production. It is necessary to distinguish between production intended for the market and that connected with self-consumption and also production for export and for the local markets.

— Profile of the agricultural practices for each product and each production system. Historical values and future forecasts. This profiles should include co-efficients related to:

TABLE 8.6—*contd.*

Labour,
Working animals,
Agricultural machinery (by type),
Fertilizers and agrochemicals,
Water for cattle,
Water for irrigation,
Fixed mechanical energy,
Transport (local),
Required surface—ploughed—sown—harvested,
Yields,
Generation of agricultural wastes.

In a complementary manner, it will be necessary to know over-all data for each region or zone about:

— Total surface devoted to annual or permanent cultivation
— Total surface devoted to grazing
— Surface covered with spontaneous and reforested woods
— Total register and horsepower of agricultural machines
— Number of working animals
— Surface irrigated by pumping

(e_3) Mining

— Evolution of the physical production of mining products which cover a high percentage of the total production, distinguishing between that intended for local markets and for export.
— Profiles of energy inputs, present and future, for each product and type of exploitation by product unit.

These profiles will include:

— Labour
— Working animals
— Fixed motors
— Mobile machinery
— Lighting
— Process heat
— Transport (local)

(e_4) Industry

According to the internal subdivision adopted each group requires:

— Evolution of the physical production of the main industrial products, so as to cover a high percentage of the added value or of the total Gross Production.

With regard to subdivision of the industrial sector, it is considered useful to take into account, in addition to the classical subdivision by branches, a more significant

TABLE 8.6—*contd.*

regrouping from the development process point of view and style of development allowed for in the different scenarios (for example durable and non-durable goods, intermediate goods, capital goods, etc.). This grouping must be coherent with the structure of the macro-economic model to be used.

— Profiles of energy inputs, present and future, for each product and type of process, by product unit.
These profiles will include:
Total energy consumption,
Mechanical energy,
Lighting,
Specific electricity,
Combustible wastes generated,

(e_5) Construction

— Evolution of housing construction (m^2) and building for the service sector (m^2) and for the industrial sector (m^2),

— Evolution of Investment in Infrastructure (as an indicator of the volume of construction connected to infrastructure development)

— Profiles of energy inputs, present and future, for each type of construction, by unit of surface or of investment.

These profiles will include:

— Fixed machinery
— Mobile machinery
— Lighting
— Transport (local)

(e_6) Transport

— Evolution of the motor vehicle register, broken down by type (diesel, gasoline) for public and private passenger transport.
— Average distance travelled by type of vehicle.
— Passengers/kilometers transported by means and by type.
— Evolution of freight transport vehicles register, by means and by type.
— Average distance travelled by means and by type.
— Tons/kilometer transported, by means and by type.

In this sector the informal type of transport, both by land and by water must be taken very particularly into account, supplying information on the variable considered most suitable in each case (existence of animals for transport, average distance travelled etc.)

TABLE 8.6—*contd.*

(e_7) Services

This sector is highly diversified and in each case its internal structure must be defined. On this basis the main explanatory variables must be chosen for each activity (commerce, education, public health, tourism, government, armed forces, etc.)

It will probably be difficult to make a selection of sufficiently homogeneous activities and on which there is segregated information which would enable an important percentage of the sector's total to be covered.

In this case it will be treated in aggregate form by means of overall coefficients of energy input by unit of added value or of personnel employed.

The analysis of personal or informal type services presents a special problem, as it is very difficult to find a reasonable, explanatory variable for them (it could be personnel employed) and the respective energy coefficient.

From the energy point of view it is important to consider independently within this sector, certain activities such as:

— Street lighting
— Public water supply
— Government services (including consumption by the armed forces which may constitute an important percentage of liquid fuel requirements).

The National Statistical System presently contains most of the necessary information in primary form, either systematically prepared or otherwise, in accordance with the following details:

(i) Information systematically prepared:

— Population (a) and Social Variables (b):

National Institute of Statistics and Census: 10 year census of population and dwellings, Permanent Home Polling and Provincial Statistical Directorates,

Ministries of Education and Social Welfare, Statistics for FAO and ECLA.

— Macro-Economic Variables (c):

Central Bank of the Argentina Republic in Publications on Gross National Product and Quarterly Poll for establishing the Product,

Federal Investment Council, Regional Estimate of GNP (in preparation).

(ii) Information not systematically prepared.

— Industry (e_4), Mining (e_3) and Services (e_7) INDEC, National Economic Census (ten-yearly), Central Bank, Quarterly Poll for establishing the Product, Bureau of Industry and Mines.

— Rural Production (e_2) partially
INDEP National Agrarian Poll

(iii) Non-systematic primary information

— Construction (e_5)
Provincial Statistics Directorates
Chamber of Construction

— Rural Production (e_6) partially
Agricultural Bureau

In what follows the manner of putting into improvements to be introduced into the information, both energy and non-energy, will be analysed, to allow for in depth planning.

FEASIBLE IMPROVEMENTS TO BE INTRODUCED INTO THE INFORMATION SYSTEM

This section will deal with changes to be implemented (short terms) and which may eventually be implemented (medium term) for this purpose.

Changes to be put into effect in the short term

The information Sector of the E.B. is implementing standards jointly with the main State enterprises to overcome the problems arising at the level of the "breakdown" of final sectorial energy consumption.

Fuels

Earlier we indicated the problem arising from the marketing habits of the different companies for the majority of their products, which affect the quality of information given by these companies.

However, since a capital-intensive industry is involved, it is logical to recognize that the basic information has a fully justified cost. Therefore, every effort should be made to integrate the distributor into this task; in other words demanding it if the contracts with his respective enterprises allow it, or reaching an agreement with them over the way it can be done.

It is possible that this difficulty of obtaining data on the final use by zones can be solved in a short period by taking advantage of the possibilities offered by computing systems.

Therefore in the near future it will be feasible to have available sales figures of all the fuels with a degree of discrimination which will make regional studies possible, since by making additional efforts the information can become available at district level. It is also expected that classifications according to the UIIC Code will become possible. Consequently, the studies which are presently made at overall country level could be undertaken at regional level, if circumstances so require.

In connection with terminology, some progress has been made as regards fuels, on similar lines to the proposal concerning electrical energy, without creating a new terminology but rather adopting the terms already in use in International organisations.

Electrical energy

As pointed out earlier, the different tariff structures which do not take into account the end use of energy but instead have consumption scales and different user classification system, impede and delay the processing of the information received. This has led to the need to schedule consultations with those public service companies which have a suitable organisation for achieving the desired uniformity of data by the use of a common form. This would meet the double purpose of facilitating the process commented on and of adapting and/or incorporating other headings which present requirements demand, for example the classification of consumptions in accordance with the UIIC Code.

Once this stage has been completed and in order to achieve continuity of efficient information, it is intended to appoint a permanent delegate in the respective companies whose mission will be to advise other suppliers within his jurisdiction in the case of provincial enterprises.

Since the production of electrical energy is one of the indices mostly intimately connected to the evolution of the economy, the data supplied by the companies on the consumptions of the different sectors of industrial activity add a valuable element of knowledge to the existing information. Moreover the self-generation data is to be published next year in the *Electrical Bulletin* as an amplification and complement, due to the high degree of reliability achieved in generation by the self production sector. This represents a substantial volume of energy, without which it is not possible to know the real behaviour of the sector.

This information is presently obtained monthly and on the basis of a sampling designed for the purpose. With it the number of despatch and reception documents can be reduced as much as possible.

Other information

Apart from these improvements, the EB is in a position to put into effect a Data Bank for which it can count on the possible assistance of local companies and cooperation with other countries.

This possibility enables other types of changes to be visualized in the energy information system in addition to those already mentioned :

(i) The EB for lack of physical capacity, does not process a series of information which is available but is filed.

(ii) The preparation of statistics referring to the non-commercial sources. Although consumption of these sources may not be important in Argentina, if a regional analysis were made, these sources may, in given sectors, be significant.

These sources which are not recorded either in the apparent consumption category or in the energy balances, are neither evaluated at reserves level (for example, firewood) nor at consumption level (wind, biomass, vegetable wastes, animal energy, and so on.)

(iii) Preparation of the sub-matrixes (7), (8), (9) and (10) of the Integral Energy Balance shown in Fig. 8.3. For the purpose, primary information from other sectors (transportation) will have to be processed, surveys carried out with designs already established, also by other organisations (Urban residential, Industrial and Mining) and further surveys undertaken for the rural areas and the Service sector.

In the case of the Transport sector, primary information regarding the transportation of passengers exists, filed in the Transport Bureau. A Survey at national level is being prepared for freight transport.

For the Residential and Industrial Mining sectors two organisations carry out periodic surveys. INDEC, conducts the Permanent Home Poll whose mechanized processing enables questions to be added to the original form for example on types of equipment in the different energy uses and consumptions. The second is the quarterly survey made by the Central Bank, through the Department of Industrial Activity. This has a significant coverage and by taking advantage of its sampling design, the questions corresponding to energy equipment, use and consumption could be added.

Finally, the cases for the Services sector and the rural areas are the only ones which have no background data in other organisations, so it would be necessary to carry out a survey handled by the EB

Since more than 80 per cent of energy consumption is concentrated in three sectors: Industrial, Transport and Residential, for which primary information or sampling designs already exist, it is evident that the costs of preparing matrixes of useful energy consumption by source and by use in each sector, are not high.

These three extentions would adapt the information system to a more analytical form of energy planning. A large part of the information to be prepared would serve as a basis for the analytical examination of the demand which forms part of the scheme referred to earlier.

Changes to be implemented and medium term prospect

The organisation of a network for exchange of energy information would have ample advantages for the institutions involved and for the users in general.

A formal energy information network does not exist in Argentina and therefore each of the entities compiles and files its own information, resorting when necessary to others for information it does not possess. The only organisation which compiles multisectorial energy information is the Energy Bureau.

It should also be pointed out that in Argentina the main energy companies are state enterprises, as are the organisation compiling the energy and associated non-engery information.

Consequently the initiative for improving the energy information system should emanate from the State or if it emanates from its enterprises, they should at least have its support. If this objective is to be achieved, the State, the enterprises and the organisations which make up the system, must be fully aware of the importance of exchange of information and of the implementation in the not-too-distant future of a data transmission network at national level. This network, named Arpac, has arisen out of the need for communication between the different groups located at considerable distances from each other. Thus, for example, data can be transmitted between the computer (central process unit located in the computer center of a company) and its terminals located in the plants generating the information.

It is possible to transmit digital signals over the telephone network but for this the digital signals have to be transformed to analog signals. This adaptation is done by using modems which transform the digital pulses into analog signals which can be transmitted within the width of a telephone channel band and vice versa. The Arpac network will soon operate throughout the country; this will be a great step forward for data exchange.

It is obvious that to transmit data it is useful to have a computer where the information is organised. This organization can be simply a file or can have more complex features like those of the Data Bases and the Data Banks. The fundamental difference between a Data Base and a Data Bank is that the first provides bibliographic references of the documents and the latter provides the information directly (statistics, laws, and so on.)

There are already many Data Bases which have been created in a number of countries, some of them having world coverage, but these bases only compile the most important information published in each country. This is one of the reasons why most countries are faced with the need to create their own Data Bases. Apart from the problem of creating the data base, the countries must have qualified personnel for handling it. In order to obtain optimum results from all the scientific and technical possibilities which exist in the bases, it is necessary to train the personnel connected with these bases conscientiously.

ACKNOWLEDGEMENT

The present study has been done with the help of Engineer Hector Piegari from the General Directorate of Energy Coordination and Information of the Energy Bureau and Licenciada Stella M. Torres from the Institute of Economic Energy (IDEE).

REFERENCES

1. F. Blamoutier, *Energie, L'offre et la demande d' information*, Collection Energie et Societe, CNRS.
2. *National Energy Balance—Republic of Peru*, Ministry of Energy and Mines, UNDP, December 1977.
3. *Integrated Energy-Economic Planning*, LJ. Legisa-IDEE—APDC—UNDP Workshop on Training issues for Energy Planning and Management—Kuala Lumpur, 26-30 July, 1982.
4. Methodology for the evaluation of energy requirements in the developing countries, Working paper prepared by IDEE—(Second Draft)—February, 1982.
5. Mundo Informatico, Vol. II, No. 35.

DISCUSSION

NIKHIL DESAI

This is not a question but a suggestion. I am quite impressed by OLADE's activities in generating regional energy data and establishing contact with governments and other organizations in Latin American countries and I would like to make a suggestion, that probably other regional organizations can follow their example, and try to establish both systems of collecting the data and providing assistance to member countries in order to generate those data.

JHIRAD

I know that you have an interest in Argentina for energy economic modelling. My question is, is the drive towards a better energy data system in some way governed by the need to have a particular kind of information for the modelling efforts that you are carrying out ? Are you also collecting information in Argentina with a view to incorporating this information in some energy-economic macro model of Argentina?

ALTOMONTE

Yes, the data are there. But I am not sure if the macroeconomic modelling is relevant to Argentina, we use the model of Argonne. We cannot account for the energy in the macro economic model because the system of information was not prepared in such a way it could be incorporated in the macro economic model.

JHIRAD

Dr Mubayi's comment was, that they should have used the Brookhaven model and not Argonne's.

NINE

EVOLUTION AND PERSPECTIVES OF THE BRAZILIAN ENERGY INFORMATION SYSTEM

JOAO LIZARDO R.H. DE ARAUJO

INTRODUCTION

BRAZILIAN energy data collecting and processing systems following the national energy planning process, have evolved from sectorial supply-oriented systems towards greater integration and comprehensiveness. Early false starts generated frustration, but overall there is a tendency to more organic systems with better trained people. Of particular interest are an Energy Matrix as part of the Brazilian Interindustry Matrix, census data on Energy, and a decentralized, disaggregated and consistent structure for the National Energy Balance.

As most developing countries, Brazil has only very recently developed information systems. In many cases, they suffer from serious deficiencies. Energy data, in particular, have serious gaps. These stem, on the one hand, from the fact only very recently have decision makers become aware of the need for integrated energy planning, and therefore, of comprehensive data on energy production and consumption; and on the other hand, from the role played by non-commercial sources, specially firewood.

To these two specific problems we may add institutional, political organizational and technical ones. All these are responsible to some extent for frustrations and retrocessions in the process. But it is not our purpose to analyze here Brazilian Energy Policy; we shall, therefore, refer to these problems only as far as they affected directly the evolution of Brazilian Energy Data Systems.

Lastly, when talking of National Energy Data Systems one has to be selective in order to give a clear picture; we have thus concentrated on the evolution of integrated energy information systems, omitting discussion of sub-systems, except when relevant to the evolution of the whole; and on systems created for specific programmes.

THE EARLY YEARS

During the fifties and sixties two essentially independent information systems evolved, corresponding to the oil and electricity sub-sectors. The information gathered reflected the responsibilities allotted to PETROBRAS and CNP for oil, and ELETROBRAS for electricity. Thus, far more effort was dedicated to production data than to consumption, and virtually none to integration of energy data in a comprehensive picture. The dominant role played by hydroelectricity in Brazil helped this attitude.

Nevertheless, this scheme was hardly compatible with the creation of a Ministry of Mines and Energy in 1960; and during the sixties the Ministry tried to set up a comprehensive system. The culminating effort of this phase was the creation of a task group for the elaboration of the Brazilian Energy Matrix in 1969-70.

This was an endeavour of unprecedented magnitude. Its aim was double: on the one hand, profiting from the Census to be taken in 1970* it was to gather very disaggregated data on energy production and consumption, distinguishing urban and rural sectors and taking into account non-commercial energy; on the other, to elaborate scenarios and plans based on the census sample of energy data.

The result was not an unmitigated success; while the sample

*This census, for the first time, collected data for the construction of an input-output matrix for the Brazilian economy.

was carried through and the first disaggregated energy matrix of Brazil was constructed, the scenario studies were only completed in 1973, when the first oil crisis rendered them obsolete. This study never reached publication, and only the 1970 matrix ever saw the light of day as a report of limited circulation.

This work absorbed scores of high-level technicians from all over the energy sector, working together for this single task, but there were no institutional mechanisms capable of ensuring continuity. Thus, upon completion, the task group was dispersed; the frustration with the results contributed somewhat to this.

Nevertheless, as we shall, this study was not fruitless; it prepared the ground for less ambitious, more systematic action.

AFTER THE FIRST OIL SHOCK

For all its shortcomings, the 1970 Energy Matrix had shown the possibility of integration; and the oil shock of '73 showed clearly the need for an integrated comprehensive National Energy Balance. Thus, in 1976 the Committee for the National Energy Balance (COBEN) was created, coordinated by the Ministry of Energy and comprised of representatives from organisations and enterprises of the energy producing sector.*

From 1976 to 1980 this committee prepared very simple energy balances, based on the data collected by the member institutions. As these data referred basically to the supply side, the published energy balances were little more than statistics on the quantities of primary energy produced and consumed, broken down by source, supplemented by data on electricity generation and consumption of oil products. This was accompanied by projections made by the team.

In terms of method and scope, this was a big step backwards in relation to the Energy Matrix; but it was more organic than that initiative, in the sense that the participating institutions were more deeply involved.

Thus, during this period the sole integrated data on energy production and consumption were given by energy source, not

*See the appendix for a list of the institutions comprising COBEN.

by consuming sector, with one important exception to which we will return. It is true, that the National Oil Council (CNP) and PETROBRAS had data on sectorial consumption of fluid fuels, DNAEE and ELETROBRAS had a clear picture of electricity consumers, and coal was monitored by CAEEB, ELETROBRAS and CONSIDER (this last also accounted for most of charcoal consumed).

Nevertheless, two blatant gaps remained: one, firewood production and consumption were unknown, and crude estimates gave only a rough order of magnitude. The sole non-commercial energy to be measured with some accuracy was bagasse, as the IAA (Institute for Sugar and Alcohol) monitored closely the quantities of sugarcane processed by refineries and stills.Two, comprehensive analyses of consumption by end-use sectors were lacking, specially residential, agricultural, services and transport. This problem was compounded by the fact that these activities fell under other ministries, and there was little coordination of efforts at that level.

This can be illustrated by two important studies on the residential sector done in 1974 and 1975. The IBGE (Barzilian Institute for Geography and Statistics)* launched the PNAD (National Home Sampling Plan) and the ENDEF (National Survey of Family Expenses) aiming to establish the standard of living of Brazilian families. Even though the emphasis was on nutrition and general conditions of life, energy consumption data were collected. Nevertheless, these took the form of expenses in cruzeiros, even when energy was not purchased (as was the case for most firewood consumption in the rural area). In this case, physical quantities were converted into cruzeiros using local prices. Since the price of firewood varies widely among regions, and the local prices were not retained, it is difficult to reproduce the physical quantities of firewood consumed by families; even so, useful approximations could be obtained.

*IBGE is the organisation responsible for the census, and generally surveys most socio-economic activities at county, regional, and national levels.

(In 1979 the PNAD was repeated; this time greater attention was given to energy aspects and, in parallel with the sample, local prices for firewood and charcoal were collected so that the physical consumption could be reproduced. The results of this study have not been published as yet).

The need to integrate efforts was thus keenly felt by all workers in the field; and starting from 1979 the Ministry of Energy sponsored workshops aiming for interaction among a wider audience than the COBEN. (The first two workshops were organized by COPPE and held at Itaipaiva, RJ).

Also in 1979, the new administration introduced the Brazilian Energy Model. This was not really a model, but rather a set of energy substitution goals, however, it also contained disaggregation of energy consumption by sector. This disaggregation raised some questions on the methodology and data; and, pending a clearer methodological definition, the 1980 National Energy Balance returned to its previous format.

RECENT DEVELOPMENTS

After 1979 two developments are worth noting: the creation of an energy group within the IBGE and the restructuring of the National Energy Balance. Let us analyse them by turn.

The IBGE energy group was formally created in 1979, having as two of its functions : to gather detailed data on energy consumption and to construct a Brazilian Energy matrix which would be regularly updated. The Energy Matrix was to be a part of the Brazilian Inter-sector Matrices resulting from the economic censuses, starting with the 1975 matrix. For the first time, besides input-output matrices relating to monetary quantities, two submatrices were constructed for the Energy and the steel sectors, relating also to physical quantities. The 1975 matrix is practically ready for publication, and the update should come from two sources: the economic censuses, held every five years, and the annual economic surveys.

The other task initiated by the energy group was the inclusion in the 1980 census of a thorough investigation of residential energy consumption, covering sixteen energy sources and five end-uses (including individual transport for a sample of the population, besides other economic and demographic information. At present the data obtained are being tested and discussed with researchers in the energy field.

The significance of this development lies in the large infrastructure and experience of IBGE in collecting and analyzing data. Its contribution towards obtaining comprehensive, disaggregated, reliable energy data will be invaluable.

The restructuring of the National Energy Balance is another significant development. Already in 1980 COBEN started to revise the methodology, and reached a format compatible with one proposed by OLADE. The first Energy Balance to identify production, transformation and consumption centres in a consistent way was published in 1981, and covered the years 1976 to 1980. No attempt was made to project these balances into the future.

This year a second balance was published, with minor methodological revisions and covering the period 1976 to 1981. This is being followed now by an attempt to decentralize the construction of Energy Balances to each state of the Federation, which in turn is likely to have a large influence on the construction of an integrated Energy Data System.

In fact, the requirements and features for the Energy Balances and those of the IBGE work are supplementary. The NEB demands agility and often has to rely on secondary and indirect information corrected whenever possible by information from IBGE or other primary sources. Also crucial is consistency especially when one tries to decompose the NEB to the level of Federation Units under the responsibility of state-level teams.* This has led to the design of a computerized information system, with the double purpose of compiling relevant information in a readily accessible form and providing a standard guideline for the newly-formed teams forming state level Energy Balances.

This information system is still under construction in: order to achieve operational realisability, in a first phase it will contain information disaggregated only to the level used in the Energy Balance.

*At the end of September a short course on Energy Balances for state government technicians was organized by SUDENE and CHESF in Recife, and sponsored by the MME.

This was wise, as it considerably simplified design and allowed efforts to be concentrated on standardizing procedures for the use of state teams; there were other reasons for doing it too, in particular the difficulty of obtaining the relevant primary data.

PROBLEMS AND PERSPECTIVES

These developments should not make us forget the problems remaining. First, the integration of data even within COBEN is imperfect. The design of the information system was assigned to ELETROBRAS, since it already had a similar system for the electrical sector; but of its own this creates institutional problems, which was one of the reasons for the gradual strategy agreed upon. Also, energy consumption in agriculture and transport is still imperfectly known, and will likely require specific studies.

Another problem lies in inter-ministerial collaboration : although a specific body, the National Energy Committee, has been created, containing representatives of all ministries concerned and chaired by the Vice-President of the Republic. The problem of institutional attrition has not been wholly solved.

Finally, there is lack of coordination even among specific energy programmes, specially when developed by distinct ministries; as a result, information relevant to the assessment of the programme may be withheld from other institutions or simply not collected at all.

Even so, there seems to be a tendency for better and more integrated collection and analysis of energy data; this comes both from the larger and better-trained group of technicians involved *in an organic way* with the task (which stands in a sharp contrast to the 1970 Energy Matrix experience) and from the need for a clearer, more accurate energy picture to aid policy decisions.

REFERENCES

Balanco Energetico Nacional (National Energy Balance) MME, years 1976, 1977, 1978, 1980, 1981, 1982.

Modelo Energetico Brasileiro (Brazilian Energy Model) MME, 1979.

ELETROBRAS, Energy Sources Department, Energy Division Reports 3 and 4 on Project monitoring, Information File (1982).

ELETROBRAS Balanco Energético National—1982, Informacao sobre Energia electrica, report 005/82.

CODESE, Projeto SC. 01.01—Informacoes Estatisticas—Relatorio final, MME, 1980.

Carmen Garcia, Estatisticas Demograficas Continuas e Uso Domiciliar de Energia : O caso da PNAD, IBGE/DESDE, Grupo de energia, April 1982.

Roberto Ramos, Matriz de Energia 1975, IBGE/DESDE, Grupo de Energia, April 1982.

Victor Gouvea, Informacoes sobre o consumo domiciliar de energia nas estatisticas censitarias : a pesquisa de avaliaeao do censo demogràfice de 1980, IBGE/DESDE, Grupo de Energia, April 1982.

Oliveria, J. Araujo, L. Pinguelli—Cenarios Energeticos para o Brasil, COPPE, 1979.

APPENDIX

A—Composition of the Committee for the National Energy Balance

(COBEN)

Presidency : First Secretary of MME
Coordination : Secretary for Technology of MME (2 representatives)
Executive Secretary : Companhia Auxiliar de Empresas Eletricas Brasileiras. — CAEEB (2 representatives)

Members

1. General Secretary (appointed by MME).
2. Representatives each of—

(i) National Oil Council—CNP.
(ii) National Department of Waters and Electric Energy—DNAEE.
(iii) National Department of Mineral Production—DNPM.
(iv) National Committee for Nuclear Energy—CNEN.
(v) Centrais Eletricas Brasileiras—ELETROBRAS.
(vi) Petroleo Brasileiro S. A—PETROBRAS.
(vii) Empresas Nucleares Brasileiras S. A—NUCLEBRAS.

B—Organisations mentioned: names and functions

(i) MME : Ministry of Mines and Energy.

COBEN : Committee of organisations under the Ministry of Mines and Energy (MME), charged with preparing the National Energy Balances.

CNP (Conselho Nacional de Petroleo) : a normative agency, responsible for regulatory aspects of pricing policy, consumption quotas, and monitoring of fuels—specially oil and its derivatives, coal and alcohol.

PETROBRAS : State enterprise charged with the exploration, production and import of petroleum and its products (all state monopoly), the buying and mixing of alcohol; it is also active in oil shale exploration and production, and participates in distribution through its subsidiaries.

DNAEE : (Departamento Nacional de Aguas e Energia Eletrica) : a normative agency, charged with all regulatory aspect of electricity production and consumption.

ELETROBRAS : State enterprise functioning as a holding company, charged with the execution of electricity policy.

CAEEB : (Companhia Auxiliar de Empresas Eletricas Brasileiras) Originally concerned with thermo-electricity, it supervises steam coal production and utilization.

CNEN (Comissao Nacional de Energia Nuclear) : normative organ charged with regulatory aspects of nuclear energy policy.

NUCLEBRAS : State enterprise charged with uranium production, enrichment and reprocessing, as well as with the construction of nuclear plants.

DNPM (Departamento Nacional de Producao Mineral): normative agency charged with the assessment and supervision of the production of mineral resources.

(ii) MIC : Ministry of Industry and Trade; it is charged with the execution of several sectorial programmes in renewable energy, specially alcohol.

IIA (Instituto do Acucar e do Alcool) : agency charged with the supervising and the distribution of quotas for sugar and alcohol production.

(iii) CENAL (Comissao Executiva de Plano Nacional do Alcool) : executive agency supervising the National Alcohol Plan.

CONSIDER (Conselho National de Siderurgia) : normative agency regulating the steel industry.

(iv) SEPLAN (Secretaria de Planejamento) ; ministry of planning, it is ultimately responsible for all economic planning.

IBGE (Instituto Brasileiro de Geografia e Estatistica); bureau in charge of the census, also surveys geographic and economic data.

(v) CNE (Comissao Nacional de Energia) : Inter-ministerial body charged with the coordination of energy policy.

DISCUSSION

LIMSUMALEE

I would like to know how you collect data on bagasse as an energy commodity. I know that in Brazil you produce sugarcane and transform the sugarcane to alcohol and the bagasse from sugarcane you use as a fuel for boilers. I would like to know how you collect accurate data.

ARAUJO

First of all, to figure the amount of bagasse produced is a swift calculation. Bagasse is about 25 per cent of the weight of the sugarcane crushed. The amount consumed is a matter of estimation. Because traditionally bagasse was seen mostly as an annoyance, the boilers were forced to consume all of it. Now it is beginning to become an interesting fuel. Theoretically you need only 60 per cent of the bagasse produced to produce alcohol. The quantity for sugar is somewhat more. But even so, that shows you can have an excess of bagasse which you could use to produce electricity or to sell as cattle feed. Some of it is sold as cattle feed. Now what one does is just to take the whole bagasse produced and assume that it is also consumed.

MUBAYI

My impression of the Brazilian Energy Model, that was being developed and the methodology that was being adopted for projections and for integrating of supply and demand for the years 1985 to 1990, was that you had not included much of linkages to the economy and between the supply sector and the demand sector. Has there been any change in the last few years?

ARAUJO

Sure. I am glad that you asked that because I have not included them in the paper. COBEN has also created an energy model group. There has been an enormous study. As for modelling they have even proposed a method of gradual model building which I think would be very interesting if it had been carried out. Unfortunately, or may be fortunately, it is a matter of personal viewpoint, the government have bought the MARKAL Model. So they had to stop what they were doing and concentrate on how to use, analyse, and criticize, the MARKAL Model. But even then, I think, some interesting things were coming out, because after the analysis they were proposing substantial changes in the model so that it would be suitable for the Brazilian case.

As far, the linkage with economy, you know it is the weakest spot, the least understood. There are some people who propose using econometrics for this procedure. The Institute for Applied Economic Research in Rio de Janeiro is doing the work along those lines. My own feeling is that it is better to use econometrics just as much as it is needed but not to trust it too much, but instead rely on a careful and qualitative scenarios analysis by discussion with people who are familiar with the economy and are looking forward to see what comes out of it.

MUBAYI

One more question I have is on the bagasse utilization and some of the environmental problems caused by the alcohol programme. Now you have almost 40 per cent or so theoretically of bagasse as surplus. Is there any consideration of using it as a feed stock, for other industries like paper, or something like that.

ARAUJO

Yes, there have been considerations. There are several possibilities. You might pellet it and sell it as industrial fuel. You might use it as a feed stock for paper and pulp industries. Most of the additional use outside burning it in the sugarcane industry was as cattle feed. Now the second question. Vinasse is a highly polluting agent that has a very high biochemical demand for oxygen. So it demands treatment. When the alcohol programme was launched no distillery was prepared to treat large quantities of vinasse. They just dumped it into the river. It was clear that most of the life in the river was killed. So beginning in 1979, a law was enacted prohibiting any throwing of vinasse into the river. The alternatives for treatment are that you may have lagoons as for urban waste and refuse or aeration or you may concentrate vinasse and then use it directly as a fertilizer for sugarcane although the only problem is that not every soil accepts it well, because it is very acidic. So in some regions, like Rio de Janeiro, the soil quickly mineralizes when you put vinasse into it. Another possibility, which seems to be gathering more and more favour, is of using it with anerobic digestion to produce biogas. In that way you can have a supplementary energy source for the distilleries. Now this dislodge can be used very well as fertilizer as it is no longer acidic. It can be just neutralized and put into the soil. These are the possibilities. But these methods are not monitored very closely and we do know at least in some cases distilleries are still throwing vinasse into the river. How extensive that phenomenon is, I am not sure. It is not widespread, but it is significant enough.

MOOLCHANDANI

Professor, are you utilizing alcohol as a mixture in both petrol and diesel engines or only in petrol engines and to what extent?

ARAUJO

In the first phase, 1975 to 1979, anhydrous alcohol was used exclusively as a mixture in gasoline up to a limit of 20 per cent. It did not have very much affinity with gasoline engines because it increases octane. The octane content in alcohol is 120 to 130. So our gasoline is very poor. It increases the performance of motors. Starting in 1979, we started the programme of auto vehicles that run on straight alcohol. We have now 500,000 automobiles that run on straight alcohol. These displaced so much gasoline that we had a surplus of gasoline. We did not know what to do with this. We tried to export it. The international market was not favourable. We mixed it with diesel up to a point that diesel got completely changed. Now diesel in Brazil is flammable, one should not smoke near it. Even so due to the very high taxation on gasoline, we have a very significant dieselization of our trade and passenger traffic so that diesel now replaces oil imports. Therefore, starting this year, we are also mixing alcohol with diesel up to 7 per cent mixture, so that alcohol is slowly spreading throughout the transportation system.

JHIRAD

I will take the licence to spend more time discussing some of the energy data problems that we encountered in the Brazilian situation in an extensive study which we did there and refer only obliquely to some of the Dominican Republic work with which I was not personally involved. The study that we carried out in Brazil was part of the series of studies which were supported by the US Department of Energy and US Trade and Development Programme on assessing in some way the potential for using renewable energy sources and energy efficient technologies within a country's energy system. The Brazilian study was one of these but there were other studies carried out in Thailand, Indonesia, Argentina, and Columbia. We tried to adopt a very consistent methodology in assessing the potential for certain renewable technologies and energy efficient technologies. This work was done mainly in 1981. At that time, you know, there was lot of funding for these problems as it coincided with the year when the UN Conference on New and Renewable Sources was held in Nairobi.

We hoped that the particular lens that we used to view data might be able to answer the questions, what kind of data do we need and for what purpose? So we had very specific questions in mind that we posed when we did this investigation; the kind of data gaps and deficiencies that emerged in the course of my discussions were data gaps for answering these questions. The gaps that Dr. Araujo has pointed out are certainly the ones that we encountered. Ideally, we wanted to know everything in order to assess the role of renewables in a country's system. We would like to know in great detail the end-use structure of the energy system, broken down by fuel type, at a very disaggregated sector level; number of households that use various kinds of fuel for hot water purposes, for cooking, for lighting, etc. In other words we would like to have a complete reference energy system for the country beginning with fuel supply and all the losses and conversions, in transmission and transportation right through to the end-use, alongwith efficiency of end-use devices. We would like to have this for commercial and non-commercial use as well at the national and state level. Ideally speaking this would be marvellous; we could take it technology by technology and look at the end-use and figure out how that technology could be matched with other competing technologies to meet the same needs.

Now as Dr. Araujo has pointed out the petroleum data was in fact quite good. We were able to get a good sectoral breakdown of petroleum products by state, and by sector, such as industry, commercial, residential. We had some problems in the rural areas. I will give you a specific example. We were told that there was great interest in solar crop drying because the use of fuel oil for crop drying was being prohibited. So the question was: how much diesel oil is used for crop drying in a particular state? We could not get a clear answer to that question except in the State of Parana where they could tell us exactly how much diesel oil was used for crop drying. Then we could say, well, if you use solar crop dryers, in a particular installation, this is the amount of energy that you could expect to save. These are the economies of crop dryers compared to using fuel oil. This was an example where in order to give a fairly good answer to the question of using a particular renewable in a particular application we needed to know the end-use data, which generally we did not have. Then ideally speaking, if we had the whole reference energy system, we could say, all right, if you displace so many barrels of oil at the point of end-use, this is what it means in terms of supply. But we did not have a complete reference energy system so we were not able to make that extra calculation. In the transportation sector there were certainly deficiencies.

We focused specifically on the industrial sector because of the tremendous emphasis given by the Brazilian Government to reducing the use of fuel oil in the industrial sector, both through substitution and through energy efficient technology. So what we really wanted to know at the state level and at the national level is how much of each product is used for each industry, in steel, in chemical, in refining and so on. We wanted to know how much electricity was used in petroleum for processing and also for drives and we wanted this breakdown for industrial sub-sectors. We could find a lot of data. I am not talking now for data needed for technical potential assessment. We could then say, well, we have so much heat produced below 100°C, below 200°C, below 300°C, and so much petroleum and electricity are used in the industrial sector. So we could make an estimate of what the technical potential for a given technology might be. Of course, we did not have the detailed temperature breakdown. But we had to calculate it from the US data, assuming that US cement plants and Brazilian cement plants were about the same. Well, these are some of the gaps in assessing the technical needs for waste heat recovery of solar energy. We did not look at alcohol or large scale hydro.

We also found that the wind resources data generally were collected at airports which are not the best places to collect wind data because they are located in places which are not windy. So we found that wind data is subject to tremendous variations and we scanned the literature and found a lot of statements of potentials about wind in pumping and irrigation application in the South of Brazil but we really did not find the kind of data that could be used to make a good assessment. That is the data need for renewable sources that we felt could be corrected.

We found very good solar energy data specially in Sao Paolo which probably had the best data out of the many states. We were confining ourselves to the four southern states.

Then there is the question of economic and financial data. Once having decided that certain technical potential was possible, we then asked the question how much of this could be financially feasible, and the more difficult question, how much of this could be economically feasible. To assess the financial feasibility we need a whole set of technology data, what are the cost and performances of the relevant systems, reliability of the systems, and so forth in order to do this in some rigorous way. We did invite a group of six energy planners from Brazil to visit various installations in the US where this technology was in use and talked with the plant managers to see whether such technology could be used and if so under what conditions. There was a whole set of technology data which we felt Brazilian energy planners were lacking and incidentally the US energy planners also were lacking. Now we are talking about renewables and energy efficiency. Let us be specific. What technology are we really talking about? Whether it is solar ponds or high temperature collectors? Or low temperature collectors? 20 KW wind machines? 40 KW wind machines? Where are these functioning? What problems have been encountered? In order to meet this technology data gaps we had this tour which lasted for about a month and we went round the various installations, and tried to fill some of the gaps. That is the question of technology.

Then I said there were financial and economic data that were needed in order to really make decisions because the purpose of gathering data is to make intelligent decisions. We did some financial analysis at the micro project level of heat pump and industrial processing installations or heat recovery system. We asked the question how would this system work in the absence of a subsidy, and with a subsidy given at the same rate or equivalent to the subsidy that is given to the coal and alcohol programmes in Brazil. The answer was very interesting and the main result is that if heat pump and waste heat recovery in solar industrial process heat were given the same benefit, it would be an extremely attractive investment. With the inflation running at 100 per cent, the indexing was only 55 per cent.

I would expect that, that was not revolutionary. The point is, do they make economic sense from the national viewpoint? We do not have the kind of national economic energy model that you would have to use to answer that question. In other words, we can say that heat pumps looked very good for the textile and food industry and they could perhaps substitute 5 to 10 per cent fuel oil and give a rate of return which is very high. You recover the capital in a year and a half. But the question is how does it look from the government's viewpoint in terms of investment in this technology versus competing investments, in terms of foreign exchange requirements, technology transfer questions and so forth. In order to answer the question at the national planner's level, we simply did not have the analytical tools to really say we should be shifting investment from alcohol into heat recovery, or from coal into solar. That kind of decision we could not make on the basis of our data, and our analytical tools. So we actually pointed out not only a data gap but also an analysis gap. Most of this exercise was really quite straightforward because it is possible to get data on technology, it is possible to identify resource gaps, it is possible to get financial analysis in the way that the World and the Inter-American Development Bank and other development banks expect, and we had the data to do that with the subsidy, but the area where we felt that there was really a large gap was the question of investment priority for Brazil. As Dr. Araujo pointed out the Brazilian Energy Model is really not a model. It is simply a set of targets for substitution and conservation and there are no investment numbers attached to those various targets. So that even though we could identify technical potential to say we could save so many thousands of barrels of oil or diesel in the industrial sector and for substitutions of electricity in the residential sector by a judicious combination of technologies, we had no base on which to attach investment priorities. So it seems to me there is a gap, which is mainly an analysis gap.

ARAUJO

Well, to comment on what you said about the data of alcohol. It really has some very substantial gaps. For instance, in alcohol if you take the subsidy which is given in terms of rates of interest and then use the figure of inflation, it depends on the data you start with, whether you get it at $50 a barrel equivalent up to $80 a barrel. That makes a difference. Now the other gap in data is on the charcoal costs. Charcoal, even if we plant the wood, is a very competitive fuel in Brazil. But it so happens that the problem of reforestation for energy purpose has been on the seesaw so the tendency is for a greater consumption of native wood.

Much of the planning is really done without good analysis. The fuel oil substitution promises to be one such. In 1980, the government just put down a 10 per cent reduction on oil production. So the industry had to adopt, or slow down their production. One another was the increase in the fuel wood and charcoal production through native wood and the fuel oil had been replaced quite extensively both by charcoal, by coal and by electricity. Now for coal too we have very ambiguous data on the economics because it is also heavily subsidised.

SRIVASTAVA

David, if I understood you right, you were able to get almost all types of information except for the investment priorities, but couldn't you make your model a little more flexible on, say, goal programming concepts, so that whenever the Brazilian Government decided on some priorities they can work it.

JHIRAD

On the question of a model used to determine investment priorities, we did not use such a model for Brazil. May be my colleague, Dr. Mubayi, could say what kind of information you need to construct such a model, since he has been working in that area. A couple of things that were raised about the information about the economies of some of the substitution devices whether of charcoal, or bagasse, were vital in order to make correct decisions on competing technologies. Capital cost on competing technologies have to be fed into some kind of broad economic model but we did not actually go to that stage. But Dr. Mubayi who has been constructing an energy economic model at Brookhaven may say a couple of wise words on the subject.

MUBAYI

David presented the different levels of analytical data needs. There are several problems that arise in looking comprehensively at the energy economy interaction, especially as regards investment planning. These are largely country specific and have to be dealt with in those terms. The two issues are the goals of your model; and secondly, what are the various types of information that should be fed into the structure to achieve the goal for which you are doing the exercise. Now construction of national energy economy models has taken place in several countries. These are major tasks because the energy sub-sector is a very important and a large component of the economy as a whole. For example, a substantial fraction of investments goes into the power system. To understand what is the relative priority of investment in the power system, as opposed to investing somewhere else in the economy, one really needs a national economic model, macro model, to start with and the energy sub-sector model which is actively associated with a macro model. We have to be able to both understand what will be the evolution of demand and the least expensive way of meeting that demand with a certain mix of energy products. This is quite a challenging exercise and in the study that Dr. Jhirad mentioned, there was not the scope to take all this effort. But in order to really prioritize the investment as he put it, one does need this kind of analytical exercise which has not really been done to any significant extent in most countries.

One of the major problems is that investment planning or just planning in the energy sector historically was done almost in complete isolation. Each sub-sector is a kind of kingdom unto itself. Oil, coal, electric, power, and so forth. I think it has been now recognized that some sort of an integrated approach has to be followed because these things no longer develop in isolation from each other. As to precisely how these interactions should be put in terms of an analytical structure, there are various models available. Dr. Araujo referred to one MARKAL Model which is a dynamic programming model that has been used in Europe and the countries belonging to the International Energy Agency. We have a model ourselves which we are hoping to develop for developing countries. It is a very substantial and challenging task. It requires a very large assembly of economic information of energy sector information at a very disaggregated level in order to understand how substitution can take place, where the gaps are, where the technology rigidities are, what the progress of technology is, what will be the pattern of likely demand for different types of energy depending on the evolution of the economy and its various sub-sectors. But I think this is in fact quite a necessary task because for the energy sector long lead times and very large investments are required in order to meet projected or evolving demands. Therefore, this kind of exercise needs to be coordinated at a fairly high level, so that a directive from the national standpoint can be given to involve rather a large number of people whose cooperation is necessary for making it a success.

JHIRAD

I just want to add one very serious gap in data which is quite generic in industrialized countries and developing countries. It has not come up in the sessions. That is with regard to the potential for revolutionary break-throughs in energy technology. In Brazil there was a great deal of interest in certain technologies such as amorphous silicon cells for solar photovoltaic conversion in a few cells, and in hydrogen production even their hydro resources. There was a great deal of interest in collaboration with advanced countries in some of these areas. One should not take too static a view towards technology such as this. Also to plan one should have a dynamic view, the fact that such technology can change very rapidly over a short period of time. Micro computers are an example of that. This kind of change in the cost and performance and widespread availability is also likely to happen in the solar photovoltaic field and particularly in some of the new technologies such as amorphous silicon technology. It is very important for planners in developing countries to familiarize themselves at this stage, with the development in these technologies because they are likely to progress very fast and much of our discussions about electric power supply might well be obsolete five years from now if we have silicon cells costing 25 cents a peak watt. One can, of course, question whether that can really happen. But it is an event of non-zero probability. In fact in some of the technical papers that I read of the US-Japan Symposium on Semi-conductors, many are already investing in Japan for amorphous silicon facilities where they are projecting costs of 25 to 50 cents a peak watt if they have 100 MW or so for the market. We should all be aware of the potentials for a change and monitor them at a very early stage.

TEN

ENERGY POLICY PLANNING DATA SYSTEMS IN THE DOMINICAN REPUBLIC

ANTONIO J. D'ERRICO, ANDRES DOERNBERG, ELISA GARCIA, ROBERT NATHANS AND MICHAEL OWENS

INTRODUCTION

THE Institute for Energy Research (IER) is an organized research unit of the State University of New York at Stony Brook. IER works with U.S. and international agencies and developing country counterpart groups to develop new analytical methods for the evaluation of energy alternatives, and to educate individuals from developing countries in the use of such techniques to support the efforts of these countries to increase their energy planning and management capabilities. The IER has worked with individuals from over fifty developing countries, including the Dominican Republic.

The authors of this paper were involved in a National Energy Assessment Project which came about as an outgrowth of discussions held between the Comision Nacional de Politica Energetica (CNPE) of the Dominican Republic and the Mission of the United States Agency for International Development in 1979. The overall goal of that project was to provide policy makers in the Dominican Republic with a body of data and supporting analysis capable of offering a comprehensive picture of the energy situation as it existed then and was likely to develop over the next several decades. Out of this project grew the need for some type of systematic approach to the storage, retrieval and presentation of the data. This led to a project to install a computerized energy information system at the CNPE for this purpose. It entailed the selection of an appropriate system, training of CNPE staff in its use, support for the collection and analysis of data, and the installation of the computer hardware and software.

ORGANISATIONAL OVERVIEW OF ENERGY PLANNING IN THE DOMINICAN REPUBLIC

Energy Planning in the Dominican Republic (D.R.) has historically been undertaken by various entities in government and the private sector on an as and when needed basis. Due to economic development, a growing need for oil, and the rising international oil prices in the 1970s, the need for a more centralized approach to energy planning was recognized. There were a number of groups collecting and analyzing their own data, but no single organization to oversee and to ensure that one entity's information was consistent with another's or that their respective plans were complementary. The following are some of the agencies involved in energy planning in the D.R. along with the role they play in an integrated system :

Oficina Nacional de Planificacion (ONOPLAN) is the Presidential office that compiles the nation's economic data and sets national priorities and development goals. It has published a document which analyzes possibilities for the energy sector and is concerned with the effects of energy prices and energy infrastructure investment on the national economy. In publishing its "Plan Triennial" (Three-Year Plan), it allocates the percentages of public investment that will go to energy projects.

Corporacion Dominicana de Electricidad (CDE) is the public utility which, in addition to generating and distributing electricity, forecasts the demand for electricity and makes plans as to how that demand will be met.

Instituto Nacional de Recursos Hidraulicos (INDRHI) is the agency devoted to hydropower development, planning, construction and management. Its planning activities centre around establishing small hydroelectric plants in rural areas that are not connected to the grid, as well as multi-million dollar projects to substitute hydropower for oil in electrical generation.

Oficina Nacional de Estadistica (ONE) is the clearing house for all Dominican statistical data and is, therefore, responsible for petroleum import data as well as production, sales and export data for the country's industrial sector.

Secretaria de Industria y Comercio (SIC) is the major planning group for the commercial and industrial sectors and sets the government's priorities in these sectors.

Oficina Nacional de Presupuesto (ONOPRES) is the national office which, along with ONOPLAN, determines the financial allocations the government will give to energy projects

Oficina Nacional de Transporte Terrestre (ONATRATE) does the planning for the public transport sector, along with FENATRADO, which studies freight transport. Between these two agencies, the bulk of the transport sector is covered.

Secretaria Estatal de Agricultura (SEA) and *Consejo Estatal de Azucar* (CEA) are the agencies in charge of agricultural planning, the latter concerned with sugar exclusively. SEA also concerns itself with the production of appropriate technologies for use of renewable energy (e.g., windmills for pumping) and also undertakes biomass and alcohol studies.

Subsecretariado Tecnica de la Presidencia is a Presidential office which makes policy recommendations involving technologies to the President.

There are also some research groups in the Dominican Republic that provide technical assistance for industrial development and the productive use of energy (INDOTEC, INTEC and CODIA). These groups provide another side of the integrated energy system.

Lastly, there are the financial organizations, such as the Central Bank, which are an important part of the energy planning system, because it is with the Bank's loans that the bulk of the energy investment is made.

Outlined above are the many facets of the Dominican Republic's energy data collection, analyzing and planning activities that existed before the formation of the *Comision Nacional de Politica Energetica* (CNPE). All of the agencies listed still remain players in current national energy planning, but (CNPE) is the centralized data collector and disseminator.

The CNPE is divided into sub-groups that include Conventional Energy, Renewable Resources, Hydropower and most important Energy and Economic Data. The last group is the heart of the CNPE, providing support for all of its other functions as well as information for policy makers. The Data section is charged with the monumental task of data collection, analysis and dissemination.

The first project undertaken by the CNPE in 1979 was a national energy assessment, the overall goal of which has already been stated. The first step in any assessment is data collection, which in this case has become an ongoing activity. Over the past several years in the Dominican Republic, a number of the aforementioned agencies have conducted studies which when taken together gave a comprehensive view of the energy resources in the country, so now the collection efforts emphasize the demand sectors. Data capable of providing the following were sought :

1. A detailed profile of the current energy situation in the major sectors of energy consumption;
2. A basis for projecting changes in the patterns of energy end-uses over the next one to two decades; and
3. Estimates of the extent to which energy could be conserved, especially through increases in the efficiency of energy end-use, and the extent to which other fuels could be substituted for oil-based products.

Data collection efforts were based on the following sequential process :

1. Secondary data from existing economic, demographic,

social, and technical sources, including all previous sectoral energy studies, were reviewed.
2. Data from fuel suppliers, distributors and the electric utility were compiled along with fuel import statistics.
3. Energy surveys of specific sectors and interviews with consumers were conducted to fill gaps in the data base.

The specific approach and the level of detail sought varied considerably from one sector to another depending upon the size and diversity of the sector, the magnitude of electricity or oil products consumed in the sector, the sensitivity of consumers to changes in either form or amount of fuel consumed, and the magnitude of the effort needed to increase the precision of the data base. For example, the direct use of petroleum in agriculture (including sugar) is less than one per cent of total energy consumption and its use is dispersed among numerous farms and processing units; energy consumption in this case was estimated from farm production statistics combined with survey-based information on average petroleum use per unit of crop production output. On the other hand, most of the industrial consumption of oil takes place in a few specific factories. Therefore, each major consuming facility in the industrial sector was visited to solicit information.

A variety of models exist for relating macroeconomic performance to future energy demands. Almost all of these, however, require a rich data base, in the absence of which estimates must be made of the missing parameters and the manner in which they change in the future. The procedure adopted, therefore, does not attempt to forecast energy demand from future levels of macroeconomic activity, but rather arrives at a set of sectoral demands that are compatible with a given level of GDP growth. This interactive procedure consists of three steps :

1. Target performance was determined for each sector, based on resources available for that sector, anticipated growth in key sub-sectors, the degree of planned public investment, and historical growth. Sectoral projections were made after discussions with Dominican Economic and Planning Agencies, the Central Bank, the World

Bank, ECLA, UNDP, and private sources as well as the Ministries of Industry, Transportation and Agriculture.

2. Projected GDP growth was then adjusted in the light of the principal growth constraint that the Dominican Republic is likely to face: limited investment financing, insufficient resources to rapidly develop all sectors at once, fluctuations in global economic activity, and most importantly, shortages in foreign exchange which would result if all sector growth targets were realized. Overall growth targets were scaled down to be consistent with reasonable long-term foreign debt position, thereby yielding a set of preliminary macroeconomic projections.
3. These preliminary macroeconomic projections were then discussed with each sector specialist on the project team to determine their consistency with the microeconomic (sub-sectoral) insights gained by those specialists. The micro projections were then adjusted to yield a consistent set of final GDP and sectoral projections, which then formed the basis for the 1990 and 2000 reference year projections and for the economic sensitivity analyses.

The data collection for the assessment was a four-month effort for five full time members of the Energy Commission and consultants, supplemented for weeks at a time by outside experts in the various sectors. Such an effort cannot be repeated periodically for the purpose of keeping the new Dominican energy data base current. From this evolved the need for an Energy Information System for the Dominican Republic, which is an ongoing integrated approach to data collection, organization and analysis. CNPE, with support from USAID, conducted a survey for a computerized system capable of supporting this comprehensive approach. Selected was the Energy Development Information System (EDIS), a bi-lingual, interactive energy information system designed for use by policymakers and analysists in developing countries. EDIS was developed under a grant from the al Dir'iyyah Institute of Geneva, Switzerland by the Institute for Energy Research at the State University of New York at Stony Brook.

The objective of instituting an energy information system is to allow for the regular collection of information, which is done through the use of instruments (questionnaires, energy audit forms) designed by and for the energy data group at the CNPE for the express purpose of compiling monthly or yearly statistics from the same sources used in the assessment. The data group has four full-time staff members devoted to the regular compilation and analysis of the data entered into the EDIS system.

ANALYTICAL FRAMEWORK

The Developing Countries Energy Planning Framework[1] used by EDIS and the Dominican Republic assessment was developed at the National Centre for the Analysis of Energy Systems at Brookhaven National Laboratory and the Institute for Energy Research at the State University of New York at Stony Brook. In brief it is a national energy production, distribution and end-use system viewed as a set of linkages tracing the steps between individual resources and energy demand sectors. The steps correspond to the processes and technologies used to transport, distribute, convert and store energy—in other words, the process by which fuel resources are converted to usable fuels.

The References Energy System (RES) is a schematic representation of a nation's energy situation in a given year. A sample overview is graphically depicted in Fig 10.1. Resources categories are displayed on the left-hand side, demand catagories are displayed on the right, and along the top the activities needed to convert energy to the forms required to satisfy the country's energy demands. Included under each general activity category are the specific technological processes which perform the functions of that activity within the country. The process of energy balancing involves an interactive check and adjustment between the entries on the supply side and specific estimates of energy demand. This allows for the identification of errors, the reconciliation of information from different sources or perspectives, and a check whether certain assumptions made are reasonable.

This analytical framework is also useful in undertaking

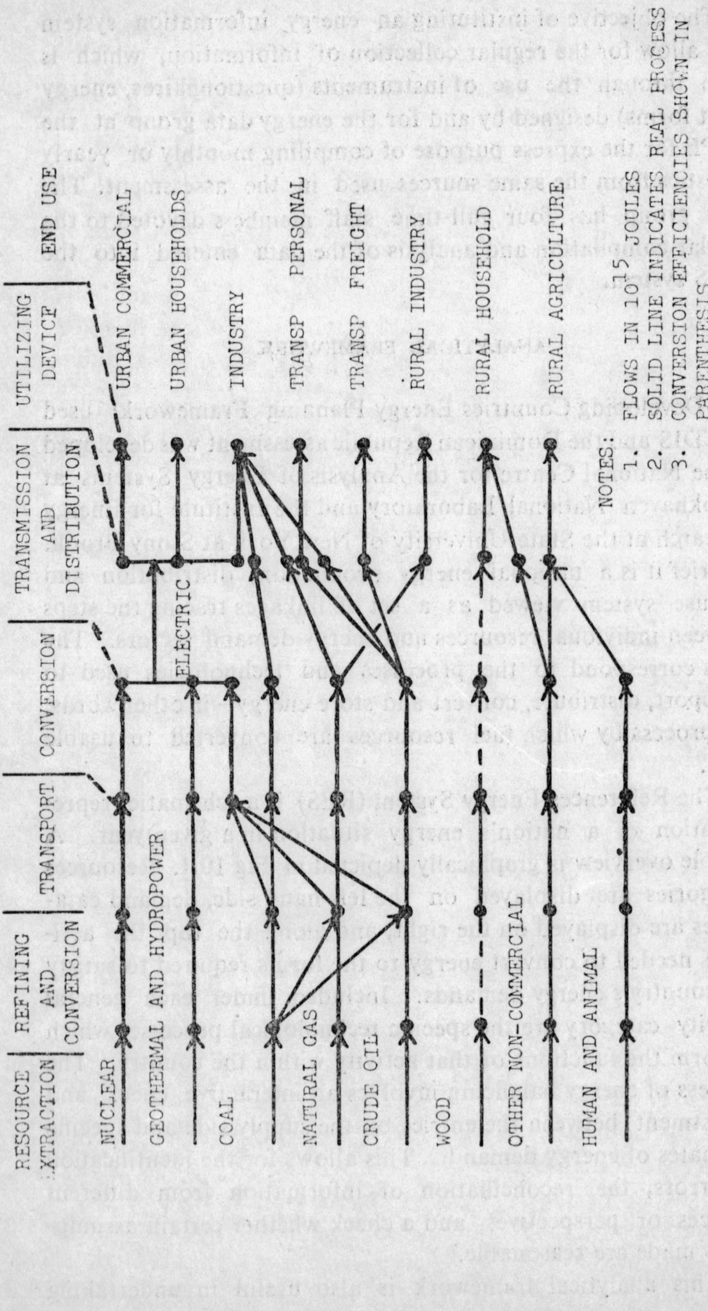

FIG. 10.1 *LDC Reference Energy System (Illustrative)*

certain necessary steps in conducting national energy assessments. Such assessments precede the formulation of detailed national energy plans and project implementation for the energy sector. Policy options can be examined in an integrated manner. Since the network representation of the energy system contained in the RES is process and function oriented, it is possible to examine the physical/economic transformations that often accompany economic development in terms of their effects on fuel demands. The RES is constructed around descriptions of specific economic activities or social functions, such as cooking in rural households, or the transportation of people or goods over certain distances. In addition, the RES shows the technical activities of the energy system through which fuels are delivered to these end-use activities. Therefore, in national assessments of the energy system, policy makers can use EDIS to examine expected patterns of demand and supply options, as well as the effects of selected policy interventions. A schematic view of the role of EDIS in such an assessment process is shown in Fig. 10.2.

In understanding the process described in Fig. 10.2 it is important to observe that its purpose is to determine the scope of desirable policies, rather than to predict the future. Future scenarios are built (a) to take into account the wide range of possible future developments over the time-scale of interest, and (b) to analyze a wide set of programmes and policies that may be of interest to decision-makers. These policies might include those designed to reduce energy demand in a group of end-uses, to improve conversion efficiencies, or to accelerate the use of new technologies and renewable resources, to mention a few examples.

Since the development of the national energy system should serve national economic and social objectives, the assessment process must begin with a recognition of these goals, and their related impacts on population growth, income distribution, patterns of urbanization, industrial and agricultural development. To examine these impacts, it is clear that a relatively extensive set of information is required on current energy use, population patterns, and levels of economic activity throughout the economy. If national economic models are available in which such changes in economic and demographic structure

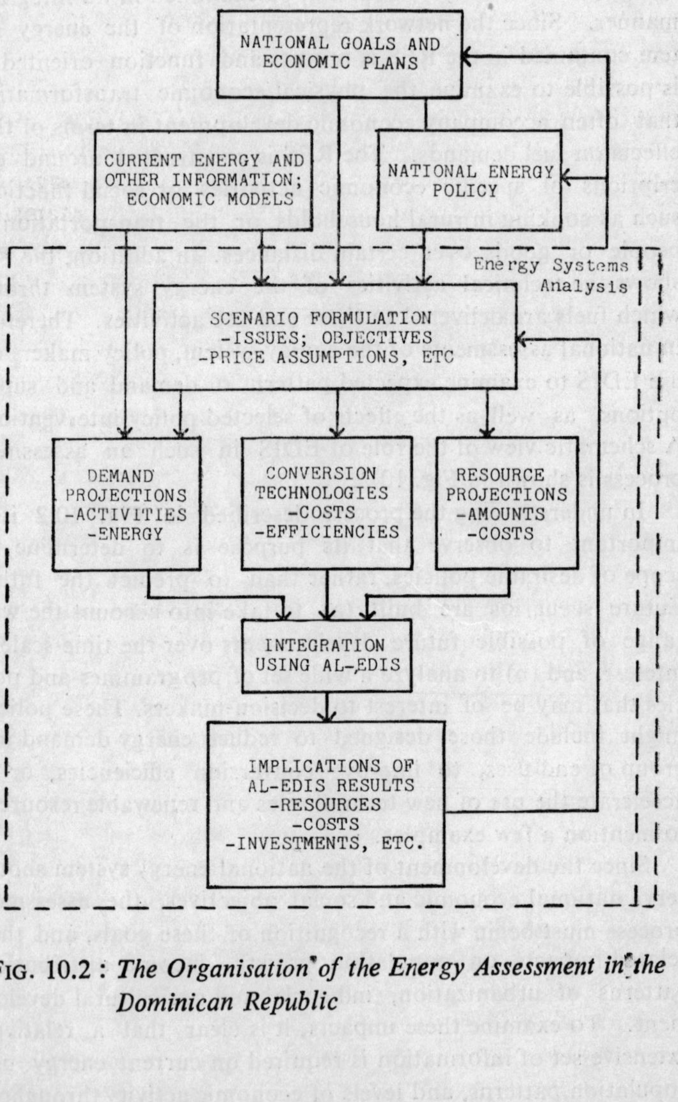

FIG. 10.2 : *The Organisation of the Energy Assessment in the Dominican Republic*

are explicitly analyzed, they can be very useful in the assessment process. (However, it should be pointed out that in formulating the relevant conditions under which the future direction of the energy system must be analyzed, many factors other than those that can be easily quantified will have to be considered in order to establish the priorities of an energy policy.)

The first step in the analytical process is to formulate a scenario of the future evolution of the overall system. It is this scenario which ties the analysis together. As part of the definition of the scenario, one must identify the issues to be addressed and the corresponding assumptions, for example, the price and availability of imported oil. The usual starting point for analysis is a scenario in which the assumption is made that no major policy intervention occurs. Such a scenario is very useful in analyzing the energy implications of current trends, and to highlight the major problems that need to be addressed in plans and policies.

The construction of a scenario requires projections and analysis in three areas: energy demand, conversion technologies, and resource production. EDIS imposes no specific set of projection models for this purpose. These are exogenous to the system.

Energy demand : In most cases it is preferable to project future demand in two steps. First, the levels of energy consuming activities (such as automobile vehicle-kilometres to be travelled, amount of cooking to be performed, tons of steel to be produced, and so on.) are projected. Then assumptions are made regarding the devices or technologies used to provide these energy services. For example, the number of kilometres per litre of gasoline for the vehicle fleet is specified, as are the devices and fuels used for cooking, along with their relative efficiencies, the processes used for steel making, and so forth. This step leads to a complete specification of the relative efficiency of energy use. It provides the basis for analyzing policies involving technology/fuel substitution or energy conservation activities.

Where economic models are available, they can provide a basis for tying specific activity levels to macro-economic parameters such as GNP or disposable income, and they can be used to provide consistency between activity levels in different sectors. For example, in an advanced industrial economy, input-output models can be used to relate the amount of glass produced to the activity levels in the construction and automobile manufacturing sectors.

The level of detail required, by region, industry income group, and so on depends on the specific policy issues to be addressed. Generally, it is convenient to select a number of specific years in the future for which detailed analysis can be carried out. The years 1985 and 2000 are often selected, although for most issues 1985 is becoming too near and 2000 not distant enough.

Conversion technologies: As already mentioned, assumptions must be made regarding the specific technologies used to convert energy fuels into services at the point of final use energy. This specification includes the efficiency of the technology (and thus the fuel-use per unit of activity) and its associated capital and operating costs, including their domestic and imported components. Similar specifications are required for the technologies used to convert energy from one form to another throughout the system, such as electricity generation and oil refining.

Resource production: The third major area of analysis for scenario development is resources. A projection is required of possible future levels of resource production and the cost of those resources. Both renewable resources such as wood, dung and agricultural wastes, and non-renewable resources such as oil, natural gas, coal and uranium, should be included in the evaluation. In most countries insufficient information is available on energy resources. Resource projections are, therefore, often very speculative. In such cases, it is much more useful to specify a range of possible future resource production levels than to choose a single figure.

In the next phase of scenario construction, the information developed on demand, technologies and supply must be integrated. It is here that the EDIS system plays a central role.

Once the scenario (which may correspond to business as usual conditions) is formulated, one can derive the implications in terms of total resource consumption, import/export requirements, total energy costs, investment requirements. An analysis of such a "reference case" then generally raises a number of issues or problems. These can then be explored through further scenario analysis.

The EDIS system allows for the use of numerous alternative formulations to construct these new scenarios by changing the assumptions of the reference future cases. One might want to investigate the effect of a complex new set of assumptions regarding, for example, GDP or population growth rates, migration patterns or fuel prices. The general scheme can also be used to analyze the energy implications of major alternative national development patterns, for example, a shift from a traditional pattern of centralized, urbanized industrialization to a more decentralized, rural development pattern. Or one might want to investigate alternative ways of satisfying the demand projected in the reference case by new technologies or resources. In the latter event, a common way to proceed is to identify the characteristics of the new technology and the rate at which it could be implemented. These assumptions will result in an altered RES calculated by EDIS and a new set of implications is derived.

Quantification of the RES requires the specification of a complete set of final end-use energy demands and, for each link in the network, a set of parameters determining its energy flow. These parameters are the conversion efficiency and the share in the total output of a process of each component technology. The energy resource requirements for a given pattern of sectoral energy demand can then be readily computed using EDIS, which computes a demand-driven energy balance for the year in question.*

Once a "base case" is constructed to represent a particular historical configuration of the nation's energy system, projected reference cases for the future can be built. The many hundreds of required computational steps of each such reference case are carried out on the computer, which functions as a large adding machine with accounting capabilities.

*Only direct energy demands are included in this computation. But the RES approach can be extended to account for 'intermediate' fuel requirements by linking the pattern of direct energy demands to an energy input-output table. This application will be made available to users of EDIS in the future.

In EDIS, the data required to construct an RES is stored in a structured data base, depicted in Fig. 10.3. Future reference case projections require that the user "create" new files in the data base for each projection year. (This is accomplished on a terminal, through which data entries are made.)

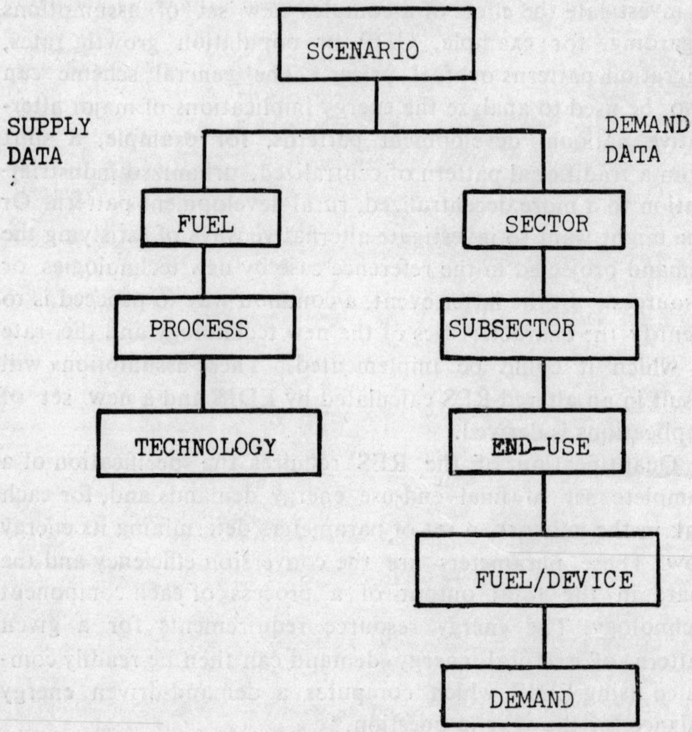

FIG. 10.3 : *EDIS—Conceptualization of data base*

EDIS provides the user with the capability of defining future energy demands in alternative ways, and to trace the impact of alternative demand assumptions on the national energy system. In Table 10.1, alternative ways of expressing the demand for energy in each sector are shown, for example the demand for fuels in each sub-sector. In transportation, the usual measure of activity is distance, expressed in terms of passenger or tonne-kilometres. Therefore, the user of EDIS, as a first step in developing 'response' scenarios to aid policy-making, must specify a set of physical activity levels for the future year in question using plausible assumptions about development.

TABLE 10.1 : **Representative measures of demand activity**

Sector	Typical physical units	Alternative physical units
Industry	Output (Tonnes, value of Sales, etc.)	Capacity
Transportation	Passenger—kilometres Tonne—kilometres Vessel—kilometres	Number of Vehicles
Agriculture	Output (Tonnes, value of sales, etc.)	Area of Land Number of Processing Plants
Rural or Urban Households	Number of Households	Population
Commercial	Number of Units (Office, Restaurants, etc.)	Value of Sales Value Added Floor space

DATA REQUIREMENTS OF EDIS

The construction of an RES for any country requires energy data at a disaggregated level, and classified in a structured format. The detail with which an RES is constructed will depend on the availability of data. Thus, EDIS is flexible since it can accommodate variations in data availability. In what follows, the general types of data required by the RES method are discussed. The problem of structuring this data for use in EDIS is the subject of the section on estimation techniques.

Demand data requirements

Since an RES analysis is demand-driven, data requirements on the demand side need emphasis. Furthermore, this type of data is often sparse for most developing countries. In order to follow the discussion concerning estimation techniques, the user should become familiar with the terms defined below.

Basis (*B*)
The physical count of all the potential (or actual) energy consuming entities for a selected sub-sector classification. Units of measurement will vary among sub-sectors of demand. For example, one of the subdivisions of the industrial sector is cement production. In this sub-sector the basis selected might be, for example, the total cement produced in a year. Similarly, for energy consumption in low income urban households, the basis selected might be the total number of rural households in this income group.

Saturation (*s*)
The fraction of the basis for which a particular end-use activity actually results in energy consumption. For example, if 80 per cent of all low-income urban households in a country own lighting devices, then the saturation for the end-use lighting would be 0.80.

Fuel Demand (*D*)
The quantity of a given fuel required as input to end-use device. (Electricity, for this purpose, is treated as a primary fuel.) In statistical sources these demands are usually reported in physical units specific to the fuel in question, such as cubic metres of wood, litres of kerosene, etc. For the purposes of comparing fuel demands, these must be converted to common energy or fuel equivalent units (that is, joules or barrels of oil equivalent).

Relative Effectiveness (*e*)
In most cases corresponds to the technical efficiency of an end-use device. However, utilization factors such as insulation levels or practices related to the use of a fuel or device may require that these coefficients be adjusted prior to their use in comparing energy demands among end-uses.

Basic Energy Demand (BED)
Represents the level of effective energy required to fulfil a particular end-use function. For a given end-use activity, the basic energy demand is expressed as:

$$BED = \sum_i e_i D_i$$

Where e_i and D_i are, respectively, the relative effectiveness and fuel demand (in energy units), of the ith fuel/device combination.

Saturated Basic Energy Demand (SBED)
Represents the basic energy demand assuming a saturation of 100 per cent for the end-use activity.

$$SBED = \frac{BED}{s}$$

Unit Basic Energy Demand (UBED)
Represents the basic energy demand per energy-consuming unit or entity for the end-use activity in question, assuming 100 per cent saturation. Therefore,

$$UBED = \frac{SBED}{B}$$

Fuel Fraction (f)
Represents that percentage of the saturated basic energy demand that is satisfied by a particular fuel/device combination. Thus,

$$f_i = \frac{e_i D_i}{SBED}$$

f_i represents the fuel fraction of the ith fuel/device combination. It should be noted that:

$$\sum_i f_i = s$$

Supply data requirements
The RES has been described as showing the interconnected

processes by which fuels are transformed and delivered to various energy-using activities. Each process takes a single fuel as input and transforms it (in time, space, or form) into a particular fuel form utilized as an input to the next set of sequential processes. Each process may consist of a number of technologies. Processes are linked together in network fashion to form an energy supply system. Any number of processes may be linked to enter or leave another process as long as consistency is maintained with regard to the input and output fuels.

In order to quantify an RES, one data item is needed for each process and three data items are needed for each technology of each process. These are defined below:

Process market share
The supply of each major fuel form is shown in an RES to consist of interconnected processes, sequentially arranged to represent the transformation of energy resources to usable fuels. Since more than one process may enter (feed) into a given process, it is necessary to specify the per cent contribution of each of these entering processes. For example, if an oil refinery process is supplied by transportation and coal liquefaction processes, and the transportation process contributes 70 million barrels and the liquefaction process contributes 30 million barrels, then the process market share of transportation and liquefaction are 70 per cent and 30 per cent respectively.

Technology market share
Each process in the system may be composed of more than one technology. The technology market share is defined as the per cent contribution of the technology to the total output of the process. For example, if the transportation process is composed of the pipeline, barge and tanker technologies, and their outputs are 35, 17.5, and 17.5 million barrels respectively, then their technology market shares are 50 per cent, 25 per cent and 25 per cent respectively.

Technology constraint

The constraint of each technology is defined as the maximum output (in energy units) of that technology. A mechanism is provided to allow for unlimited output.

Technology efficiency

The efficiency of a technology is defined as the ratio of energy content of the output fuel form to the input fuel form.

Estimation techniques

In the RES approach fuel demands are often broken down by sector, sub-sector and end-use. Each of these categories of demand represents a level of detail.

At the most detailed level, fuel demand is broken down by end-uses (such as cooking in households, process heating in industry, or irrigation in agriculture). Within each such end-use, fuel use may be further broken down by devices.

Data available in more aggregate forms on the demand for fuels must frequently be transformed to allow this type of disaggregation. These transformation include a variety of steps to derive the *basis* for energy demand in a sub-sector, the *saturation* levels within end-uses, and the *fuel demands* for each end-use or technology, and so on. This frequently requires the use of working assumptions to disaggregate available data, or to extrapolate data from available information.

No single set of procedures can be followed to accomplish this, since the choice of a procedure is guided by the available forms of data. But examples of procedures used in the Dominican Republic will be helpful in clarifying what is often called for. The data used in the following three examples were acquired from the energy assessment which is described in another section of this paper.

EXAMPLE 1 : **Demand for charcoal in upper income households**

In the Dominican Republic, as a result of the energy assessments, the following data were available :

Upper Income Households	239,600
Percentage of Above Households Using LPG as a Fuel	63%
Percentage of Above Households Using Charcoal as a Fuel	37%
Total LPG Consumption	352.9 KBOE
Percentage of LPG Used for Cooking	95%

The following data were obtained from general technical sources :

Average Efficiency of LPG Stove	60%
Average Efficiency of Charcoal Open Hearth	15%

Unit Basic Energy Demand can be calculated to be:

$$\frac{(\text{LPG consumption})\ (\%\ \text{LPG used to cook})\ (\text{efficiency})}{(\text{total households})\ (\%\ \text{households using LPG})}$$

$$=\frac{(352.9\ \text{KBOE})\ (9.5)\ (.60)}{(239{,}600)\ (.63)}$$

$$=1.33\ \text{BOE/Household}$$

This calculation is a key step in extrapolating the charcoal fuel demand. The unit basic energy demand is assumed to be constant for all makes of cooking in the upper income sub-sector. Using the relationship:

$$\text{Charcoal Demand} = \frac{(\text{UBED})\ (\text{No. of households using charcoal})}{(\text{efficiency of charcoal use})}$$

Then charcoal fuel demand is calculated to be

$$\frac{(\text{UBED})\ (\text{total households})\ (\%\ \text{using charcoal to cook})}{(\text{efficiency of charcoal})}$$

$$=\frac{(1.33\ \text{BOE/Household})\ (239{,}600\ \text{Households})\ (.37)}{(.15)}$$

$$=786\ \text{KBOE}$$

It should be noted that in this example the charcoal fuel demand was extrapolated from a few known magnitudes and a few assumptions. Changing these would obviously alter the computed demand for wood. Therefore, it is advisable to document these assumptions to allow for updating of the estimate as better data become available.

Table 10.2 shows a completed fuel mix table for this end-use.

TABLE 10.2 : **Fuel mix table**

Sector : Urban Household
Sub-sector : Upper Income
Basis : 239,600 Households
End-Use : Cooking
Unit Basic Energy demand : 1.33 BOE/Household

Technology	Fuel	Fuel fraction	Efficiency	Fuel demand
Stove	LPG	.63	. 6	335.3 KBOE
Open Hearth	Charcoal	.37	.15	786.0 KBOE
		Saturation=1.0	Total	1,121.3 KBOE

EXAMPLE 2 : **The demand for transportation fuels**

The RES approach disaggregates fuel demands for transportation into sub-sectors and end-uses, as it does in the case of other sectors. Available information to estimate passenger automobile fuel demands in the Dominican Republic includes the following:

Number of Passenger Car Registrations	61,677
Average Annual Mileage per Car (km/year)	13,300
Proportion of Gasoline/Diesel Cars	91%/9%
Unit Basic Energy Demand	137.2×10^{-6} BOE/ Veh-km
Average Engine Efficiency	20%

The *basis* is computed to be 820.3×10^{6} vehicle-km (61,677 vehicles$\times$13,300 km) which multiplied by the unit basic energy demand yields the *basic energy demand* 112.5 KBOE for this end-use.

The gasoline fuel demand is calculated to be

$$\frac{\text{(Basic Energy Demand) (fuel fraction gasoline)}}{\text{(efficiency of engine)}}$$

$$=\frac{(112.5 \text{ KBOE}) (.91)}{(.20)}$$

$$=511.9 \text{ KBOE}$$

The diesel fuel demand is calculated to be

$$\frac{\text{(Basic Energy Demand) (fuel fraction diesel)}}{\text{(efficiency of engine)}}$$

$$=\frac{(112.5 \text{ KBOE}) (.09)}{(.20)}$$

$$=50.6 \text{ KBOE}$$

This calculation has assumed that all registered cars are part of the active fleet (that is, an assumed *saturation* of 100 per cent is used. Table 10.3 shows a completed fuel mix table for this end-use.

TABLE 10.3 : **Fuel mix table**

Sector : Transportation
Sub-sector : Passenger Auto
Basis : 820.1×10^6 Veh-km
End-Use : Locomotion
Unit Basic Energy Demand : 212.45×10^{-6} Veh-km

Technology	Fuel	Fuel fraction	Efficiency	Fuel demand
Engine	Gasoline	.91	.2	511.9 KBOE
Engine	Diesel	.09	.2	50.6 KBOE
	Saturation=1.0		Total	562.5 KBOE

EXAMPLE 3 : **Industrial energy demands**

The following data were available for the process heat end-use of the sugar industry:

Output : 1.06×10^6 *Tonnes*

Fuel	*Conversion Efficiency*	*Consumption*
Fuel Oil	60%	69.6 KBOE
Bagasse	30%	4606.0 KBOE
Wood	20%	7.34 KBOE

Basic energy demand is calculated to be
=(Fuel Oil consumption) (eff.)+(Bagasse consumption) (eff.)
+(Wood consumption) (eff.)

$$=(69.6)\ (.6)+(4606.0)\ (.3)+(7.34)\ (.2)$$
$$=1425 \text{ KBOE}$$

Fuel fractions are calculated to be

$$\frac{(\text{Fuel consumption})\ (\text{efficiency})}{(\text{Basic Energy Demand})}$$

Fuel oil fuel fraction

$$=\frac{(69.6 \text{ KBOE})\ (.6)}{(1425 \text{ KBOE})}=.029$$

Bagasse fuel fraction

$$=\frac{(4606.0 \text{ KBOE})\ (.3)}{(1425 \text{ KBOE})}=.970$$

Wood fuel fraction

$$=\frac{(7.34 \text{ KBOE})\ (.2)}{(1425 \text{ KBOE})}=.001$$

Unit basic energy demand is calculated to be

$$\frac{(\text{BED})}{\text{Basis}}=\frac{(1425 \text{ KBOE})}{(1.06 \times 10^6 \text{ Tonnes})}=1.34 \text{ BOE/Tonne}$$

Table 10.4 shows a completed fuel mix table for this end-use.

TABLE 10.4 : **Fuel mix table**

Sector : Industry
Sub-sector : Sugar
Basis : 1.06×10^6 Tonnes
End-Use : Process Heat
Unit Basic Energy Demand : 1.34 BOE/Tonne

Technology	Fuel	Fuel fraction	Efficiency	Fuel demand
Burner	Fuel Oil	.029	.6	69.6 KBOE
Burner	Bagasse	.970	.3	4606.0 KBOE
Burner	Wood	.001	.2	7.34 KBOE
		Saturation=1.0	Total	4682.9 KBOE

APPLICATIONS OF EDIS

Application of EDIS has been initiated with the installation of a base case scenario and several future cases. As a first step, the linkages that transfer energy flows through the network as it exists now were determined. Over one hundred energy forms to meet the energy demand categories were specified. These demands are primarily supplied by petroleum, electricity, charcoal/firewood and bagasse. Electricity in turn is generated from petroleum, hydropower and purchased from industrial producers. Petroleum products are either produced at the refinery (from imported crude) or from direct imports. Charcoal is produced in kilns from domestically produced woodfuel.

As a next step, new linkages that respond to potential new energy flow patterns, to be implemented during the next two decades, were entered into the computer. Principal among energy resources are the use of coal, either imported or domestic, and the use of direct solar heat, windpower, and biomass. Vastly augmented uses of hydropower are planned, a linkage in the EDIS system that is part of the existing base year. For demand categories, new linkages corresponding to new energy resources used directly (such as solar heat) had to be added. Other possible new linkages, including electricity and biomass fuels to meet process heat applications, decentralized options such as wind and photovoltaics and the like, were incorporated for specific scenarios. It is worth noting at this stage that by virtue of being a small island nation, fewer alternatives are open to the Dominican Republic than to other nations. It would be entirely feasible in many applications of EDIS that the linkages which determine the existing energy flows of a nation (or a region) require much more significant revisions than those indicated above when alternative strategies are considered.

Four alternative scenarios were formulated for the Dominican Republic during the course of the energy planning activities undertaken by the Commission Nacional de Politica Energetica (CNPE). The ultimate objective of all the alternative strategies was the same: to meet the energy requirements of an expanding economy in the most cost-effective fashion by replacing petroleum with less costly energy forms or with economically beneficial technologies that utilize available domestic resources. The alternatives that emerged were as follows :

1. *No-Action Case* : a future energy configuration which continues existing trends, taking into account higher petroleum prices and planned investments in hydropower.
2. *Conservation Case* : an examination of the potential savings in all categories of energy demand from investments in more efficient end-use devices, mandating appliance standards and other policies that lead towards reduction in energy demand levels.
3. *Coal-Hydropower Case* : an alternative in which first the consumption of electricity is increased greatly by substituting electricity for other energy forms at the end-user level, where the electricity is obtained from accelerated hydropower development and coal-fired power stations. Secondly, coal substitutes for oil products used in as many large scale process heat applications as possible.
4. *Renewable Case* : a case in which small-scale renewable resources are implemented in rural/industrial applications, and large-scale ones involving bagasse drying and

electric generation from wood lots and solar ponds are implemented throughout the system.

Each case provided an excellent opportunity to demonstrate the capabilities of EDIS. Once the appropriate starting point is selected (most likely to be the No-Action Case) for the year desired, it is introduced into the EDIS data base. The years 1985, 1990 and 2000 were the reference years used in this study. From this starting point, each modification of the No-Action Case as called for by the alternative cases formulated above becomes a new balanced RES performed by the EDIS system. At all times, a balancing system is automatically ensuring that energy demands are being met and total resource consumption is being calculated. EDIS plots the total consumption of each resource for as many reference years as the system has been given.

The introduction of a new energy resource or energy form typically requires at least two pieces of information: (1) the level of yearly production of the additional energy, and (2) the category or categories of end-users that will be able to utilize this new energy resource in its proper form. The limits of the capability for substitution of one resource for another are perfectly illustrated by EDIS. A massive electrification programme, which was explained in the Coal-Hydro Case, has the potential of vastly increasing production of electricity. The potential end-uses of electricity, however, cannot expand nearly as fast, and as a result, petroleum demand in the year 2000 is still nearly 20 million barrels (about half for vehicles), only 25 per cent lower than the 26.7 million barrels projected for the Base Case.

Another case prepared by the CNPE took a different perspective: availability of petroleum in the 1995-2000 scenario was limited to no more than the 14.7 million barrels consumed in 1978. The EDIS system for the No-Action Case had to be balanced by first implementing all or part of the actions contemplated in the alternative cases: apply conservation, develop hydropower, use of renewables, use of imported coal, etc. As expected, these changes could not achieve a balanced system. At these stages, other actions had to be taken, such as modifying the levels of demand. With over 100 end-uses specified in the system, the choice of which end-uses to cut petroleum from becomes a policy issue. In many countries, these choices are being made often nowadays by energy planning institutions. EDIS, with an integral configuration of the energy system, will provide quick responses by quantifying the impact of policy choices as they are formulated.

Listed below are four examples of the type of questions that can be answered by using EDIS.

1. Because of balance of payment problems, it is necessary for the government to restrict oil imports in 1983 to 75 per cent of the 1981 totals. How could the remaining oil be rationed to minimize disruption of the Dominican economy?
2. The Corporation Dominica de Electricidad is interested in stimulating on-site generation of electricity using non-petroleum based fuels. What are the potentials for having such sources substitute for additional central station capacity being planned?
3. Because of the sharply rising prices of charcoal and fuel wood, the government finds itself in the position of having to provide low income and rural households with petroleum-based fuels at affordable prices. What is the total amount of individual fuels (kerosene, LPG, etc.) that will be needed as replacements and what are the total subsidics that will be required on the part of the government?
4. The price of sugar, the main export crop of the Dominican Republic, begins to take an upward turn which leads to a stimulation of the Dominican economy, thereby allowing the GDP to grow in the years 1985 to 1988 at a rate of 4 per cent to 5 per cent per year. What consequences does this have in terms of increased oil imports, demands on the central station electric generating plants in the distribution systems, and the in-country refining facilities?

FINALRE MARKS

Implementation of any information system, such as EDIS, in

a developing country can run into a number of problems. These include training of staff, transposition and updating of data, educating outside agencies to become users instead of just providers of data, and the installation of hardware and software which entails transportation, insurance, export and import issues. Because of the generosity of al Dir'iyyah Institute, there is absolutely no charge for the EDIS software. However, it would be unrealistic to think that proper installation of EDIS does not require an outlay of resources of both manpower and money.

The cost of hardware to support EDIS properly is U.S. $60,000. Several man-months of training, both in the methodological approach and the use of the computer, are also necessary. Support in terms of the start-up efforts, data transposition and collection, and on-going assistance must also be included in the implementation cost.

Despite the difficulties associated with a first installation, the project in the Dominican Republic to date has been a success. The tasks remaining will be completed early in 1983. We have also responded to several other countries, such as Indonesia, Thailand and Kenya, that have expressed an interest in EDIS.

REFERENCE

1. An Analytical Framework for the Assessment of Energy Sources and Technology Options for Developing Countries, BNL 50800, Upton, New York, February 1978.

ELEVEN

ENERGY STATISTICS IN INDIA

K.C. SEAL

IN INDIA, energy data are collected by different agencies and are scattered. The government departments and the administrative bodies collect limited information on energy statistics, making an overall view difficult. During the Fifth Conference of Central and State Statistical Organisations, energy statistics was included in the agenda in order to focus attention on the information system available for different sources of energy like coal, power, oil, etc., and to point out gaps which stand in the way of preparing energy balances. The recommendations included setting up of a statistical cell in the Ministry of Energy which would be a coordinating agency for energy statistics; improvement of the data base for non-commercial energy; standardisation of concepts and definitions and conducting comprehensive surveys to assess the energy flow in and out of rural areas. So far no attempt has been made to bring in the data users and producers on to a common platform to discuss the data systems. The present status of energy statistics is described below.

FORMS OF ENERGY

Two forms of energy are distinguished : primary and secondary. The primary form s of energy are those which are used in the form in which they are available in nature, that is, withou t conversion or transformation. Examples are coal, lignite, crude petroleum, natural gas, hydel electricity and nuclear power. Secondary forms of energy signify those energy commodities which are utilised after transformation. They are briquettes, refined petroleum products, cokes, manufactured gases and thermal electricity.

SOURCES OF ENERGY

Another dichotomisation of energy by its sources is into commercial (conventional) and non-commercial. The term 'non-commercial' denotes those sources which do not normally enter commercial energy markets; these are generally obtained by private efforts of individual at a very low cost or even zero private cost. They are fuel wood, saw dust, charcoal, crop and animal residues, etc. Besides these there are other sources of energy which are renewable such as geo-thermal, solar, wind and tidal power and biomass.

SOURCES OF DATA

In India, data on energy are collected by different agencies. Some of them collect and compile information regularly, while others collect on an *ad-hoc* basis, that is, as and when the need arises. Some of the important organisations in the former category are the Coal Controller's Office, Central Electricity Authority (CEA) in the Department of Power, Economics and Statistics Division in the Ministry of Petroleum, Central Water Commission in the Ministry of Irrigation, Atomic Energy Department, Department of Science and Technology, Coal India Ltd., Oil Co-ordination Committee, Planning Commission and CSO (Annual Survey of Industries). Coal Controller, CEA and Ministry of Petroleum bring out data on commercial energy sources under their respective charges. These are no regular sources of data on non-commercial energy.

Information on energy had been collected on an *ad-hoc* basis through the 1973-74 census of Small Scale Industries. Some other *ad-hoc* information has become available through the reports of various Committees and Working Groups like the Energy Survey of India Committee (1965), Fuel Policy

Committee (1974), Working Group on Energy Policy (1979) and the Rajyadhaksha Committee on Power (1980). Besides, several research scholars have also carried out energy surveys. But these are in the nature of case studies, usually limited to small areas and cover only specific issues. Data on energy consumption are being collected at regular intervals by the National Sample Survey Organisation (follow up surveys of economic census and household consumer expenditure surveys). The latest such consumer expenditure survey was carried out during the 32nd round (1977-78) of NSSO. Excepting collection of data on the use of power (electricity/ diesel) in rural areas through National Sample Survey Organisation's 31st round (1976-77), no special survey on energy was carried out.

DATA AVAILABILITY

Department of Coal, Ministry of Energy

The Coal Controller in the Department of Coal, Ministry of Energy is the statistical authority for collection of coal statistics. The Coal Controller brings out publications entitled 'Coal Statistics' on monthly, quarterly and annual basis. They together provide figures of production, despatches of different varieties of coal (including lignite) and coke (separately for soft and hard), with breakdowns by coalfields and geographical sectors. Pithead stocks of coal and stocks at consumers, allotment and loading of wagons, despatch by rail, road and coastal shipping, exports and distribution by industries and States together with information on prices, plan targets and achievements are included in these publications.

The Economics and Statistics Division, Ministry of Petroleum

The Economics and Statistics Division in the Ministry of Petroleum is responsible for monitoring information relating to petroleum and petro-chemical statistics. It caters to all the statistical requirements of the Ministry. It brings out a comprehensive annual publication 'Indian Petroleum and Petro-chemical Statistics' and also a quarterly bulletin with the same name. Certain important information relating to production is brought out monthly for restricted circulation. Production, imports, exports, consumption of petroleum products, deliveries to international bunkers, maximum selling prices of petroleum and petroleum products are included in the annual publication. This publication includes international petroleum statistics, plan statistics as well some state-wise data.

Department of Power, Ministry of Energy

The Department of Power entirely depends on the Commercial Directorate and the Operations and Maintenance Information Directorate in the Central Electricity Authority (CEA), Ministry of Energy for all types of statistics needed for its policy and decision making purposes. It does not have a statistical cell of its own. The CEA is a statutorily co-ordinating body for Electricity Statistics. The State Electricity Boards (SEB) are mainly responsible for development and supply of electric power. A few private licensees and some municipalities, government departments and corporations are also engaged in the generation and supply of energy. The CEA co-ordinates statistics from State Electricity Boards, other public utilities, etc., for its comprehensive annual publication, 'Public Electricity Supply, All-India Statistics, General Review'. The publication is the responsibility of the Commercial Directorate of the CEA. It contains state-wise information on generation, capacity and aggregate maximum demand, installed plant capacity and generation, transmission and distribution lines, utilisation, acquisition of electric supply undertakings and electrical accidents. Progress of 'electric-supply' as an industry, financial statistics and electrical tariffs are also included in the publication. However, electricity sales are available only by broad consuming sectors. Concepts and definitions adopted for the pumpsets energised or villages electrified are not uniform for all States. Consumption data is generally available after a time lag of nearly two years. There is no other publication on electricity statistics with a lesser periodicity.

National Sample Survey Organisation (NSSO)

Household survey

The NSSO in its periodic household consumer expenditure surveys, covering both rural and urban areas, collects information on quantity and value of consumption on coke, coal, electricity, gas, kerosene, methylated spirit, firewood, dung cake, charcoal, matches and candles. However, in final tabulations, the above energy sources are clubbed under the category 'Fuel and Light', except, that detailed tabulations for each of the above sources were carried out once for data from the 18th round (1963-64). The latest survey for which results are available was conducted in 1973-74 (28th round), while the latest round for which such data were collected refers to the period 1977-78 (32nd round). No attempt is made in the National Sample Survey to collect information on purposes for which energy sources are used. For example, information on energy use for cooking, lighting, heating, cooling, and so on, are not available.

Annual Survey of Industries (ASI)

ASI is designed to obtain comprehensive data on almost all aspects of the registered factory sector which comprises factories using power and employing 10 or more workers and also those not using power and employing 20 or more workers. The survey is carried out under the statutory provisions of the Collection of Statistics Act, 1953. For the purpose of the survey, the frame of factories is divided into two categories. The first category is called 'census sector' and it consists of all factories employing 50 or more workers and using power as well as those factories employing 100 or more workers and not using power. The census sector factories are surveyed on a complete enumeration basis. The second category is called 'non-census' sector and it comprises all the remaining registered factories in the factory sector. Broadly, only fifty per cent of the non-census sector factories are covered in each survey round. In ASI, information is collected on quantity and value of consumption of coal, coke, coal gas, liquefied petroleum gas, natural gas, firewood, charcoal, aviation and motor spirits, diesel oil, furnace oil, lubricating oil, and electricity purchased. These are brought out both in quantity and value terms industry-wise in the detailed reports. However, this type of result for the census sector is available up to 1971 and for 1973-74 only. Data on value of consumption for the factory sector are also available up to 1978-79.

Economic census

Two economic censuses of all economic enterprises have been carried out in this country during the years 1977 (non-agricultural sector only) and 1980 (all sectors excluding crop production and plantations). The only information on energy collected in these censuses was the type of power used in the establishments. The follow-up surveys of the first economic census were carried out by the National Sample Survey Organisation in their 33rd (1978-79) and 34th (1979-80) rounds covering manufacturing and trade, transport and services sectors respectively. In follow-up surveys, information were collected on consumption (quantity and value) of coal, coke, diesel oil, lubricating oil and electricity purchased, wherever possible. In other cases data only on value of fuels, lubricants, etc., and electricity purchased were collected. The results are yet to be published.

Office of the Development Commissioner, Small Scale Industries

A census of small scale industries was carried out in 1973-74 by the office of the Development Commissioner, Small Scale Industries. Data on quantity and value of consumption of coal, coke, coal gas, firewood, charcoal, diesel oil and other fuels were collected in the census. However, results were tabulated in value terms. The census covered only those units which were registered with the State Directorates of Industries and excluded those that fell within the jurisdiction of specialised boards and agencies. As such, the coverage is not complete and there are some overlapping units which were covered by ASI.

Gaps and deficiencies

Following are major gaps and deficiencies in the system of collecting energy statistics and the existing data :

(i) Absence of proper co-ordination between different agencies responsible for collecting energy statistics.

(ii) Inadequate statistical staff in concerned departments.

(iii) Non-availability of basic data, such as

— Changes in the stocks of various kinds of energy.

— Consumption of energy by different sectors.

— Losses of energy in conversion, transport and distribution.
— Other solid fuels like peat, tar in addition to some non-commercial fuels.
— Information on gases.

(iv) Absence of reliable data on production and consumption of non-commercial energy.

(v) Lack of regular publications dealing with production and consumption data of the important non-commercial fuels.

(vi) Problems of conversion of non-commercial fuels into energy equivalents.

Suggestions for improvement

The present system of energy statistics is not considered to be very satisfactory. While the concerned departments in the energy sector maintain energy data relating to their sectors, there is no single agency at present which maintains comprehensive energy data on a regular basis thus making an overall perspective difficult. The first pre-requisite would be to identify a nodal agency charged with this responsibility. It has already been recommended that the Ministry of Energy should be the nodal agency for collecting and coordinating energy data. This nodal ministry should set up the necessary statistical machinery in consultation with the Central Statistical Organisation and develop a comprehensive system of energy statistics on a priority basis.

DISCUSSION

MAURICE LEVY

I would like to say a few words to summarize the session this morning. We have heard an analysis of the energy data situation in a number of countries, including India which we heard in detail for the last two days. I think all of you will probably agree that there are many similarities in these presentations as well as differences. I think many countries at the moment are in the process of completely revising their methodology and this is, of course, an important point. Several times it has been pointed out that the data are not available nationally for a number of sectors and also are not given to international organizations which will disseminate the data. Also there are problems in data handling, and computerization now in many countries which are in the process of computerizing their energy data system. Of course, that will certainly improve the energy data and at the same time improve the prospects of exchange. Many countries should be interested in exchanging data either directly or through international organisations. First of all methodology is very important, and secondly very little energy analysis can be done in isolation and you need to have plenty of data from the international scene, including on the technology prospects in order to do energy planning. I think this workshop has been convened at a very opportune time because this is a time of transition in both energy data gathering and in energy planning. I think there is a large interest in comparison and exchange of data.

TECHNICAL SESSION-6

Chairman : MAHESHWAR DAYAL

Theme Papers : B. BOWONDER
K. RAVISHANKAR
RAMESH BHATIA
DR. C.L. GUPTA

Discussant : T.K. MOULIK

TWELVE

RURAL ENERGY DATA SYSTEMS

B. BOWONDER AND K. RAVI SHANKAR

INTRODUCTION

THOUGH there are a number of studies on rural energy use in India, they vary widely with respect to the following parameters :

(i) fuels included or considered (some studies do not include animal energy, chemical fertilizers, compost, green manure, human energy whereas others do);
(ii) end use (some of the rural energy studies examine only household sector, but exclude agricultural use or use in rural industries)
(iii) unit of aggregation (some of the studies use kWh where as others use coal replacement or coal equivalent units);
(iv) level of aggregation (some studies are village level surveys whereas others are macro level analysis; and
(v) flow of energy from fuels to end-uses (except for some, most studies do not depict the flow of energy completely).

Table 12.1 shows the differences in approach adopted in various studies. A comparative analysis of the various surveys indicate the following issues :

TABLE 12.1 : **Rural energy studies**

Survey	Whether flow of energy is shown	Differences identified
National Sample Survey	No	1. Quantity of fuels used only is included
Von Oppen, Parthasarathy and Bowonder	No	1. End-use data only 2. Domestic energy use only
Pachauri and Govinda Rao	No	1. Chemical fertilizers not included
Reddy and Subramanian	Yes	1. Income-wise energy usage not examined
NCAER	No	1. Village level data not available
Gupta, Usha Rao, and Vasudevaraju	No	1. Green manure, fertilizer use not included. 2. End-use data not available
Malhotra and Lal Chaurasia	No	1. Fuelwise break up only 2. No end-use data
Makhijani and Poole	No	1. Income-wise energy use patterns examined
Maheshwari *et al.*	Yes	1. Industrial energy use not included.

(i) there is a great deal of variability in energy use in rural areas;

(ii) without detailed energy surveys village level energy data systems cannot be generated;

(iii) uniform procedures should be used so that inter-regional comparisons are possible;

(iv) Energy use patterns are dependent on variables such as :

—population density
—agro-climatic condition
—irrigation facilities available
—agricultural cropping pattern and productivity

—soil conditions
—biomass productivity
—cattle population
—seasonal biomass availability patterns
—socio economic conditions.

Rural energy data systems should contain these parameters as well since energy use is very sensitive to some of these parameters. Table 12.2 and Table 12.3 show comparative analysis of various rural energy surveys reported in India. Shirapur, is a semi-arid village with very low biomass productivity and hence availability of dung agricultural residues and firewood is uniformly poor.

(v) without detailed analysis of rural energy use no meaningful renewable energy programmes can be implemented.

COMPARATIVE ANALYSIS

The authors have conducted a rural energy survey covering three villages of different agro-climatic regions. This study was funded by the Tata Energy Research Institute (Project P-068). Detailed energy balance diagram of one of the villages surveyed by the authors is given in Table 12.4. This table also gives the details of energy flow from the resource to the end use of a village in Andhra Pradesh. The fuel wood usage in that village is low since it is an irrigated village with a good number of cattle. The full details of the survey is available in the Report of TERI Project P-068. Tables 12.2 and 12.3 give a comparative analysis of the various Indian studies. As can be seen from the Tables 12.2 and 12.3, many of the entries are not available for some of the studies. The main point which emerges from this is that there is a need for the standardization of rural energy surveys. Comparative and cross sectional analysis can be meaningful only if this is done.

To illustrate a point, the energy sources if disaggregated into owned, purchased and collected yield much more information than is available in the aggregate and only such detailed disaggregated analysis will help us in understanding

TABLE 12.2 : **End-use analysis of rural energy consumption per capita per year**

Village (and investigators)	Domestic lighting %	Domestic cooking %	Domestic Total %	Agriculture %	Industrial and Commercial %	Total energy Giga calories
Lankapally (Pachauri & Govinda Rao)	0.1	23.3	23.4	71.2	5.4	3.8
Pura (Reddy & Subramanian)	2.2	90.6	92.7	2.9	4.4	2.6
Vruli Kanchan (Bowonder & Ravishankar)	0.9	28.7	29.6	54.8	16.2	3.5
Islam Nagar (Maheshwari *et al*)	1.9	81.1	83.8	16.2	—	2.5

TABLE 12.3 : **Fuelwise energy use in rural areas**

Village (and investigators)	Fuel wood %	Dry dung %	Agricultural residues %	Kerosene %	Diesel %	Animal energy %	Human energy %	Electricity %	Total energy Giga calories
Mangaon (Makhijani & Poole)	6.9	13.8	6.9	—	—	1.4	20.7	—	3.6
Pura (Reddy & Subramanian	89.2	—	—	1.9	—	—	7.2	0.2	2.5
Shirapur (Von Oppen, Parthasarathy & Bowonder	78.4	5.8	14.9	0.8	—	—	—	—	0.5
Saram (Gupta, Usha Rao & Vasudevaraju)	71.1	6.0	2.0	12.0	8.8	—	—	—	1.3
Lankapally (Pachauri & Govinda Rao)	4.4	57.8	32.7	—	1.2	1.2	2.3	0.2	3.8
Silora (Malhotra and Kaur)	60.5	20.4	7.6	3.4	3.6	—	—	3.2	NA
Plains (NCAER)	36.8	26.7	31.7	3.50	—	—	—	0.9	—
Vruti Kanchan (Bowonder and Ravishankar)	5.9	45.1	39.0	0.1	2.7	3.9	2.7	0.5	3.5
Islam Nagar (Maheshwari *et al.*)	49.5	31.1	10.3	1.6	3.1	1.0	1.4	1.9	2.5

completely and comprehensively the determinants of rural energy use. Table 12.5 gives a comparative analysis of energy use in two other villages surveyed by the authors.

TABLE 12.4 : **Energy flow in a village in Andhra in Giga calories**

End-use	Agriculture	Commercial	Domestic		Total
			Lighting	Cooking	
Dung	937	200	—	—	1137
Agricultural residues	—	527	10	770	1307
Fuelwood	—	—	—	10	10
Charcoal	—	0.5	—	2.4	3.0
Animal energy	280	—	—	—	280
Human energy	90	—	—	—	90
Green manure	499	—	—	—	499
Electricity	—	—	—	—	—
Kerosene	—	—	25	0.14	25
Diesel	2.5	—	—	—	2.5
Coal	—	—	—	—	—
Total	1808	728	35	782	3354

TABLE 12.5 : **Rural energy use in terms of collection in Giga calories per year**

	Owned	Collected	Purchased
Dung			
Karnataka village	314	—	—
Andhra village	1461	6	—
Agricultural Residues			
Karnataka village	1	2480	36
Andhra village	16	1155	138
Fuelwood			
Karnataka	1330	1743	7818
Andhra	1.7	8.0	—

CONCLUSIONS

1. Before creating rural energy data systems it is necessary to conduct detailed analysis of energy use in various regions. Energy use patterns are location specific as mentioned earlier.
2. The rural energy data systems should show the complete energy balance of the region under study. Studies reporting only end-use or fuel use data completely neglect the flow aspects. This cross linkages between fuel and end-use will indicate the areas of thrust needed for renewables.
3. Apart from rural energy data parameters the energy data base should include other major determinants of energy use such as cropping patterns, land use patterns, socio-economic conditions, etc.

ACKNOWLEDGEMENT

The authors are grateful to the Tata Energy Research Institute for sponsoring this research study.

REFERENCES

Bowonder, B. and K. Ravishankar, Energy use in three villages, Report, ASCI, Hyderabad, 1982.

Gupta, C.L., K. Usha Rao and V.A. Vasudevaraju, Domestic energy consumption in India, *Energy 5*, 1213-1222, 1980.

Maheshwari, R.C., Energy use in Islam Nagar, *Proceedings of the National Seminar on Adoption of Renewable Sources of Energy*, ASCI, Hyderabad, 1982.

Makhijani, A. and A. Poole, *Energy and Agriculture in the Third World*, Ballinger, Cambridge, 1975.

Malhotra, K.S. and K. Kaur, Energy use in Silora Block in Rajasthan, *Proceedings of the National Seminar on Adoption of Renewable Sources of Energy*, ASCI, Hyderabad, 1982.

National Sample Survey, 18th and 23rd round, CSO, New Delhi.

NCAER, *Rural Energy Consumption in Northern India*, NCAER, New Delhi, 1978.

Von, Oppen, M. R.T. Parthasarathy and B. Bowonder, *Rural Energy Use in five Semi-arid Villages*, ASCI, Hyderabad, 1980.

Pachauri, R.K. and P. Govinda Rao, Energy flows in rural areas of Andhra. Report, submitted to Tata Energy Research Institute, ASCI, Hyderabad 1981.

Reddy, A.K. and D.K. Subramanian, *The Design of Rural Energy Centres for Pura Village*, ASTRA, IISc, Bangalore, 1978.

DISCUSSION

DAYAL

I feel very strongly that we need to make continuing efforts in the field of renewable energy because it has a very important role to play, particularly in our rural settings in the Third World. Renewable energy will make a very significant contribution towards local development and for improving the quality of life. At the same time I also feel that we have to go about this in a very scientific manner. There are problems of costs and of economic choices and priorities which are very critical and we have to consider all the various options that are available. This leads one to the obvious conclusion that one cannot simply throw over board the conventional energy systems. We need to have a very large and massive increase in the availability and production of the various conventional energy systems also, including, in my opinion, nuclear power. But it has to be done in a way which is optimal, most efficient and also matched to local requirements and local economics.

In this connection the importance of data, especially for rural energy seems to be very critical. In many cases, we find if we go simply by the traditional thinking of continuing to invest in large transmission lines and various other costly distribution systems, then perhaps we do not choose the most economical form of supplying energy, from an overall national point of view. There are extremely heavy subsidies involved in the supply of power to many rural areas and some urban centres. But the fact is that in many cases these subsidies are being borne by the states. As compared to that, a renewable energy system has to be installed by the individual consumer. The consumer has to bear the cost of the initial investment.

This in a way loads the situation somewhat unfavourably in favour of the conventional power system. But if you look at it from the point of view of national economics, you may find that the nation is investing much more in the conventional power system. This does not help to promote the renewable energy resources. This kind of situation has definitely arisen in many places and this is causing people to consider shifting the burden of subsidy from the central energy generation system and giving a somewhat lesser subsidy to the individual consumer thereby making it possible for the nation as a whole to benefit and at the same time encourage local generation of energy. But this requires a much more detailed analysis and data is absolutely basic to this so that one knows precisely what is the cost, the costs of alternatives, also the availability situation for local energy sources, and also the demand patterns that are likely to arise.

Again if we talk about the different energy sources, we do feel that there is a substantial advantage in having local energy generation from the point of view of control of the energy supply and demand by the local people who are going to benefit most from it. This has a very good potential fall out in terms of development at the grassroot level which can help to prevent migration of people from the village to the cities, enable increase in purchasing power, local industrialization, and better agriculture. But we need much more information to see what can be done at various places. For example the whole question of agricultural residues, or woodbased systems, biomass systems, need much more quantity of data. Our Commission for Additional Resources of Energy is supporting studies for gathering of some of this data. I, therefore, feel that in the context of an overall energy crisis the transition to a more diversified energy mix has to be based increasingly on renewed and renewable forms of energy but without in any way diminishing the importance of the conventional and nuclear energy sources in the short and medium terms.

MOULIK

Dr. Bowonder said that the data should be decentralized, location-specific. If you have to do this for all the villages, or a number of villages, it becomes really a problem. I do not think that any agency can handle that level of annual data for villages.

BOWONDER

This is a very valid point. At least they should have a cross-section of agro-climatic conditions, a cross-section of regional variation. For example, semi-arid zones have not been studied in detail. There are various kinds of villages in semi-arid zones. In Rajasthan, in some places the annual rainfall is less than 400 mm, or 200 mm in a place like Mithapur. If you study this in different regions, different agro-climatic regions, that may be a good thing. I am not saying that you should do this in every village. To start with a sample village can be selected in different regions but there should be a proper procedure for selecting, so that all types of villages are included in this survey.

MUBAYI

I just wanted to make a few comments to strengthen some points that Dr. Bowonder has made. What he said is very important in the case of rural energy. Perhaps India was one of the few countries where, good or bad, some attempt was made since the 1950s to gather some data in this area. Those of us who are working in this field for some time are aware of the inadequacies of the National Sample Survey, particularly of renewables and rural energy. A homogeneous classification of different areas is extremely important in two respects: in setting the pattern of demand that obtain, as well as in looking at the source of supply from the point of view of availability of these different renewable sources which depend on agro-climatic patterns.

For example, suppose we are interested in a technology like pumping of water by solar energy technology in which international agencies like the World Bank are spending lots of money. Now our problem is how do we go about selecting areas in which this technology is going to be useful. An organized classification based on agro-climatic condition is a great help in knowing at what depths the water to be pumped is available. In arid zones, water is available at great depth, so pumps of a certain capacity are required. The solar insolation is certainly good in such areas. But in effect you can only pump a small amount of water because it is available at great depth. So you might instal pumps of different sizes, different capacities, mostly for drinking water or water for livestock. Because of the pattern of demand, you will not have irrigated agriculture in an arid zone so your water demand is determined by the kind of agro-climatic condition. But unfortunately it is obvious that you just cannot have every village analyzed at the same level of detail. There are 550,000 villages in this country. An attempt should be made to group fairly homogeneous areas for the purpose of collecting this kind of data.

The other point I have is in the way the aggregation has been done in these tables; it is not clear on what basis the animate and inanimate energies are being aggregated. This is a well-known problem of aggregation and you might be knowing there may be some problem of compatibility in the way the data has been presented here.

BOWONDER

For aggregation, we have used a common conversion factor for all the data. It was not available in this form. We have used a common conversion factor similar to the one which was used in the Dominican Republic. In the detailed report all these conversion factors which we have used are given. This is only a starting point. For a renewable energy programme, you would require more disaggregated information. Even for commercial lighting, cooking, etc., you have other kinds of things, you may need a more disaggregated level of data than what is given here. If you want to have a very meaningful renewable programme, you have to match the needs with the resources of that region. But for a detailed survey those kinds of things also have to be included. Cross linkages between the fuel and end-uses have to be studied in detail.

NIKHIL DESAI

It was not clear when you said that standard procedures have to be formulated for these. Now if I take your point that the rural energy system is subjected to great variations even amongst regions, I would tend to think that a standard set of rules would not suffice for doing the survey.

BOWONDER

No argument. I am advocating for the questionnaire a standard format which can be used for comparison across regions. A similar questionnaire which includes all the factors, different end-uses, different kinds of fuel, different purposes, different efficiency, different income level, different kinds of buying all these should be there. That is all I meant—a standard survey procedure.

THIRTEEN

ENERGY DATA SYSTEM
The Role of Traditional Sources

RAMESH BHATIA

INTRODUCTION

NON-CONVENTIONAL sources of energy can be divided into two broad categories: (i) Animate energy sources and (ii) Biomass and related sources. The first category includes animal draft power and human labour which are important sources of energy in the Third World agriculture, providing up to 90 per cent of all energy used in traditional agricultural systems. The second category includes biomass and related sources such as woodfuels, firewood, shrubs, charcoal, sawdust, crop residues (straw, rice husk, bagasse, jute, cotton and tobacco sticks) and animal residues (dung cakes and biogas). In literature, the second category of sources are usually referred to as *traditional fuels*. Most of these sources are also called 'non-commercial' fuels, although charcoal and part of fuelwood are usually traded in markets.

Although a number of problems of measurement of non-conventional (traditional) sources have been recognized, it is considered very important to collect and analyze data on these sources for the following reasons :

(i) these sources supply a major portion of the total energy consumed in many developing countries, amounting to 80-90 per cent in some cases;

(ii) most of the households, especially in rural areas of developing countries, depend on these sources not only for their cooking needs but also for energy required for farm operations, transport and so on;

(iii) the poorest of these households, especially in rural areas, spend significant amounts of their time in collecting fuels for meeting their 'basic needs' which has implications for their income/welfare levels;

(iv) collection and use of these sources, especially in view of rising prices of conventional fuels, has important implications for ecological balance in some regions and requires very careful consideration;

(v) the economic viability and social acceptability of a new energy technology or intervention (for example a biogas plant or an improved stove) depends crucially on the socio-economic-ecological environment in which it is introduced. For successful dissemination of any improved device/technology it is important to know how it affects the pattern of generation and use of these traditional sources by different categories of households.

This paper is divided into three parts: Part I discusses significance of collection and analysis of data on traditional energy sources. Part II reviews, briefly, the attempts made at quantification of the role of these sources at the national, regional and village levels. It also includes a discussion of some selected village level comprehensive surveys attempted by individual scholars or research teams. Part III outlines a framework for acquiring data on the traditional sources through direct surveys as well as through indirect methods. The paper also attempts to list a few important points for detailed discussions at the International Workshop on National Energy Data Systems.

SIGNIFICANCE OF TRADITIONAL SOURCES

With the renewed emphasis on energy planning in the mid-seventies, some attempts have been made to quantify the role of traditional sources of energy in developing countries. As discussed later, most of these attempts suffer from inadequate coverage, problems of definition and measurement, and conceptual difficulties of aggregation of different forms of energy. Nevertheless, these estimates do indicate the role that these traditional sources play in developing countries and underline the significance of systematic attempts at data gathering and analysis.

At the international level, the United Nations, the Food and Agricultural Organization, the World Bank and the International Energy Agency have put out figures on production and trade of fuelwood and charcoal for various countries.[1] Although developing market economies account for 75 per cent of the fuelwood production in the world, no attempt is made to measure its contribution to total energy supplies in these countries. The World Bank Study[2] has measured woodfuels consumption in 1976 and presented projections for 1990 based on FAO statistics covering the period 1961-75. The study also discusses the significance of traditional fuels for households through case studies in Bangladesh, Thailand, Nepal and India. The study also discusses the complex interrelations among eco-agricultural and socio-economic systems in the context of choices regarding food-fodder-fuel and fertilizer.

Openshaw (1977) estimates that between 40 and 50 per cent of the world's population relies on firewood for cooking and that firewood supplies about one-fifth of the world's fuel requirements. Parikh (1978) has estimated that commercial energy sources supply less than 10 per cent of total energy in such countries as Nepal, Tanzania and Mali.[3] The majority of the developing countries rely on traditional fuels for 30 to 80 per cent of their energy use. Various studies of rural energy use in developing countries—Bangladesh, India, Nepal Pakistan—show that non-conventional sources account for over 90 per cent of the energy use and, in some cases, rural systems are closed ecosystems with very little energy input from outside.

Given the predominance of traditional sources, especially in rural areas, it is important to assess the demand-supply situation in respect of these sources as an input to energy planning. It is necessary to know how the demand for these sources varies with growth in population, changes in household size, income levels, urbanization and education. Similarly, it is necessary to make supply projections based on agricultural growth, cropping patterns, livestock situations, mechanization, afforestation schemes, etc. These estimates provide the basis for planning supplies of commercial sources such as kerosene and LPG as well as determining prices (or taxes and subsidies) for substitute fuels such as soft coke, diesel and electricity. For example, the kind of answers that a macro-level analysis would require would include (i) The extent of oil, coal or gas that would be required if traditional sources of energy are not found adequate for meeting 'basic needs' in future, (ii) The extent of conventional energy sources required if adverse environmental consequences (for example, deforestation) of extensive use of fuelwood have to be avoided, (iii) The level of conventional fuel sources required if availability of crop-residues and animal residues is reduced as a result of use of high-yielding variety seeds, changes in crop pattern and higher levels of mechanization, and (iv) The socio-economic aspects of using new technologies (fermentation, gasification, etc.) which convert traditional fuels (fuelwood, crop residues, animal dung, and so on) into conventional sources of energy, such as fuel alcohols, biogas, pyrolysis gas and electricity.

Apart from the macro-level issues, information on the processes of generation and use of traditional energy sources is required for planning new energy sources that will be more efficient and cheaper or that will provide higher quality energy services. For example, the successful adoption of a biogas programme by the farmers would depend on the pattern of ownership of energy-producing assets (cattle, orchards, land), availability of crop residues, preference for cleaner fuels, cooking habits, prices of alternative fuels, etc. Such a programme will not be successful in areas where large farmers, who can afford biogas plants, have adequate quantities of crop residues and/or dung for burning while marginal farmers and agricultural workers, who face shortages of fuels, typically, do not have adequate cattle and capital resources, for installing biogas plants. Similarly, a programme to introduce improved and more efficient cooking stove would require information on cooking habits, the role of women in collecting fuelwood, the trade-off between time spent on collecting fuelwood vis-a-vis cooking, the level of seasonal unemployment and under-employment in the region and the social customs concerning fetching of wood, cattle grazing and working in the fields.

Thus, systematic efforts at collecting information on traditional sources of energy are required so as to provide necessary data-base for analyzing both the macro-level and the micro-level issues outlined above. The next section reviews the efforts made in this direction while the following section presents a framework for measuring traditional energy sources in developing countries.

AVAILABLE DATA ON TRADITIONAL SOURCES : A REVIEW

As was mentioned earlier, available data at the international level are confined mainly to fuelwood production and trade. However, in some countries, efforts have been made to estimate the consumption of traditional (non-commercial) sources of energy at national or regional levels. These estimates are based on large scale surveys organized by the Government or research institutions. Besides, some indirect estimates have also been made on the basis of area under forests, production of grain and straw from various crops, age and sex composition of different livestock and working human population in the country. Apart from these macro-level estimates, a large number of village surveys have been completed in the recent past. Since a detailed review of all these efforts would be outside the scope of this paper, an attempt is made to review only a few studies relating to India, Pakistan, Nepal, Bangladesh and Fiji.

National/regional studies

In India, the Fuel Policy Committee (1974) and the Working Group on Energy Policy (1979) have given estimates for non-commercial sources of energy for the relevant years. These data are based on fuel consumption surveys conducted by the National Sample Survey Organization (NSSO) of the Government of India. Although data on expenditure on fuel classified by expenditure groups and states are available from the fifties, the estimates of quantities consumed are available only for 1963-64 and 1973-74. Thus, the year-wise estimates of consumption of firewood, and vegetable wastes given in the reports of the two committees referred to earlier are essentially point estimates of fuel consumption multiplied by estimated rural/urban population in each year.[4] The shares of firewood, vegetable waste and cowdung, calculated on the basis of a survey are 65 per cent, 15 per cent and 20 per cent respectively.

The Committee noted that "the different surveys[5] have yielded divergent results regarding the use per household of non-commercial fuels in the domestic sector". Such divergence is to be expected in view of the seasonal and regional variations in the quantity of fuel consumed as well as the inability in many cases of the sample population to quantify the fuel consumption accurately. As most of these quantities are not purchased but gathered free of cost from forests and grazing grounds, no record of their usage is kept. Field surveys to assess the extent of quantities used also proved to be unreliable as the vast majority of rural population who use these quantities have no sense of proper measurement. The Working Group on Energy Policy (WGEP, 1979) has also adopted the figures on per capita fuel consumption used by the Fuel Policy Committee (FPC, 1974) and suggested that[6] "A survey of available information revealed the large gaps in our knowledge of rural energy needs; in a village community, the current flows of energy within the rural system, and the exchange of energy between rural and urban areas".

Apart from the difficulties of getting estimates of quantities used, there are problems in aggregating traditional fuels with commercial fuels. The FPC as well as the WGEP have adopted the measure of Tonnes of Coal Replacement (TCR) by using data on "the amounts of each non-commercial fuel required to replace a tonne of coke and the amount of coal required to make that coke." The ratios used were : 1 ton of firewood replaces 0.65 tonnes of soft-coke or 0.95 tonnes of coal (since 1.5 tonnes of coal are required to manufacture 1 ton of soft coke); 1 tonne of dried dung replaces 0.27 tonnes of soft-coke or 0.40 tonnes of coal; 1 tonne of waste products replaces 0.63 tonnes of soft-coke or 0.95 tonnes of coal. However, no details are given about the assumptions made regarding gross energy content of these fuels and the efficiency indices of use of these fuels. As has been pointed out elsewhere[7] there are large variations in the values used for energy equivalents (gross energy contents) of traditional fuels depending on the type, moisture content, ash content, and so on. For example, NCAER (1959, 1982) has used a figure of 4700 kcal/kg for dry firewood while ASTRA (1980) has estimated a range of 4147±379 kcal/kg for dry woods.[8] This difference may be due to the fact that average values (used in NCAER studies) are for wood in the form of logs, chips, etc., whereas the values reported in ASTRA study are for twigs, branches, etc., used by villagers. Dungcakes have values ranging between 2130 and 3300 kcal/kg while the range of values for plant residues are 2690 to 3800 kcal/kg depending on the source of data. Similar differences are noted in efficiency indices of firewood (13-15%), plant residues (11-17%) and dungcakes (10-11%). This indicates the need for experimental observations of the energy equivalents of domestic fuels for different regions and seasons as well as quantification of factors explaining variations therein. Field-level experiments[9] on relative efficiencies of different combinations of fuels, utensils, stoves and food items cooked are required on a regular basis. In the absence of such a data-base, the contribution of non-commercial sources of energy in India is nothing better than informed judgments—or "guesstimates". Besides, the available estimates ignore altogether the contribution of animate power in total energy consumption.[10]

The NCAER (1978) has recently published the results of a survey in Northern India which are quite comprehensive. The objectives of the study are stated to be (i) to estimate the consumption of energy in rural Northern India to include all aspects of rural activity—agriculture, domestic, commerce, industry and transport, by source of energy, (ii) to bring out the differences in energy usage between the different environments—hilly areas, plains and desert areas, and (iii) to determine the changes in the pattern of energy consumption in rural areas in the wake of increases in agricultural incomes. About 1488 households in 100 villages were surveyed in seven states of India. A two-stage stratified design was used for the selection of households. Villages were the primary sampling units and households within the villages formed the last stage units. A total of thirty strata were formed for three regions in seven states. Relevant data on consumption of energy were collected over two seasons—summer and winter—from households and other commercial establishments through a comprehensive questionnaire. Thus, the survey covered 1488 households and 706 trading, industrial and public service establishments.

After listening all the households in the 100 selected villages three schedules were filled: Household Schedules for a total number of 1595 households, Agriculture Schedules for 971 farming households and Establishment Schedules for 800 village establishments. During winter months, an additional household schedule was used to collect food and fuel consumption data. Data were collected on consumption of: (i) coal/coke, kerosene, charcoal, firewood, vegetable waste and dungcakes for cooking, water and space heating, (ii) kerosene and electricity for lighting, (iii) animal and human labour and diesel oils and electricity used in agriculture, and (iv) consumption of electricity, diesel, furnace oil, kerosene, coal/coke, fuelwood, cowdung, vegetable wastes and charcoal in rural establishments, such as brick kilns, eating houses, oil extraction plants and grain mills.

The survey presents very useful data on share of different fuels for various rural activities. It also indicates changes in level and pattern of energy consumption in hill/desert regions or with season and/or income level. The limitations of such a large scale survey are:

(i) data on energy consumption in cooking have not been obtained directly. In the case of many households, fuel consumption for cooking has been calculated on the basis of consumption of foodgrain and the norms (average consumption) obtained from households where all cooking fuels were purchased;

(ii) no attempt has been made to distinguish between different categories of fuelwood or vegetable wastes;

(iii) there is no attempt to relate availability of non-commercial fuels with agricultural production and the supply aspects of these sources have been more or less ignored;

(iv) relationships among different categories of households with regard to energy exchange or ownership of energy assets has not been quantified; and

(v) available data has not been fully analyzed, using

econometric methods, to yield estimates of income and price elasticities and/or elasticities of substitution.

A more recent all-India survey for the year 1978-79 covering domestic fuel with special reference to kerosene has also been completed by the NCAER whose results are awaited.

The Energy Resources Cell, Government of Pakistan (1978) conducted a large-scale survey on consumption of non-commercial fuels in rural and urban areas of all the four provinces. A total of 18,297 households were covered, of which 10,748 were situated in urban areas and 7549 were located in rural areas. 4962 establishments covered in the survey included sugarmills, gur making industry, tobacco curing kilns, blacksmiths, laundries, restaurants, tandoors, *pakora/tikka* shops, and so on. The questionnaire for collecting the information was designed in cooperation with the Central Statistical Division, The United Nations Project Manager and Messrs Acres International Limited of Canada. In order to carry out the sample surveys (including collection, tabulation and analysis of the data), a group of 30 enumerators, field investigators and research investigators was recruited.

The main results of the survey can be summarized as follows:

(i) a total of 15.5 million (long) tonnes of indigenous coal. equivalent of various non-commercial fuels were consumed in Pakistan during the year 1974-75. It worked out as 47 per cent of the total energy consumed in the country;

(ii) firewood accounted for 54 per cent of the total consumption, while the shares of dungcake and bagasse were around 16 per cent each;

(iii) 82 per cent of the total non-commercial fuels were used for the domestic sector and 16 per cent for the industrial sector;

(iv) bagasse was mainly used as fuel in the boilers of sugarmills and gur (jaggery) manufacturers; and

(v) about 80 per cent of the non-commercial fuels were used in the rural areas; and

(vi) animal dung was also utilized in cities like Karachi, Lahore, Hyderabad, Lyallpur and Peshawar.

The available report does not give much detail about the following:

(i) the method of calculating Btu content or coal equivalent of different fuels;
(ii) the method of data collection regarding the non-commercial fuels such as firewood or dungcakes, especially whether any actual measurements were made or not;
(iii) the reasons why firewood is not distinguished by type as discussed in earlier sections;
(iv) the reasons why only 41 per cent of the total households surveyed were from rural areas, while the share of rural population is 75 per cent; and
(v) the method of selection of households in each block.

It may be mentioned that the results of the national/regional surveys discussed above cannot be used for any policy purposes except to get a 'rough' idea of the magnitude of use of non-conventional energy sources. Although large samples were selected (18,297 households in Pakistan and 1,500 households in North India), the data are not reliable since these are not based on direct observation and measurement of different fuels used for various activities. Besides, data on gross energy content of different fuels as well as efficiency of various end-use equipments were either not collected or have not been reported. The available data cannot be used to identify :

(i) which regions/households are facing energy shortages?
(ii) which are the activities for which energy constraint is felt the most—that is, is it cooking, irrigation or ploughing where energy shortages are critical?
(iii) what type of conservation or substitution possibilities exist and what policy measures need to be adopted?
(iv) how will energy use pattern change with changes in economic activity?

In addition to the government sponsored surveys, contributions by Pimental and Beyer (1976), Desai (1980), Tyers (1978) and Manibog (1982) have attempted to take a comprehensive view of energy balance at the macro level. Some of the main results are :

(i) although a part of energy produced by agriculture was consumed within agriculture (crop residues used for draft animals), a part was available for outside use as food, milk and cooking fuel. Net energy availability in the form of cooking fuel alone was of the order of 176 trillion kcal in 1972 which was higher than the total outside input into agriculture (127 trillion kcal) required for irrigation, fertilization, field preparation, etc.;

(ii) an analysis of data on biomass yield and its utilization for 1973-74 in India shows that, along with food, milk and industrial raw materials, agriculture provided around 74 million tonnes of fuel for non-agricultural uses;

(iii) in Bangladesh in 1974-75, a total energy equivalent of 146 trillion kcal of crop residues are produced, about 62 per cent of which are used for cattle feed, 31 per cent for cooking and about 7 per cent for food processing. These crop residues along with other fodder, produce animal labour equivalent of 25.4 trillion kcal and cattle during equivalent of 31.8 trillion kcal making Bangladesh an almost closed ecosystem from an energy standpoint.

Surveys at the village level

As a number of surveys at the village level have been conducted, it would not be possible to present a comprehensive review in this paper. Fortunately, some of these surveys have been reviewed in Bhatia (1980), National Academy of Sciences (1980), Manibog (1982) and Desai (1982). However, a few methodological issues are outlined below based on personal discussions with some of the researchers involved in these surveys.[11]

(i) in some surveys where the objective is to study the processes of generation and use of traditional energy sources, it is necessary to use the method of participant observation. Data on several types of crop residues, animal residues and woody fuels may be obtained by actual measurements of different fuels consumed for various activities used. Such a survey is also useful to show the dynamics of energy availability and use over time and to speculate how poorer villagers would be adversely affected by the introduction of HYV seeds and other development projects;

(ii) whenever, a comprehensive survey over a long duration is not possible, it may be necessary to use a judicious combination of large scale surveys with 'participant observation and measurement method', preferably, repeated over two or three seasons;

(iii) a detailed survey may be useful before a programme (for example biogas units, social forestry schemes, improved stoves) for augmentation or improved use of traditional sources is envisaged. Such a survey will help in defining the scope of the programme as well as motivate people to take interest in the programme;

(iv) in some cases 'unstructured interviews' may be more useful compared with structured questionnaires;

(v) there were complications involved in the usage as well as acquisition of non-conventional energy sources which had to be understood thoroughly within the context of the rural environment. There should be a key problem focus along the lines of deforestation, diversion of manure to fuel, etc., which had to be incorporated through examination of the land-man-animal—forest inter-relationships. Concurrently, full account had to be taken of the local people's perception of problems and priorities which reflected the energy situation in the overall rural context;

(vi) it is necessary to disaggregate woody fuels and crop residues into a number of categories on the basis of method of acquisition, energy content, moisture content and end-use;

FRAMEWORK FOR ACQUIRING DATA ON TRADITIONAL SOURCES

For energy planning, required data on traditional sources may be grouped under the following categories :

(i) pattern of energy-use for various agricultural domestic and industrial activities in different seasons along with the methods of acquisition (ownership, collection, exchange, purchase, etc.);

(ii) the relationship of pattern of energy-use with family size, land holding, cattle population, income, education, urbanization, etc., so as to predict changes over time;

(iii) possibilities of conservation (improved efficiency) and inter-fuel substitution under different ecological fuel price scenarios;

(iv) assessment of the resources position that is livestock population, forest land, uncultivated waste land, pastures pattern of crop production, pattern of processing of agro-products, etc.

(v) possible changes in supply position that is changes in forest area, grazing land, crop pattern, straw-grain ratios, livestock population (size, age), dung production per animal, efficiencies of collection of crop/animal residues, etc.

It is considered that all these data require well-planned energy surveys at the national/regional/village levels. Even though some of the existing sources may provide data on some variables (for example from agricultural and livestock censuses, farm-management surveys) a complete and accurate picture cannot be obtained without the help of pre-designed surveys. Hence, the remaining part of this paper[12] is devoted to the methodological issues relating to identification of the type of survey required, selection of villages, households and respondents and the methods of data collection and analysis. In short we try to answer questions such as *who* are the people or groups to be surveyed, *how* are these target populations to be selected, *what* type of data are to be collected, *how* is this information to be obtained and from *whom* and *who* should be included in the survey team. A short discussion of the techniques of analysis of survey data are also presented towards the end of this section.

As was discussed earlier, the scope and coverage of the survey to be conducted depends crucially on the objectives of the survey in question. If the objective is to assess the level and pattern of consumption of non-conventional fuels, a large-scale extensive survey at the national/regional level may be conducted for various households, small industries and other commercial establishments in rural and urban areas. However, a set of comprehensive surveys at the village level will be required if the objective is to assess the demand and supply of non-conventional energy sources in rural areas with special reference to improvements in the existing use-patterns and introduction of new technologies.

National/regional surveys

The first step for such a survey would be to do a critical evaluation of existing information. Surveys are rarely done in a vacuum and there is always some existing information on aggregate fuel use, agricultural production, livestock availability, and so on, that can be used as a starting point. Existing information may also be available on either the economic, social or physical aspects of rural systems and urban areas. Sources might include academic studies, agricultural censuses, farm surveys, surveys of small-scale manufacturing units, labour force surveys, general consumption level surveys, forestry data, meteorological measurements and other survey or project studies that may not necessarily be energy related, but could still be a valuable source of information. Such information must be located and evaluated not only to avoid needless repetition, but also in order to build upon existing groundwork. It may be useful to coordinate energy surveys with other socio-economic surveys so that a large amount of data on these variables is simultaneously available for the target groups. The surveys done by the National Sample Survey Organization (NSSO) of the Government of India are examples of similar exercises.

Since the target population of a survey of this type will be very large, a multi-stage, stratified sampling approach would have to be used. The following criteria may be used for stratification: rural/urban, hill/plains/desert, electrified/non-electrified, irrigated/non-irrigated, high/medium/low income levels, predominantly agricultural/industrial, etc. Sampling may be required at state, district, block, taluka, village and household or establishment levels. Given the limitations of time and resources, it may not be possible to survey a large number of households or establishments. One of the accepted procedures is to select households on a random basis. Although it may be a good procedure for general socio-economic surveys, and may even be appropriate for urban areas and industrial establishments, it may not be very advantageous for rural energy surveys. A more appropriate method would be to select the sample households in a village, using purposive sampling method in a two stage process.* In the first stage, one household is to be selected from each of the following categories: large farmer, medium farmer, small farmer, marginal farmer, agricultural worker, artisan, trader, teacher, priest, and so on. At the second stage, an attempt should be made to identify households which have energy related relationships with these households selected in the first stage. For example, if a large farmer is selected at the first stage, all those households who work for him in the fields, receive crop residues or firewood, animal dung or dungcakes from him or share ploughing, threshing or irrigation activities with him, have also to be included in the sample. Similarly, in the case of agricultural worker or artisan or trader, it is important to include households which supply non-conventional fuels to them. This procedure may have to be slightly modified to give proper representation to various categories of households in a village. However, it is considered that this method of selection of households will give a much better insight into the inter-relations among households in the production and use of non-conventional sources of energy than would be obtained if households are selected randomly. It must be kept in mind that obtaining good quality data along with a better insight into the processes of generation and use of these energy sources should be given more

*This approach has been followed in selecting households in two villages in a survey in progress in Bihar, India.

importance than a mere statistical formal requirement of random selection of households.

In addition to a proper selection of villages and households, (or establishments), it is equally important to collect the right type of information using appropriate methods of data collection of informal questionnaires and interviews or informal household interview, etc. Since these aspects are also relevant for village level surveys, a discussion of these issues is presented in the later sections.

Identifying the type of survey required at the village level

Although the economic viability and social acceptability of new energy technology depend on the socio-economic environment, it does not always follow that exogenous technologies can only be introduced after completing large-scale, comprehensive surveys in a number of villages involving numerous households. A survey should be structured so that it can be used to identify and support useful policy interventions in the system being studied, not just to understand the system. For example, in the case of household energy, source augmentation, conversion technologies and end-use devices would be likely points on which a survey would focus.

For the purposes of this paper, rural energy surveys can be broadly classified into two types. The first type is a short-duration (a few months) survey in which the focus is on identification of the relevant constraints and defining the scope and dimensions of the contemplated project (or technology application). The second type is a more comprehensive survey (over a 12-15 month period) in which the objective is to quantify the inter-relations and inter-dependencies among the major components of the energy system, such as household consumption, crop production, livestock maintenance, agro-processing, and rural transportation. The type of survey to be conducted in a particular village depends on the following factors :

(i) the types of technology (or intervention) to be introduced, tested, and disseminated. For example, installation of biogas plant(s), use of improved stoves, improved charcoal production and introduction of new water lifting devices may require only the first type of survey, while pyrolysis of agriculture residues, village woodlots programmes, improvement of animal power utilization, energy conservation for small industry, and conversion system integration require the second type;

(ii) the type and quality of existing information regarding agricultural production, land use, food habits, livestock maintenance and use. For example, short-duration surveys would be quite adequate in regions/villages where in-depth farm management, agronomic, or socio-economic surveys have been conducted. Comprehensive surveys may be necessary where no such earlier studies exist;

(iii) the nature of the energy availability constraints. For example, if energy is not available or is inadequate for water lifting for irrigation or for activities that prevent multiple cropping and higher yields, the first type of survey may be adequate. However, if the bottleneck occurs in cooking (reflected in long distances and deforestation) or threshing/chaff cutting, it may be necessary to do a comprehensive survey (type 2);

(iv) the type of household or groups affected by energy shortages or higher costs. The first survey is adequate if the bottleneck is felt by relatively wealthy people who can afford to seek alternative solutions. If agricultural labourers, artisans, marginal farmers, or particular households are facing the bottlenecks, it may be necessary to do a comprehensive survey to identify the inter-relationships and inter-dependencies of these households with wealthier ones;

(v) the need to quicken the process of action research in new energy technologies. The first type of survey is recommended to meet the objective of not unduly delaying action research needed to firm-up the techno-economic parameters of rural-oriented technologies under varying physical (temperature, rainfall, groundwater quantity and quality, soil type, and so on), economic (farm size, family size and composition, occupation of household, patterns of income and expenditure, and so on) and social conditions.

Selection of villages, households and respondents

The methodology of selection of villages and households depends crucially on the type of survey being contemplated.

For example, the following characteristics would be taken into consideration for vlllage selection in the case of short duration surveys in villages where introduction of biogas units is being considered :

Availability of animal residues, especially cattle dung, pig manure, etc.;
Relative shortage or adequacy of plant residues available for fodder and fuel;
Problem of deforestation, as demonstrated by need to walk long distances;*
Availability of adequate water supply, construction material, etc.;
Relatively mild winter, since gas production in cold climates decreases significantly;
Demand for organic manure for its social buildup properties and nitrogen content;
Availability of groundwater or surface water (canal, river, etc.) for irrigation purposes.

A community-size biogas plant may be considered for a village that has all or most of the above characteristics and in which the people are eager to participate in such a programme. However, it should be kept in mind that the villagers should participate in the determination of project size, gas allocation, pricing, collection of residues, etc.

Selection of households should be carried out according to the following criteria :

Farm size, crop pattern, and livestock ownership;
Availability of groundwater (or surface water) for irrigation near the field;
Acute shortage of cooking fuels.

*It would have to be specifically documented rather than taken in a general way.

This type of survey can be carried out over a period of 2-3 months through the method of participant observations by a person who may live in the village(s) or be there almost continuously. Data should be collected at both the village-community and household levels. Questionnaires should be prepared to obtain data at weekly and fortnightly intervals.[13] This data can be supplemented by asking other relevant questions about the rest of the period between questioning. Usually, data may have to be collected from the male head of household as well as from the female head of the household.

As a contrast to the short duration survey, the focus of the comprehensive survey of village energy system will be :

1. To quantify the inter-relations among major components of the village energy system—namely, crop production, agro-processing, livestock maintenance, household use (Figure 13.1);
2. To identify and, to the extent possible, quantify the inter-dependencies among different households in the community, e.g., the dependence of the small or marginal farmers agricultural workers or artisans on medium or large farmers for crop residues, dungcakes, and animal residues;
3. To postulate the dynamics of village energy generation and use patterns, especially the possible scenarios with simulated proposed changes (e.g., new irrigation, new crop varieties, fuel prices, new infrastructure project, etc.);
4. To document energy resources and nutrient flows *before* and *after* the introduction of new technologies (or intervention), such as biogas, village woodlots, improved stoves, kerosene lanterns, etc.;
5. To identify social, cultural, and organizational constraints that would affect the successful development, implementation, and diffusion of new technology.

A comprehensive survey may have three stages :

1. dividing the country in question into several agro-climatic regions;
2. selecting two or more villages for comparison within a

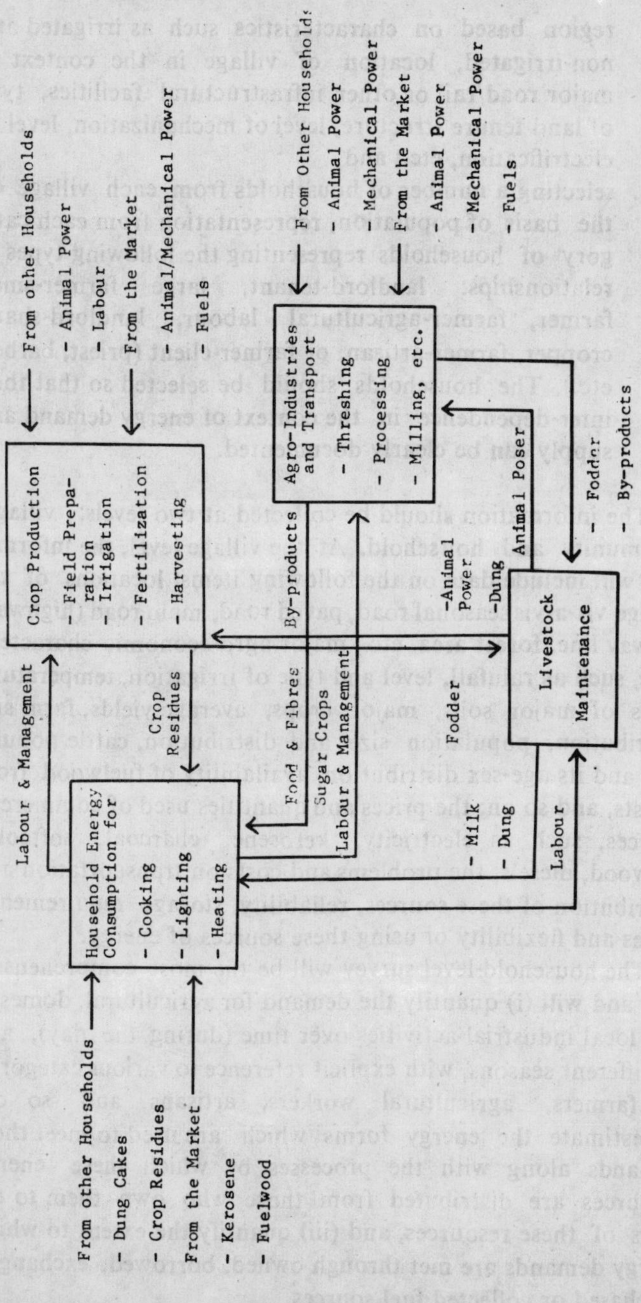

FIG. 13.1 : *Energy generation and use in rural areas*

region based on characteristics such as irrigated and non-irrigated, location of village in the context of major road/rail or other infrastructural facilities, type of land tenure structure, level of mechanization, level of electrification, etc.; and

3. selecting a number of households from each village on the basis of population, representation from each category of households representing the following types of relationships: landlord-tenant, large farmer-small farmer, farmer-agricultural labour, landlord-sharecropper, farmer-artisan, or farmer-client (priest, barber, etc.). The households should be selected so that their inter-dependence in the context of energy demand and supply can be clearly documented.

The information should be collected at two levels: village/community and household. At the village level, the information will include data on the following items: locations of the village vis-a-vis seasonal road, paved road, main road (highway) railway line, forest area, etc., major agro-economic characteristics, such as rainfall, level and type of irrigation, temperature, types of major soils, major crops, average yields, farm size distribution, population size and distribution, cattle population and its age-sex distribution, availability of fuelwood from forests, and so on; the prices and quantities used of commercial sources, such as electricity, kerosene, charcoal, softcoke, firewood, diesels; the problems and costs on transportation and distribution of these sources, reliability, storage requirements, losses and flexibility of using these sources of energy.

The household-level survey will be the most comprehensive step and will (i) quantify the demand for agricultural, domestic and local industrial activities over time (during the day), and in different seasons, with explicit reference to various categories of farmers, agricultural workers, artisans, and so on; (ii) estimate the energy forms which are used to meet these demands along with the processes by which these energy resources are distributed from those who own them to the users of these resources, and (iii) quantify the extent to which energy demands are met through owned, borrowed, exchanged purchased or collected fuel sources.

Type of data to be collected

In a comprehensive survey, four components of energy generation and consumption should be identified: crop production; animal husbandry; domestic activity; and crop processing and local industry (Fig. 13.1). Energy flows into the crop production component in the form of work (from men and animals), machine power (for ploughing, irrigation and planting), and other forms of fuels. Crop production makes energy available in the form of food, fodder, fuel (straw, husk, firewood, and so on), and crop residues including vegetable wastes. Energy flows into the animal husbandry component as fodder, fuel, and human energy for caretaking. Energy is generated in the form of work by animals (ploughing, water lifting, threshing, transporting, inputs/outputs, and so on), food, fuel (dung), and fertilizer. Domestic activity consumes energy in the form of food, fuel (for cooking, lighting, heating, and human work. The activities included are harvesting, threshing, winnowing, transporting, and other agro-based industries. For each of these components, the survey will quantify the pattern of users as well as the magnitude of the stocks and flows over time. As a background to data collection on fuel use, it will also be necessary to obtain information on the type of equipment used for various activities, for example, the type of stove used for cooking, lamp used for lighting, kerosene stove, appliances, pumpsets, tractors, electric motors and so on.

The survey should concentrate on collecting data on the amounts of energy resources produced, collected, stored and used in each of the four components of the system. Greater attention should be paid to the amounts of fuels produced and consumed, since it is this area that is most important and for which data are least available. *Annexure I* provides a tentative checklist for the method and frequency of data collection for measuring non-conventional energy used in different activities.

In data collection, the emphasis should be on :

Consistent measurements of quantities when fuel is used in non-standard units, such as bundles, bags, back-loads, etc.;

- The variety of fuels, carefully distinguishing between different items such as twigs, branches, fallen or dead trees, self-grown firewood, etc., which are usually classified under "fuelwood";
- The relationship of fuel availability to agriculture in terms of quantity and type of residues related to crop rotations, and availability of livestock and manure, etc., over different seasons;
- The pattern of use of various crop residues as fuel, fodder and construction material;
- The use of human labour (by age and sex) for collecting and processing of different fuels.

The measurements on fuel consumption will require detailed weighing, timing, and observation of activities in all sectors of the system. These measurements will answer questions about how much energy is produced and consumed in a household by socio-economic level (large farmers, medium farmers, small farmers, sharecroppers, landless labourers, and others) and by fuel type. The following fuel types should be clearly distinguished: kerosene, gasoline, light diesel oil, high speed diesel oil, electricity, biogas (gobar gas), soft coke, fuel alcohol, charcoal bagasse, rice straw, hulls and fuelwood. In the case of fuelwood or firewood, care should be taken to distinguish between different species and material usually classified as firewood. This material may include twigs, some type of crop residues, leaves, branches of dead trees, small fallen branches, fuelwood from dead trees, any specially grown form of wood, and fuelwood purchased from the market. It would be necessary to estimate the moisture content of different fuels so that calorific content of each fuel can be estimated accurately. *Annexure II* provides a tentative list of energy sources for which relevant information may have to be obtained.

The survey team should also solicit opinions of users on each fuel-type in terms of its suitability for a particular form of cooking, timing of cooking, ease of increasing/decreasing heat versatility in many uses, mobility of use, uncertainty of supply with and without storage, as lack of storage in the case of biogas against storage with kerosene/dungcakes, seasonality of fuels as in the case of wet fuelwood versus dry dungcakes, convenience of storage, etc. The user may be asked to rank each type of fuel according to the above criteria, and a weighting system may be developed to incorporate this information.

In addition to the measurements, data on the patterns of energy distribution and ownership should also be collected. This information will answer questions about where each socio-economic group gets its supplies of energy, the mechanisms (ownership, purchase, exchange, or patronage flow) through which each group obtains its supply of food, fuel, and fertilizer, and how energy is related to the land ownership and labour structure of the village.

Methods of data collection

Repeated interviews covering the previous 24 hours may be necessary to provide the basis for calculating the amount of energy expended by human beings and animals during domestic and agricultural activities. Data may have to be collected almost continuously for a period of about 15 months so that coverage of all the principal activities during different seasons of the year along with an overlapping period for consistency checks is possible. The frequency of data collection will vary for different activities and in some cases may require interviewing once each month or every two months. Data collection will be through actual observations and measurements made by a field investigator who may live in the village almost continuously during the period of the survey. It may be noted, the type of data that is required for an evaluation of a technology in the context of a complex rural energy system cannot be obtained through one or two interviews, using the recall (memory) method.

In all of these surveys, effort should be made to ensure that there is no data collection merely for its own sake. For example, it may not be necessary to repeat fuel-use measurements every day over the 15-month period. This may be done only for a few selected days in each season to reflect the variations in fuel-use over seasons, since both demand for energy as well as availability of different fuels will show seasonal variations.

The surveys should be designed so that necessary data

required for policy interventions that affect the community as a whole (and not individual households) are collected at the village level. For example, information on average energy consumption, major fuel types used, the share of fuelwood in average energy consumption, and extent of fuelwood obtained by cutting trees in nearby forests can be obtained at the village level, along with other characteristics of the village mentioned earlier. However, necessary variations in energy use from one household to another, resulting from household size, farm size, cropping pattern, level of mechanization, level of income from other sources, and so on, should not be ignored, since decisions on energy choices regarding purchasing, investment, exchange, and so on, must be made at the household level. The perception and evaluation of benefits from cleaner biogas, smokeless fuel, or efficient stoves are going to vary from household to household, depending on the level of income and education, so it is important to ensure that such differences are not ignored by using averages or by using the perceptions of relatively well-off villagers. In cases where it is found that there are significant variations among households, it may be useful to survey all the households in the selected village(s). However, care must be taken to see that interference in the activities is minimized and the survey should be organized so that the investigators are participants rather than intruders in village life.

It may be useful to discuss the advantages of using formal questionnaire-based interview method against using an informal, unstructured household interview approach.[14] The advantages of the questionnaire method usually claimed are :

(i) it is easier to administer through hired and trained investigators/enumerators;

(ii) uniform questions can be asked from all the respondents and the data can be quantified, using computers ;

(iii) pre-testing of questionnaires can be done in order to modify questions or eliminate questions; and

(iv) the biases or prejudices of investigators can be minimized as there is very little left for subjective judgment or memory of the investigator.

As against these, the experience of individual scholars conducting such surveys shows that: (i) use of questionnaires results in distractions and leads the discussion into unwanted directions, (ii) the words used in the questions themselves could be misunderstood due to differences in dialects, (iii) the answers would usually reflect the views or biases of the investigators or others present at the time of the interview, and (iv) villagers are usually reluctant to give information on income levels, and, respondents tend to either exaggerate or understate answers to some questions—depending on their perception of the biases of the investigators(s).

The use of unstructured, informal, conversational "immersion-perception" technique is considered advantageous for obtaining, not only qualitative, but also quantitative information. This method relies on the investigators gaining the confidence of the villagers initially through community leaders and by spending time at places of community gatherings. During these days of initial stay, no specific 'interview type' questions are asked and the investigator tries to avoid giving impressions about his biases or preferences or hypothesis in regard to the questions to be investigated. After initial acquaintances, house-to-house interviews are conducted instead of administering pre-dcsigned questionnaires. A list of questions to be asked is always prepared in advance, although no list of questions are held during the individual visits. The information obtained during the interview is recorded afterwards, almost immediately in the case of quantitative information. The advantages of this approach are that the investigator gains the confidence of the respondents and there are adequate opportunities of cross-referencing and cross-checking of information provided by respondents through actual observation or through interviews of other villagers. The disadvantages are that the method can be very time-consuming, the investigator's reliance on memory may not give accurate statistics and the results may show subjective biases of individuals.

It is suggested that a judicious combination of the two methods should be used in village surveys. After the initial period of "immersion-perception" techniques to gain confidence of the villagers, it may still be useful to fill out well-designed, pre-tested questionnaires. However, adequate care must be taken in wording the questions and a lot of preparation, training and patience on the part of investigators will be required for a successful survey.

If the survey is elaborate, pilot or test surveys are needed to both identify the main features of the data to be collected and pre-test the questions, data gathering procedures and measurement methods. These preliminary tests should include not only the relevance and practicability of the questionnaires and data gathering procedures, but also the processes whereby survey staff can most effectively communicate with those being surveyed. It is important that questionnaires and field procedures are tailored both to the capabilities of the enumerator and to what information the respondents can be reasonably expected to provide. In thec ase of surveys relating to non-conventional sources of energy, it would be useful to involve some technical people who can help in measuring moisture contents, energy content of each type of fuels, efficiency of utilization of different fuels in various end-use devices, and so on.

Rural energy surveys benefit when local people are involved in the design and implementation of the survey. The type and degree of involvement would depend on the specific characteristics of each project and location.[15] Experience has shown that field investigators can come from other nearby villages. In rural energy survey, a man and woman team from the local area would be an asset. Role of women, both as investigators as well as respondents, should be explicitly recognized. Motivation of the people can be increased by providing some benefits as a part of the project. Response is likely to be better if the survey can be directly and explicitly connected to the local energy situation, and if some long-term improvements can reasonably be expected. However, people should not be given expectations that are unrealistic.

Analysis of data

Analysis of data collected through a survey is one of the important components of the study. Sometimes (for example, in surveys of NCAER (1978) a detailed analysis of data is not carried out, and, hence, full benefits from survey results are not obtained.

The methods of data analysis can be broadly grouped into two :

(a) Tabular and graphical presentation of data to present an energy profile of the area;
(b) Statistical and econometric analysis of data to present inter-relationships between the energy profile and socio-economic variables.

The energy profile to the region would provide quantitative data on the level and pattern of consumption of energy. The tabular and graphic presentation of data would include details about share of different fuels in the total energy consumption in various activities (cooking, lighting, industry, and so on), the seasonal variations in the level of energy use and the types of fuels consumed, and differences in energy consumption in villages with differing levels of irrigation, electrification and economic activity. At the household level energy consumption pattern should be discussed for different categories of farmers, agricultural workers, artisans and traders. Data on per capita consumption of energy should be compared across households and seasonal variations in fuel-use should be noted. Results should be given not only in terms of the average (mean) but variance around the mean should also be discussed. The magnitude of estimation errors should be indicated, wherever possible. In the case of villages, where comprehensive surveys are undertaken, a detailed quantitative and qualitative description of the processes of generation and use of non-conventional fuels by different households should be given.

One of the difficult problems, both conceptually and empirically, is to estimate the contribution of animate energy (animal power and human labour). According to one view[16] "a rural energy survey need not cover both animate and inanimate energy. If it does, the two should not be added together; a distinction should be made between energy used for heat and work". It is suggested that measurement of animate work may have to be done in terms of labour time (measured in hours) and food intake (measured in calories). However, since some of the surveys[17] have 'measured' animate energy and aggregated it along with other sources, it may be useful to review the alternative approaches so that commonly made mistakes can be avoided. In the available studies, there are three ways in which energy contribution of draft power has been estimated. The first method[18] is to estimate energy input for draft animals, adjusted for the energy available from dung. Energy input is calculated in terms of kilocalories in animal feed which could be used directly for human consumption, and those feeds grown as fodder on land which could have been used for other crops. The second method, which is invariably used in the farm-level energy-use studies conducted by agricultural universities and the ICAR, is to estimate the work output of draft animals in terms of horsepower. In most of these studies, it is assumed that a pair of bullocks develops one horsepower. Using this assumption of 'engineering equivalence' a tractor of 35 horsepower is 'equivalent' to 35 pairs of bullocks.[19] If economic equivalence in terms of work performance or replacement ratios is considered, the equivalent number of bullocks would be much less (say 2-3 pairs per tractor). This would indicate that the contribution of animals would be much higher than what is generally assumed in these studies.

Besides, there is also the question of the method used for estimating energy equivalent of the work of animals. One view[20] is that it is the energy equivalent of the work done by them that mattered and not the energy required to support them. However, some others[21] feel that for a comparison with the other power sources, an estimate of energy input is desirable. However, if the contribution of mechanical power is estimated in terms of horsepower-hours, the consistent procedure for work animals would be to estimate their contribution in horsepower-hours. Also, if in the case of mechanical power, it is the energy equivalent of inputs (electricity, diesel, etc.) used to operate these mechanical equipments, the consistent procedure would be to estimate the energy input into work animals, corrected for the energy input 'embodied' in the stock of bullocks.

In the third method,[22] the contribution of animal power has been estimated in terms of diesel equivalents that is, the quantities of high speed diesel and light diesel oil that would be required if all the bullocks are replaced by tractors and diesel engines for providing draft power and irrigation. It has been argued that the energy contribution of bullocks is not the same as energy input of feed utilized by them for work. Instead, their contribution to the energy budget should be estimated as the amount of fossil fuels (say diesel) which would be required in the absence of animal power and so on, the additional energy which would be required if tractors and diesel engines were employed to perform the same work that is currently being done.

It would be necessary in the survey results to display separately the contribution of animal and human labour in terms of number of hours spent in each activity.

Simple correlation and regression methods should be used for statistically analyzing differences in level and pattern of fuel use. Energy consumption in different households can be explained in terms of family size and composition, level of income, farm size, type of irrigation, levels of mechanization and ownership of fuel sources. Econometric methods may be used to estimate income and price elasticities of demand for different fuel sources. This type of econometric analysis supplemented by a cross-sectional analysis of different villages can be used to present different energy-use scenarios for the future. If computers are to be used for preliminary consistency checks on data and for tabulation and statistical analysis, it is important that computer programmers are associated with the designing of questionnaire, etc.

After the completion of the survey, it is important to ensure that its results are used by the relevant agencies. Survey results should be made available to the following groups:

—Policy makers and international financing agencies who may be interested in using these results in defining or implementing their programmes and projects;

—International agencies who may be acting as central repositories of data, making survey methods and findings accessible to all countries;

—People in the surveyed area who should also receive information on decisions that have been taken on the basis of these results;

—Other potential users such as researchers, educational institutions, government agencies not involved in the survey, and even other countries.

SUMMARY AND CONCLUSIONS

This paper has reviewed the attempts made at quantification of traditional energy sources at various levels. It is found that the available studies are rather inadequate in providing a data-base which can be meaningful for energy planning in developing countries. Most of the studies suffer from inadequate coverage and problems of definition, measurement and aggregation of animate energy and different forms of fuels. The available information is not sufficient to indicate in which regions and/or activities energy shortages are being felt and are acting as constraints to development or leading to ecological imbalances. Besides, very few studies have described the processes of generation and use of these traditional sources by different households which are important for projecting demand-supplies as well as for evaluation of new technologies and programmes. Hence, it is felt that a thorough discussion of methodologies of energy surveys is required as a basis for planning.

This paper points out that the scope of a survey would differ according to whether the objective is to define general policy problems at the national or regional level or to aid specific project design and implementation. It is important to outline clearly the objectives and scope of a survey—since survey parameters such as type of information to be collected, type of respondents and investigators, timing, frequency and methodology (what, who, where, when and how) depend on the policy issues in question.

Large scale surveys using mainly the questionnaire method for measurement of non-conventional sources of energy at regional or national level are not very appropriate, especially in the case of rural energy systems. A judicious combination of large-scale surveys with 'participant observation and measurement method' should be adopted for rural areas. A very good example of this type of survey is one conducted by John Arnold (1980) in Indonesia. With suitable modifications in the methodology and a repetition of surveyor visits in two or three seasons, very good insights can be obtained in the energy profiles of various rural households.

Another type of survey required, especially for rural areas, is specific purpose short duration surveys where the objective is to collect enough information to define the project parameters and/or to explore possible impacts of the project on the local people. Examples of such studies include those by M.N. Islam (1979) for Bangladesh, Manibog (1979) for the Philippines, Reddy and and Subramaniam (1979) and by the Government of Uttar Pradesh in India. Here again, it is crucial to make a difference between the scope and coverage of a survey required for installation of biogas plants or a survey required for studying the impact of social forestry projects. It is argued that the latter type of projects which have relatively more extensive type of impacts would require more comprehensive surveys.

A comprehensive survey designed for understanding the processes of generation and use of energy by various households in a rural community is essential if programmes (such as social forestry) which have complex interactions with the rural economy in terms of food, fodder, fuel and fertilizer choices are to be implemented without adverse impacts on the rural poor. Examples of such surveys conducted or in progress are those by Briscoe (1979) for Bangladesh, by Bajracharya (1979) for Nepal and by Bhatia, Rogers, Briscoe and Jha in North Bihar, India. This type of survey requires, preferably, an interdisciplinary team and needs very careful documentation of the quantitative and qualitative information. It is necessary for a survey of this type that the investigator stays in the village during (almost) the entire duration of the survey and uses a judicious combination of house-to-house interviews, structured questionnaires and actual observation and measurement.

For any country concerned with policy decisions relating to non-conventional energy sources, it is necessary that a combination of the three type of surveys outlined should be undertaken. The actual number of each type of survey and the timing of such surveys would vary from country to country.

There are problems in measurement of non-conventional energy sources since standard units are not used, records of collection and use are rarely kept and there is absolutely no field information about energy contents of various fuels used or efficiency ratings of end-use devices. Therefore, repetitive actual observation and measurement of different fuels (sufficiently disaggregated) over different seasons should be undertaken to overcome these difficulties. An interchange of information among different countries in this regard would be a good starting point. The following is a summary of questions that need to be thought about and answered before surveys of traditional energy use are mounted.

1. What is the role of large-scale, quick surveys in the measurement of non-conventional energy sources—especially in rural areas?
2. What is an optimum mix of quick, short duration and comprehensive surveys (as defined in the paper) for a country? What criteria should be used in selecting the number, timing and duration of such surveys?
3. How detailed disaggregation of various fuel sources is necessary? What are the advantages and disadvantages of collecting information on say, ten types of fuelwoods and seven types of crop residues?
4. How should one incorporate the contribution of animate energy sources? Should animate sources be aggregated with fuel sources? If so, what should be the principle of aggregation?
5. What are the merits and demerits of using structured questionnaires, vis-a-vis informal house-to-house interviews? Is it possible to device an appropriate combination of the two methods of obtaining information? What are the considerations in using random sampling in selection of blocks, villages and households? Should purposive sampling replace random sampling at each level?
6. How should local participation in the survey be secured? What are the considerations in involving teachers, village leaders and other influential people in the survey effort?
7. What are the training, personnel and management requirements of conducting a successful survey of each

type? What type of technical expertise is required for estimating potential ofn on-conventional sources (such as micro-hydel sites or windmill sites or biogas plants or appropriate species for social forestry) or for measuring energy content of fuels?

8. Are there some standard methodologies which can be used for analyzing data from surveys in different countries? How should one go about constructing data bases on non-conventional energy sources in developing countries? How should computers be utilized for surveys?
9. How should one ensure that the methodology and results of all the surveys are available to people who are interested in them? What type of international agency should be charged with collection and dissemination of information on energy surveys to all the countries?

ANNEXURE 1

METHOD AND FREQUENCY OF DATA COLLECTION FOR MEASURING NON-CONVENTIONAL ENERGY USED IN DIFFERENT ACTIVITIES

	Interview**		
	Fortnightly (for the previous day)	Seasonal (once in 2-3 months or for each crop)	Actual observation and measure-ment**
	1	2	3
Type of Activity of Data			
A. *Crop Production*			
Human Labour		√	
Animal Labour		√	
Diesel Fuel		√	
Electricity		√	
Biogas		√	
Fertilizer		√	
B. *Livestock Maintenance*			
Human Labour Used	√		
Fodder (Green/Dry) . . . . . . Eaten	√		
Grains/Oil Seeds	√		
Dung Production	√		√
Dung Utilization	√		√
C. *Household Fuel Use*			
Cooking	√		√
Heating	√		√
Lighting		√	
Space Heating		√	
D. *Agro-Processing*			
Energy used, by type,* for			
Threshing		√	
Grinding/milling		√	
Transport		√	

ANNEXURE I—contd.

	1	2	3
E. *Local Industry*			
Energy used, by type,* for			
Pottery		√	√
Smithy		√	√
Brickmaking		√	√
Sugarcane Processing		√	
Oil Processing/Pressing			
Others			
F. *Residue Production*			
(by type)*	√		√
Residue Collection and Storage	√		√

*The details of the type of fuels and other energy sources used in various activities are given in Annexure II.

**Measurement should be done in local units such as maunds, seers, headloads, baskets, backloads, bundles, etc. Average weight as well as its variance should be estimated for non-standard units of measurement.

ANNEXURE II

LIST OF CONVENTIONAL AND NON-CONVENTIONAL ENERGY SOURCES AND THEIR ENERGY CONTENT

	Moisture Content by Season	Gross Energy in kcal/kg (Over Dried)	Efficiency of Utilization (%)		Net Energy Available (kcal/kg)	
			End-Use Device I	End-Use Device II	Device I	Device II
	1	2	3	4	5	6
A. *Non-conventional* (traditional sources)						
FIREWOOD						
Wood from felling trees						
Dead Wood						
Twigs and branches						
Dhencha (doinshah)*						
Charcoal						
Sawdust						
Wood Shavings						
CROP RESIDUES						
Nara**						
Kher**						
Wheat Straw						
Maize Straw						
Jute Sticks						
Cotton Sticks						
Residues of :						
Oilseeds						
Vegetable oils						
Coconut husk						
Shrubs/weeds						
Bagasse						
Bamboo						

ANNEXURE II—contd.

	1	2	3	4	5	6
DUNG CAKES (or Ghoita**)						
B. *Conventional*						
Coal/coke						
Kerosene						
Electricity						
Diesel						
C. *Other Sources*						
Biogas						
Pyrolysis Gas						

*Fuelwood specially grown around the field.

**Nara is the tough straw which remains in the field after paddy is harvested. Kher is more tender straw which is carried from the field with the harvested paddy and which remains after the threshing of the grain. Ghoita refers to fuel made by packing cowdung around the jute sticks.

NOTES AND REFERENCES

1. U.N. *Year Book of World Energy Statistics*, 1979, pp. 1154-84.
2. Prospects for Traditional and Non-Conventional Energy Sources in Developing Countries, World Bank Staff Paper No. 346, July 1979.
3. The significance of non-conventional energy sources (excluding animate sources) in developing countries can also be seen by estimates given by Knowland and Ulinski (1979). According to them, in 1976, the people in developing countries would have burnt 1300 million cubic metres of fuelwood, 400 million ton mountain of dung and a similar mass of crop residues. In oil equivalent, this would equal about 2.7 billion barrels of oil worth, at least, 90 billion dollars at current oil prices. As a comparison, the total commercial energy used in all the developing countries was around 4.1 billion barrels of oil equivalent. However, in rural areas of the developing countries, non-conventional sources are predominant and these are used at a relatively low level of efficiency.
4. For example, in the Fuel Policy Committee, it is noted that according to the NSS (18th Round), per capita consumption in rural areas was higher than in urban areas (0.34 tonnes as against 0.29 tonnes per annum in coal replacement terms). However, taking into account the trends in urbanization and the effect of urban incomes on energy use, the Committee assumed an average of 0.38 tonne of fuel consumption per head in rural areas and 0.40 tonne per head in urban areas in coal replacement terms.
5. The surveys referred to were: (i) the National Council of Applied Economic Research Survey covering the cities of Bombay, Calcutta and Delhi in 1958 and a sample survey of rural households all over India in 1962. (ii) The National Sample Survey, Eighteenth Round, conducted between February 1963-64 and January 1964 covering over 4000 urban and over 21000 rural households.
6. Report of the Working Group on Energy Policy, 1979, p. 88.
7. See Desai (1982), pp. 13-19, Bhatia (1977), Appendix A.
8. This discussion is based on Desai (1982).
9. It is reported that the mean efficiency for Ungra village from the ASTRA survey is only 5 per cent as against assumed efficiencies of 10-15 per cent. Some of the difference may be explained by the vessels, which in Ungra are earthen.
10. As has been estimated by Revelle (1975) and reported in Bhatia (1977a), the contribution of animal power in energy use in rural India varies between 12-24 per cent depending on the methodology used for estimation.
11. Some of the authors are Briscoe (1979) for Bangladesh, Manibog (1979) for Philippines, Islam (1978) for Bangladesh, Arnold (1980)

for Indonesia, Bajracharya (1979) for Nepal, Reddy and Subramaniam (1979), Gupta, Rao and Vasudevaraju (1980), Gupta and Rao (1981) for India.

12. This part draws heavily on another paper by the author on Energy Survey Methodologies, see Bhatia (1980).
13. These questionnaires can be based on modification of existing work reported in Briscoe (1979), Arnold (1980), Bajracharya (1979), Manibog (1979), Gupta, Rao and Vasudevaraju (1980).
14. An excellent discussion of the various issues involved as well as details of an actual experience in the selected villages of Philippines are presented by Manibog (1979).
15. There are numerous ways of involving local people. For details see papers by Mary Elmendorf (1980) and Irene Tinker (1980).
16. Desai (1982), p. 43.
17. Surveys by ICAR (Indian Council of Agricultural Research) at five locations on Energy Requirements of Intensive Agricultural Systems and those conducted by Agricultural Universities.
18. This method has been used by Revelle (1976) and Pimentel and Beyer (1978).
19. For further discussion, see Bhatia (1976, 1977a).
20. Desai (1978, 1980).
21. Pimentel and Beyer (1978), Revelle (1976).
22. See Bhatia (1977a) in estimating the contribution of animal power in total energy use in rural India.

BIBLIOGRAPHY

Arnold, John, A revised methodology for energy demand surveys, Draft, January 1980.

ASTRA, Rural energy consumption patterns: A field study, Centre for the Application of Science and Technology to Rural Areas (ASTRA), Bangalore (Mimeo), 1980.

Bajracharya, Deepak, Rural energy utilization patterns: Some Observations from Pangama, Nepal (mimeo), Sussex, IDS, January 1979.

Bhatia, Ramesh, Energy requirements of different farm systems, Rapporteer's Report, *Indian Journal of Agricultural Economics*, July-September 1976.

Bhatia, Ramesh, Economic appraisal of biogas units in India: Framework for social benefit cost analysis, *Economic and Political Weekly*, August 1977.

Bhatia, Ramesh, Energy and rural development in India: Some issues in W. Lockeretz (Ed.), *Agriculture and Energy*, Academic Press, 1977.

Bhatia, Ramesh, Energy survey methodologies: A framework for measuring non-conventional sources in developing countries, Paper presented at the ESCAP workshop on *Energy Statistics at Karachi*, October 1980.

Bhatia, Ramesh, An analytical framework for socio-economic assessment of technological and policy alternative, paper prepared for East-West Centre, Honolulu, Hawaii, draft, January 1980 (1980a).

Bhatia, Ramesh and Miriam Niamir, Renewable energy resources: The community biogas plant, paper presented at a seminar at Harvard University, November 1979 (draft).

Briscoe, John, The political economy of energy use in rural Bangladesh, Harvard University, Environmental Systems Programme, August 1979.

DeLucia, Russell, Rural energy survey requirements: Some comments and perspectives, Workshop on Energy Survey Methodologies, Jekyll Islands, January 1980.

Desai, Ashok V., Development and Energy Consumption, *Oxford Bulletin of Economics and Statistics*, August 1978.

Desai, Ashok V., *India's Energy Economy: Facts and their Interpretation*, Centre for Monitoring Indian Economy, Bombay, 1980.

Desai, Ashok V., Rural Energy Survey in India, NCAER, Mimeo, November 1982.

Elmendorff, Mary, Finding out about energy needs and resources: The human dimensions, discussion paper for the workshop on *Energy Survey Methodologies*, Jekyll Island, January 1980.

Government of Pakistan, Energy Resources Cell, Ministry of Natural Resources, Islamabad, Energy Data Book: Part 6 on Non-Commercial Energy Resources Survey (1974-75); 1978.

Government of India, Report on National Sample Surveys: 18th and 28th Round, 1964 and 1974.

Government of India, Report of the Final Policy Committee, 1974.

Government of India, Report of the Working Group on Energy Policy, 1979.

Government of Uttar Pradesh (State Planning Institute), Lucknow, India, Bench-Mark Survey Report on Community Biogas Pilot Project in Fateh Singh Ka Purwa, Etawah (mimeo), February 1978.

Gupta, C.L.; K. Usha Rao and V.A. Vasudevaraju, Domestic energy consumption in India (Pondicherry Region), *Energy*, 5, 1980, pp. 1213-1222.

Gupta, C.L. and K. Usha Rao, Energy inputs for irrigated farming with mixed cropping: A micro-level survey of Pondicherry Region, *Urja*, May 1981, pp. 1-6.

Islam, M.N., Study of problems and prospects of biogas technology as a mechanism for rural development: Study in Pilot Area of Bangladesh, Bangladesh University of Engineering and Technology Dacca, November 1978.

Knowland, Bill and Carol Ulinski, Traditional fuels: Present data, past experience and possible strategies, U.S. AID, September 1979.

Manibog, Fernando R., An economic evaluation of renewable energy systems for Philippines Rural Area, draft, September 1979.

Manibog, Fernando R., Bangladesh: Rural and renewable energy issues and prospects, *World Bank Energy Department*, paper No. 5, April 1982.

Nathans, Robert and Romir Chaterjee, An approach to assessment of rural energy delivery systems in developing countries, Institute for Energy Research, Stony Brook, May 1979.

National Council of Applied Economic Research (NCAER), New Delhi, India, Survey of Rural Energy Consumption in Northern India, February 1978.

NCAER, New Delhi, India, Domestic fuel consumption in rural India, 1965.

NCAER, New Delhi, India, Energy consumption in greater Bombay, 1976.

Openshaw, K., Woodfuel—A time for reassessment, *East African Journal*, January 1977.

Parikh, Jyoti K., Energy and Development, P.U. Report No. PUN 43, World Bank, August 1978.

Pimentel, David and Nelson Beyer with Vinod Mubayi, Energy in Indian agriculture, Brookhaven National Lab. and Cornell University, NY (Mimeo), 1976.

Reddy, Amulya Kumar N., and D.K. Subramaniam, The Design of Rural Energy Centres, *Proceedings of Indian Academy of Sciences*, September 1979, pp. 395-415.

Revelle, Roger, Energy Use in Rural India, *Science*, p. 192, 1976.

Revelle, Roger, Energy sources for rural development, *Energy*, Vol. 4, pp. 969-987, 1979.

Siwatibau, Suliana, Survey of domestic rural energy use and potential in Fiji, The University of South Pacific, Report to the Fiji Government and IDRC, Canada, no data.

Stout, B.A., Energy for world agriculture, *FAO*, Rome, 1979.

Tinker, Irene, Energy Survey Methods: Role of Women, paper presented at the Workshop on *Energy Survey Methodologies*, Jekyll Islands, January 1980.

Tyers, Rodney, Optimal resource allocation in transitional agriculture: Case studies in Bangladesh, Ph.D. dissertation, Harvard University, 1978.

White, Wendy D., Annotated Bibliography of Energy Surveys in Developing Countries, paper presented to the International Workshop on *Energy Survey Methodologies* held at Jekyll Islands, January 1980 (reported in press).

DISCUSSION

MAHESHWAR DAYAL

I want to raise one question here that comes up as a result of the table that has been prepared particularly with regard to agro industries and transport. We must keep in mind one aspect of the whole exercise, that is, finally we want development from the present status to some improved status. Now, should not our surveys, therefore, include some assessment of what newer technologies or newer sources can do to change the situation? In other words, the surveys perhaps should not only be aligned to meeting the existing needs. Existing needs are in fact are at a very low subsistence level in many cases. Our whole objective must be to improve the conditions and quality of life. So it would be very helpful for us policy-makers if we can also have some idea of what can be done with the newer things coming in. We should also see what can be done, what types of industry can come up by a spiral of increased energy availability, and increased use of that energy, to increase the economic availability, and the purchasing power. I just felt this also might be given some thought, because that might change somewhat the type of answer that you may get. If you are only looking for a system that could meet essentially cooking needs, then you get one type of answer. If you are trying also to bring in some generation of economic benefit and purchasing power, then you may get a slightly different answer. The quality and level of energy that you have to supply will also change.

FOURTEEN

DATA FOR RENEWABLE ENERGY IN INDIA

C.L. GUPTA

BACKGROUND

WHEN we started collecting energy data, our primary purpose was to plan our research programme and to know if there is a problem, and if so, what it is, how significant it is and whether renewables can make some impact. When we looked around we found that most of the energy data system models were either too disaggregated to give any meaningful clues on the first look (mostly done by technologists or by those economists who were concentrating on small areas intensively) or they were too aggregated like, e.g., giving how much we consume per capita per year in each state. This too was not relevant for our purpose for they were stating average figures and some of them also were giving standard deviations. To find a reasonable mean, as Prof. Bhatia was suggesting, we must decide what do we want this data for. Since my predecessors have already concentrated on the survey methodologies as well as some of the specific results of these surveys, I will not go into them in any detail. I will primarily concentrate on the specific data requirements for renewables, leaving aside all the traditional sources, which have been covered by Bhatia.[1]

SCOPE

Amongst renewables, we will consider solar, wind and biogas from the point of view of design, i.e., how to match the basic demand data with the requirements of systems design and how to define criteria for integration of different energy sources in different climatic zones either amongst themselves or with the conventional sources. Of the special characteristics which mark the topic which I am going to speak on, the first one is the Indian situation itself and second is the nature of renewables. If one looks at data even in a restricted sector such as domestic surveys[2] and even in it, if one looks at just cooking, there is an enormous diversity between villages, suburban and urban areas, between families in an area, from region to region, and between various income groups. Thus it is quite a problem where to choose the reference level for design of systems. This results from a large host of particular minutae. For example, if one takes just cooking stoves, which we have been concentrating upon, heat efficiency depends on the type of pots used, the type of cooking practice, fuel used, the combustion chamber used and that too in a highly coupled manner. Second part is what concerns the renewables themselves: all renewable systems are required to use variable design point methods. For example, for any conventional or commercial source of energy like coal or oil, it does not matter whether one uses it in the day or in the night. What matters is in what kind of boilers one uses them, what temperatures one expects but otherwise the performance is more or less independent of the cyclic variations in the process as well as the diurnal and annual time cycles. This is not true for the renewables.

FEASIBILITY DATA REQUIREMENTS

These data provide a check list whether one should go into renewables at all, either with respect to the source or with respect to the task. For initial feasibility, if one looks at the sources side, there are thumb rules (not theoretically substantiated) but fairly reasonable in the final analysis. Some of these are :

—One does not consider solar systems at the first go, if

the sunshine hours are less than 2000 per year. In such a case, one has to have detailed calculations before one decides that a solar system should go in.

—Similarly for wind systems, the mean hourly annual wind speed should be greater than 12 km per hour.

—For biogas, there has been a restriction for economies of operation, that single inputs should not be less than the equivalent of four animals' waste, that is something of the order of 40 kg per day at 18 per cent solid content.

In relation to the primary source data, one would be surprised to know that the detailed data are either not available in the right form or not available at all. However, for India, we have had a well staffed and well managed meteorological network and we have one of the most reliable data bases for renewable energy sources throughout the world. However, it is not sufficiently detailed and till two years ago, this data was not available easily.

Now with the support of CASE and with the dedicated work of Mani *et al.* we have three very good quality Handbooks [3,4,5] for solar radiation and for wind data in their primary form. The raw data as collected by the Meteorological Department, is processed in terms of means and statistical variation parameters and is useful for limited applications. The problem here is how to process this data for specific applications in a sufficiently generalized manner so that only reasonable time and effort are required for computation and presentation. Fortunately, such methods are known to an extent and they are being developed further[6] in response to the demand generated by increasing use of renewables. On the demand side, the special data characteristics required for rural as well as for industrial energy systems using renewables, is the need for daily load profiles in addition to aggregated magnitudes. For example, it matters a lot for a solar water heating system, if one takes a bath in the evenings or in the mornings; both in terms of costs and design options. So we must collect the demand data on time basis, that is, when does the load occur and how much. For example, for irrigation for a given crop, one can irrigate in the day or night. If one has to irrigate in the mornings and evenings one will need some kind of storage or standby system whereas to irrigate in the day, one has to have a fine matching of variations in the energy source with demand patterns. When we want to collect this data, the problem has been for how many places to collect and whether the data is specific or transferable for a given agro-climatic region. For the agricultural data, we classified it on land holdings basis and for the domestic data we classified it on the income level basis and used this data from a limited number of sample surveys to predict the total energy profiles for a community: rural as well as urban and suburban communities from the population figures in the Census data. By 100 per cent survey for even a single village, one can prove such norms for a region. I think it would be all right for design purposes to use such data, for which basic norms can be updated from time to time.

COMPONENTS PERFORMANCE DATA

If data is not required for just wanting to know how much energy is used or for international comparison but for designing a system, what do we do? The first thing we should have is the components performance data. There has been a long debate as to what kind of component performance data are required.

There are hundreds of collectors all over the world and there is an international standard test procedure but surprisingly when one gets to design, this standard curve cannot be used directly because it is based on instantaneous data and is primarily used for comparative collector ratings. However, once we have these performance curves and luckily for Indian solar collectors, these performance curves are available—one has to use them to design. Again one has to fix the design value of solar radiation intensity. I will come back to that later. Similarly in the performance of the wind system, the performance equation coefficients are available for most of the rotors for Indian designs but again the problem there is of the cut-in and design speeds. Because the cut-in-design speed is related to cut-out and the mean design wind speed and all three depend upon the power spectrum of the wind at the place. Similarly for the bio-gas, one has to decide the inputs mix ratio between available amounts of agricultural residues and animal wastes so that C/N ratio is between 20 and 30.

CLIMATIC DATA PARAMETERS

For climatic data, one has to find one or two critical parameters in relation to which one can consider the system because one cannot rigorously consider dozens of parameters. For the solar systems, the entire solar design data can be related to three parameters, (1) Monthly mean global daily radiation ($\bar{H}$), (2) The mean monthly day-time ambient temperature ($\bar{t}_a$), and (3) Monthly Clearness Index ($\bar{K}_T$). With these three parameters and certain correlations for radiation statistics—which are now available for Indian data[6] complete solar system design is possible for any location or use. For wind systems, the main problem is how to characterize the power spectrum: if you take eye readings, it is too much of a bother in terms of data collection and processing and if one takes integrated readings, they are not good enough. If one takes continuous recording, these instruments cost thousands of rupees and the network cannot be dense enough. Also for complete reliability, captive power or a reliable grid supply of power is absolutely essential. So whether we can use integrated observations on simple integrating instruments, or hourly observations to predict a complete source profile, depends upon availability of finance and skills and the accuracy desired. For a given agro-climatic region, it is possible to do a similar analysis of wind in terms of the Weibull distribution. For example, for Ladakh, where a Dutch Team came and took spot readings for seven days, values of constants were guessed from similar agroclimatic conditions and checked with the spot observations. Using the wind statistics, they could predict for design purposes a complete annual power profile without having recording instruments and without waiting to have ten-year data. But the problem is the auxiliary data: for example, for solar energy systems one needs maximum wind velocity so that the systems are properly anchored and do not vibrate or flutter or fly away. The village water supply system based on solar distillation at Awania was severely damaged in a cyclone but it was a very heavy cyclone. To have to design for a 200 Kmph wind, the whole system would have been uneconomic. May be, depending upon frequency of such storms, partial replacement costs are cheaper than very heavy capital costs. For all the wind mills, gustiness is the main problem and for that one has to very much depend upon local or spot observations. No reliable prediction methods have been yet finalized to my knowledge nor are measurements easy.

DATA FOR SYSTEM OPERATIONS

Once the systems have been designed, their economics of operation and overall configuration depend much upon peak capacity and use factors. For example for solar pumping systems with liquid output, thermal or electrical storage may not be needed if pumped storage is provided and peak capacity is reduced. For air systems, however, no heating up time is required and radiation transients are immediately reflected. But for power systems, where PCM or hot liquid thermal storage is common and system start-up and shut-off times are considerable, the continuity of radiation data has to be known for determining operating performance. In such cases, frequency analysis for exceeding thresholds for a given period of time has to be done. For life cycle costs, operating life and replacements costs are needed and these are at the present a rather weak link in the renewables data chain. To sum up, the crux of the matter is not the availability of most sophisticated data but rather the most sophisticated processing and analysis of the nearest, reliable and easily available data is required. It is not possible to have too sophisticated data for we have neither the personnel nor have we the instruments in the numbers required. This is true for all developing countries as large as India.

DATA FOR SYSTEM INTEGRATION

To use the electrical terminology, we do need not only the peak load but also the peak to base ratio and also the time when the peak occurs. To give an example, if we have lighting loads within the time when the monsoon regional winds are available, it is possible to have an integrated system using solar energy in the day and wind from three o'clock onwards. But if in a place, wind drops after 7 or 8 P.M. and the use hours are afterwards or if industrial loads are in the night, one just cannot have this configuration. Then one has to consider storage or auxiliary systems or one has to have a different integrated system. All this micro-analysis is necessary and this cannot be helped. In fact, one of the reasons why renewable systems are not popular in the country is because very few want to do this micro-analysis. Wherever this micro analysis is not necessary, like in solar water heating systems, they are going on and they have been put in hotels, hospitals, factories and households.

DATA FOR SYSTEM ECONOMICS

Apart from sources matching with load profiles, the integration is mainly governed by economics. One must calculate what are the pay back periods in terms of capital costs and life cycle costing and what is the net energy saving. Again to give an idea, if one uses an aluminium collector the energy pay back period is much longer than if one uses a steel collector. Similarly the ratio of the energy pay back periods is 1 : 2, if one operates the collector in a thermo syphon mode where one does not use energy, as compared to an active mode where one uses pumps. So how one operates the system and what materials one uses in the system also determine cost and energy pay back periods. These have not been largely determined in our country because we are still struggling to cut down the economic pay back period from 6 and 7 years to 4 years as most industry people do not like to consider more than 5 years. As far as rural people are concerned the problem is where they need the energy. The only renewable source really viable is biogas and may be producer gas, if technology is developed well land social forestry takes roots.

However, for new sectors like rural industrialisation, costlier systems like solar pond and photovoltaics may be considered in the next 5 to 10 years. For example, it is possible to have 10 to 50 kW rural power stations either based on solar thermal concept using ponds and organic Rankine cycles or biogas based duel fuel engines or bio-mass based producer gas engines. This is an area where we can make an impact. But a clear and decisive will is lacking so far.

CONCLUSIONS

I think, it is now imperative that we give a real professional basis to the implementation of renewables both for rural energy systems and industrial energy systems. And what is necessary is not so much sophisticated data but a reliable energy data base and good processing methods based on clear understanding.

REFERENCES

1. R. Bhatia, *Traditional Energy Sources*, Chapter 13, this book.
2. C.L. Gupta *et al.*, Domestic Energy Consumption in India (Pondicherry Region), *Energy*, 5, pp. 1213, 1980.
3. A. Mani, *Handbook of Solar Radiation Data of India*, Allied, New Delhi, 1980.
4. A. Mani *et al.*, *Solar Radiation for India*, Allied, New Delhi, 1981.
5. A. Mani *et al.*, *Wind Energy data for India* (under publication).
6. C.L. Gupta *et al.*, *Radiation and Wind Statistical Corelation for India* (under publication)

DISCUSSION

SHAH

Could you say how far the 50 kW local power plants based on the three modes which you have suggested, are practical?

C.L. GUPTA

There is a discussion of this in my paper on the rural energy system at the back.[1]

GHATE

A windmill has been put up in Ghazipur in U.P. which is one of the lowest wind speed parts of the country. They got data from the Varanasi Airport and although it was useful in a very general sense when it came to putting up the windmill at Ghazipur which is about 100 km away, they found that one needed very site specific data. The presence of a few tall trees in the vicinity caused completely different results. Not just the wind, but for most technologies, you really do not know until you actually start implementing a project what kind of data you are going to need and as you start implementing it, you generate new data requirements which then feed back into the design of the project. This might represent a more realistic approach than starting off knowing exactly what you want to do.

C.L. GUPTA

There must first be the feasibility data. At Ghazipur the rotor was specifically designed for cut-in wind speed of 6 km/h. But the unfortunate fact is that the same rotor is now being advocated for use all over India. It will work. There is no question. We will have some wind mills to show, whether they are the optimal wind mills for the Madras Coast or the Sunderbans, one cannot say. Second simplistic approach is to put anemo-meters everywhere. But people may not read them. The anemo-meter data needs processing, and this is done only at the airport. So the thing to do is, like Dr. Kirk Smith did, to have some kind of battery assisted anemo-meter, and take spot observations day and night. I know that the Dutch team did this from 4 o'clock in the morning till 10 o'clock in the night in the cold of Ladakh. Once you can correlate this with that month's averages at Varanasi, you can get a power spectrum month by month. But you have to use the Weibull distribution.

DAYAL

If you talk about implementation there is a very genuine problem of time. If we have to wait for every possible village to be surveyed and within the village every site to be surveyed, then we will never get started.

C.L. GUPTA

I am not saying that we should not start. In fact I have been waiting and trying my level best for 20 years for it to start. The first level is to study the operational problem: whether the wind mills can be put up, whether they fall down or crack up. This we have done and we are doing. I am thinking of the second stage, that is, if you take a wind mill specifically designed for low cut in wind speed, which was typical of Ghazipur, it will also have a low cut-out speed, and the wind energy that you get is proportionate to the third power. So you might be losing all the higher power wind. So, may be, for example, in areas like ours where the average wind speed is 12 km/hr, we should have either a better rotor design or we should be conscious of the fact that we are using a wind mill which is safe and tried but which is not giving the maximum output.

DAYAL

Optimization will take time. That has to be noted.

MUBAYI

I want to ask Dr. Gupta one question. In the villages that you surveyed, you have got all the demand side data and the various availability data for the renewable sources. What is your conclusion for this group of villages, as to what are the feasible set of technologies both from the logistic stand point and from an economic stand point of meeting those demands?

C.L. GUPTA

You have really hit the nail on the head. To be very honest with you, in the present situation of incomes, the matching is very poor. But as I said, cooking, fertilizers, and irrigation are the three major areas which are consuming the maximum energy. When I found that solar cooking was out of the question for our type of system but that 90 per cent of energy was being consumed for cooking, we started a project on improved cooking stoves, in addition to working on solar cookers, which was really not my area. It has been effective and we have spread the first level design, like the Ghazipur design, in 100 households and found out their reaction, and as Dr. Bhatia has said, there was increased time (required for cooking). Incidentally, this increased time was brought to my notice not by our people but by the villagers. I have recently been to an International Workshop on Stove Testing and we found that this was a problem in many places.

The second thing we found was that for solar irrigation, it is very important that we improve the irrigation efficiency first, both of the distribution and of the pumping systems. Sometimes the hydraulic efficiencies are as low as 30 per cent and if the energy costs are high, there is no point in having solar energy for a pumping system which has a 30 per cent efficiency and a distribution system which has 50 per cent efficiency. I think the solution lies in improving them.

For fertilizers, we should have some kind of integrated

farming system, and wherever possible to use bio-gas fertilizers as well. The problem there is, I found from the limited survey we had, that a farmer was not willing to use more than 25 per cent of his land for fodder. That limited the number of animals, and the water requirements are heavy. So working backwards on that basis, one can calculate for integrated farming, the size of land one can afford, and what percentage of the manure one can get. We have worked this out not only for the farming system but also for the residential community system.

Similarly in the solar water heating system, the problem is not the sun, the problem is usually the water, at least in our climate. All solar water heating systems designed and sold commercially assume there is a water supply. There is no water supply in most of the villages. I found to my great surprise even in South India where there is so much heat, people do use hot water, but at the moment they heat water from the residual fire of the inefficient cooking stove, so they do not spend any extra energy. So if you are going to make an efficient cooking stove, you have to provide the incremental energy either in solar form or in some other form. But then you have to be very careful what kind of system you give them. Very very careful. It is a constant learning experience. You feel overtaken, you reach the frontiers of your knowledge very soon.

DAYAL

Perhaps we can ask Dr. Moulik to continue the discussion.

MOULIK

Having done the data collection and acquisition myself, I could give some anecdotes of field experiences. Before I do that, I would like to respond to Mr. Shah who raised the question of power plants. For GIDCO I had calculated the costs of power plants for bagasse wind and solar systems. I found that they are more costly than what was supplied by the Rural Electrification Corporation, which itself is a losing concern.

They are not economically and financially viable at the present level of technology and investment in Gujarat.

With all the bagasse in Sangli in Maharashtra it may be possible to feed power into the central grid as in the Hawaii model. With the supply position there, there is a possibility of this.

I will give you another anecdote. The man who is in the middle of action, Dr. Dayal here, one day asked me abruptly, "Tell me, what is the impact of bio-gas on the landless, cattle-less people?" At that time I gave some answers off hand. Then when I went back to the macro level data, of dung consumption and also looked into data for 20 villages of Bihar and Punjab. I found a lot of variations. I found that the aggregated macro-level data does not apply to at least these 20 villages in Bihar and Punjab. There are a lot of regional variations. And my calculation showed clearly that to invest money to collect gobar or the dung for a farmer will be much more uneconomical than to give it away to the people. This is the situation at least in Gujarat. Punjab does not use much of dung in any way.

Someone once asked me, "Tell me, what is the market segment where I can sell the solar cooker?" Now we are talking from the manufacturer's point of view. There, because of my knowledge about the data which I was gathering for these 20 villages plus the Gujarat villages, I could immediately answer, "You do not have much of a market in rural areas. If you really want to sell, here we could sell 700 pieces of pumps per year, but the market for solar cookers is in the urban and semi-urban centres".

Many of the renewable energies including bio-gas, have direct impacts on the life style, which might not be easily acceptable. Ramesh mentioned about the time taken for cooking with the stove. I happened to be in a village where some commercial agencies were selling and trying to popularize the stove, and they were saying, "Lady, it will save you a lot of time which you can spend with your children". The lady said, "I do not want that. I have 12 children, I have had enough of them, and do not want to save time". This gives me a very clear-cut message, that the motivations to go for this, is different for different kinds of people. Someone at

Jyoti once told me that people would buy bio-gas plant at any price, the way the world prices were going. But the commercial agencies going into the field suddenly found them not selling at all. Hence market segment data are required very badly. I fully agree with Dr Ghate that this kind of generalized data is all right for the initial stages. But it is very site-specific data in each one of these cases of renewable energy that will give us the proper feedback.

I will mention two other instances and then come to the substantive issues. The Punjab Government gave 35 licences for the paper and pulp industry and Mr Brar, ex-Governor of Punjab asked me: "now that I have a lot of wood stock and paper mills are coming up, could you tell me the kind of pricing and supply system I should work out?" If you take all this for paper and pulp or use as energy form, where will you get the fodder for your livestock? This will hurt the whole of the Punjab very badly. No work is done on this. Some people were saying that the cotton stalk which is being used now could be used as well. In Punjab the cotton stalk is a big problem. Transport costs make it uneconomical to go and market cotton stalk and it stays there for six months. It is voluminous. In any case people who use cotton stalk feel that is the best use for it. There is no point in replacing it by any other. This we must learn.

What Dr Dayal mentioned is a very important point. If you are talking in terms of static systems, at the present level of consumption, I do not think you will get a real picture for the future of the renewable energy programme. When you are talking about whether to completely substitute or replace, at least my calculations show that there is no chance of completely substituting or replacing so called commercial energy. Fertilizer, you cannot; in certain cases electricity, you cannot. But the most important point is that it is to be looked upon as a dynamic growth situation. When I looked into it with the data for 20 villages in Bihar and Punjab, I found that early on the energy inputs will give you more in terms of energy in the agricultural system. The scenario has changed completely. For an incremental input of energy into the system you are not going to get an equivalent amount because there is a diminishing return. You are not going to get a similar amount of energy from the output system.

Now my calculations in Punjab and other villages are as follows: if you want to have five per cent growth in agriculture, you have to put in at least 25 per cent extra energy input. This is what we have calculated from the data for the last two years. In Bihar, if you want to have five per cent growth in agriculture in the present difficult situation you have to put in around eight to ten per cent more. Dr Hopper of the World Bank used this data to say that you require more unless there is genetic engineering.

But coming to the necessity of the data, I would like to talk about the three things which are required:

(i) I think you need one set of data to plan what kind of energy mix you should go in for in a particular region, including commercial energy. But you also need to know how these renewable energies are performing, which means a data system which will flow as a follow-up measure, and as a management information system. With the bio-gas system, you did not know what would happen, as also with wind energy, and with solar energy. In the textile industry, you spend about three to four lakhs to study the feasibility of solar energy technology, but no one says that it saves about 23 kg of coal. For a textile industry 23 kg of coal in a boiler system is nothing. So commercial people would not go for that;

(ii) The second thing is most important: a good analysis of the existing systems in terms of economic viability. You can go on testing feasibility and other things besides but ultimately it comes to the price signal. Because the problem is a national concern, the question is how to translate it in terms of the consumer. Besides the problem of subsidising and trying to do the pilot testing comes the question of what should be the organizational role, who would collect what? There are certain data which the manufacturing or the supply agencies will collect. Which are the agencies? Who will do what kind of job? Because government will do only certain kinds of jobs, but certainly it cannot supplement everything—the organizational and the institutional parts are most important.

(iii) Lastly I want to mention about an evaluation I did for social forestry, which is thought to be of one of the major sources of fuel energy. I was shocked, and so were many in Gujarat. Within four to five years, two per cent of good irrigated agricultural land has been replaced by eucalyptus plantations. The consequences of this are serious. None of the eucalyptus plantations are going to be used for the fuel needs of the rural people; the wood is going to the pulp, as well as the building industry. One must think of the kind of data we would need to monitor the social forestry or energy forestry system as any other energy system.

DAYAL

Questions?

KHURANA

I wanted to ask one question of Dr Gupta. Are projections available which give for the near future years what proportion of the total energy will be supplied by renewables? the order of magnitude of costs involved? and the most typical types of areas of applications in which these are likely to succeed?

C.L. GUPTA

About the fraction it can give in the near future, it depends upon a number of things. Somehow I do not make these large scale projections, because I am working in the hardware and trying to put up as many systems as I can. I think Dr Dayal would be able to answer that question better.

For the second part, for the water heating system, the pay back period is of the order of 4 to 5 years with the government incentives, which is fairly reasonable. I can give you the capital cost which is something of the order of rupees 25 to 30 per litre. One can also work out the cost per joule for air heating systems, which have no storage need, and are economical. But for the industrial system, where you want 24 hours processing or reliability, air systems supplement the existing systems. They are not replacements. In that case the economics depends upon what quantum of renewable fraction you put in that particular process, it has to be calculated for each case. There can be no generalization.

One of the things which I would like to add here, particularly for people who deal with industry, don't get taken in by some solar fellow telling you that "I can give you 80°C here." What you need is, if you have a supplementary existing system like oil-fired, coal-fired, or air-fired, try to get the maximum solar fraction at a reasonable cost per unit input. Because for coal or oil, it does not matter, whether you heat from 50°C to 100°C or 80°C to 100°C, but in solar it matters. So I think it is important without becoming too technical in terms of the second law efficiency, energy quality, etc., that when we are taking a solar input, we must take the maximum solar fraction, consistent with cost pay back period, and a system in which there is net energy saving. We are investing so much energy in making the solar systems themselves. There is no point actually. It will only shift the energy mix. It will not save energy on a national basis. It think all these should be considered at some point.

For example, I would very much wish the industry people who have plenty of manufacturing capabilities, to engage a consultant and manufacture their equipment on site. Even though it means a little more load, this will save in cost, and in maintenance. Also it does not make sense in a country like this that we should be transporting large solar systems, unless it is done for demonstration purposes, use bio-mass to pack them, oil and coal to transport them, where we are not able to transport coal for the power system.

MAHESHWAR DAYAL

Dr Gupta has asked me to answer a question partly. A short while ago, we did a very detailed quantitative study for the type of contribution which renewable energy can make. This question was in the minds of many policy-makers. We found that there were of course several constraints in the whole business. The rate and the amount of contribution of renewable energy system will depend a lot on our fiscal, economic, technical and other policies, and also on the rate of technological development. It is an exercise in scientific crystal-gazing. The quantitative figures that we came to was that by the year 2000, provided we do give the necessary priorities and the necessary sense of direction, to meet the increased energy demand which will be of the order of almost double the total energy consumption in the year 1978-79, we would need almost 49 per cent by renewable sources. This would include 12 per cent from hydro which, of course, is a renewable source, having a conventional as well as a non-conventional component. There are conventional hydros which are known and which could be expanded. We could then give additional stress on non-conventional hydros, smaller hydros, and so on. Then 16 per cent would come from the traditional renewable forms like social forestry and other forms, but done more scientifically, so as to give us a truly renewable source, because currently some of the renewable sources are being used in a manner which in fact makes them non-renewable.

For example, wood is a renewable source of energy, but if you take the macro picture in India we are cutting wood much faster than the rate at which trees are growing. At the moment on an overall basis it is not renewable but we could make it more renewable by development of scientific methods, and new technology. Then we found that almost 21 per cent could be supplied from other sources for uses which otherwise at present are based on oil, coal, and electricity. This is a sort of projection that we came to. There are several assumptions in this but it does give you one type of figure to which we came, that the contribution in the long run, say in about 20 years, is very significant. It is not something peripheral. It can be significant provided the necessary decision is taken now and the necessary priorities are given now. Forty-nine per cent is fairly substantial. We also felt that this kind of contribution will increase gradually. In the first five to ten years, it will be obviously slower as time is taken for these new technologies to develop and cost to decline, but later on it could grow more.

But I think for all these things we need the necessary data and, as it has been mentioned earlier, we need the economic viability of these systems because we cannot obviously do it only on the basis of just subsidising these types of energy sources. We need really proper data on a rational basis. Given these data, several things can come up.

KHURANA

Most interesting. Could you touch upon the type of application, because if you say 49 per cent it means that it is almost equal to the energy consumption level in 1978-79. It is very large. Therefore, it will be interesting to hear a little of the main areas of application in the next 15 or 20 years.

DAYAL

I think they are there in the papers. As I mentioned, for hydro power, the application is in the form of electricity, which comes up to about 12 per cent of this total. Sixteen per cent is in the form of fuel wood; its use is going to continue, we cannot really cut it down very much beyond that. Nor should we, for the simple reason, there will be a large percentage of our population which will simply not be able to afford other forms. The other forms which are of very much interest and are likely to get developed very soon are the applications of bio-energy in the form of bio-gas, bio-mass, both direct as well as in a converted form. These could be used for cooking, heating, lighting and also for localized power (electrical and mechanical) for some purposes. Then solar systems, we thought, will have a greater impact in those forms of application where you need energy in the form of heat, particularly at low and medium temperatures. It was very interesting, we found even today almost 20 to 25 per cent of even our conventional sources of energy, are being used to supply heat which is required at less than 200°C. Now all this represents a very interesting and potential market which can be met relatively soon where we do not need to concentrate sunlight too much. There is a cost wherever you have to concentrate. Finally we do feel that some of the solar electrical devices, particularly portable types, will come down to meet a fair amount of these requirements.

SENGUPTA

Could I address a question? I do not really know whom to address, Dr Bhatia, or Dr Moulik, or Dr Gupta. This is about some difficulty which we have been facing and whether they could offer some suggestion. One is with regard to this question of aggregation. I will take a specific case of Karnataka State. There are 19 districts, 175 talukas or sub-districts, 27,000 villages. Now on what basis do we take a sample for any survey and particularly for rural energy consumption. Do we go on the basis of geography or rainfall, or the amount of incident solar energy, and so on. We started off with a district, where we sampled two geographically similar villages. When we started going into the details, we found that the data were very different, and it would have been wrong to have come to the same conclusions about the two villages on the basis of one sample. That is one point, one difficulty which we have been facing with regard to the question of 'some' place, and having made a detailed survey of two villages in fair detail, which are close by, we find totally disparate data.

The second question is of how to go and do the survey and this is particularly addressed to Dr Bhatia. We found that it depends on the personnel. I had a totally different experience when I went with three or four project assistants of mine, stayed there, and went into details. So far, with the help of the planning secretariat, we have covered 29 villages where the services of students on vacation were made use of. When we started analyzing the data that flowed back, frankly speaking, we were not very sure whether they had been taken at the village, or cooked up elsewhere. The moment you start to look at the data, you have an uncanny feeling that things are not all right. It is not physically possible for any investigator to go round and look into these places for himself. I would seek advice from those experienced in the line, because the result of the question depends on the person who goes, talks, in what manner he talks to the people, and how he fills in these questions. These are the two major difficulties which we are facing. I would very much like some response and benefit from the experience of the people who have been involved in this.

C.L. GUPTA

I can give you the response of an amateur. The first time you go to a village and ask them questions, they think you are crazy and a government man wanting to trouble them afterwards. We found it convenient to send two investigators at a time, prepare a good questionnaire with the help of economists, because we did not know how to prepare a proper questionnaire, and take the help of a local school child who makes your task easier. We spent two hours in each house, one person writing and the other informally talking. Then we followed this up, by going a second time to those very houses. They are being asked a lot of questions; they feel nobody is doing anything, and feel very sore about it. Sometimes I feel if I were a villager and everybody came and asked me questions, I might ask them to get out.

SENGUPTA

That is not precisely my point, as to how to approach the villager. All that I am trying to say is that if you want to disperse widely and get the information, it is not possible physically for you or just two or three handpicked people to cover all the ground and you have to spread it over a large number of people. That is when you start finding that a few are doing the job well, while others are not.

C.L. GUPTA

You have 10 or 20 teams of people for that, you have to train them and motivate them. It is a human problem.

SMITH

I am of the view that perhaps the information is going backwards. Unless you expect to operate all these wind mills, plantations, and so on, it is the people in the village that have to know about them, not you. They already know how many twigs, different species of wood they use, but that is not important to them. What is important to them is what are the characteristics of these technologies, how they interact with their lives, how can they get them financed, how do they act, what is the risk in terms of reference they can understand. After all it is the household that is the basic economic unit, you do not try to control every household, instead you get the information to the household so that they can make their own decision. What seems to be leading to an extreme is the planning system where down to the last cooking detail is planned in Delhi.

BHATIA

I would just make my position clear. The purpose of the survey is not only one way, the purpose is really to know how the different households would be affected, how they view the different technologies, etc. You can, of course, go and install them without caring about how they affect the different people. But the objective is to evaluate the social profitability and acceptability, and economic viability of the project from the household point of view because all our experience is that if you do not worry about this, then the technology is not going to be accepted. This has repeatedly happened in the past. We generate this information partly from the planning point of view. We are interested in this because we are worried if the households do not accept them for their own considerations, financial risk, their own time of cooking, etc., then this particular programme is a failure. It is the concern for the household, the way they view all these things, that we want to have more information. Otherwise we can continue in the way we have been continuing in the past—just go and put up these things and they do not work. In the design stage we do not take care of these facts. The concern for the household is the primary. aim. The planning is to be done from Delhi for that household.

RAMASWAMY

If you can discuss bio-mass, which is the excreta of animal, you should also think about the animal itself. Why not look at the animal first, and the excreta afterwards? Secondly, I also want to say that this animal energy cannot be replaced by all the 13 renewable sources of energy otherwise being discussed. For cultivation you cannot use solar energy, you cannot use hydro energy in 200 million locations. You cannot bring in wind energy. You cannot bring in bio-mass energy. Therefore, there is no other way for the next 30 to 60 years to replace animal energy being used for cultivation.

On animal power, animal energy, animal wastage, under-utilization, mis-utilization, losses, ignorance, there is no data at all. Even after 50 years of existence the FAO which produces an elegant handbook does not have a column for draught animal power. Out of the eighty-eight countries, only six countries have some data on animal power.

The draught animals alone are worth 100 million dollars. Replacement is going to cost you 250 million dollars in capital equipment and tractors. It is an appropriate source of energy for cultivation. Fifty-two per cent of the area in developing countries is being cultivated by animals, 22 per cent by tractors, 28 per cent by manual labour. Even in the advanced countries the figure for animals is 11 per cent. Animals are not only ploughing for us, but they are also used in advanced countries as well. To begin with all these are Ramaswamy's guesses, which can be partially wrong, or even grossly or totally wrong. Yet, they are worth looking into.

The efficiency of the draught animal system is hardly five per cent. In India we are using animals only 80 to 90 days in a year. That is 110 million animals are used only for a maximum of 100 days in a year. China uses them for 300 days. There is a potential to increase the use. Philippines, Thailand, Indonesia and countries not knowing this, are using them only for 50 days in a year. They are not using them for carting. China uses them for carting as well as for ploughing during the season. But in some countries where they do not use animals for carting, they come in a pick-up truck with a buffalo behind, and use the buffalo for ploughing because tractors cannot be used in all places.

Moreover the beating of animals in India alone is costing us at least 30 billion rupees by way of damaged skins. The skin is ruined by the time the animal is dead. Consider slaughter. Official statistics show 1.79 million, but the actual figure is 40 million. Government counts only those which are slaughtered in official slaughter houses where only five per cent are killed. The rest are killed elsewhere. India needs double the energy input, says the National Commission on Agriculture. Where are we going to get this double energy input? After all we, the Third World countries, are already importing 80 million tonnes of foodgrains now. The minimum forecast of import by A.D. 2000 is 100 million tonnes and a reasonable estimate is that we are going to import 150 million tonnes of foodgrains. But where is the energy going to come from? Even assuming Dr Moulik's figure for every five per cent increase in production, you need 25 per cent increase in energy, where are we going to get all this energy from? Even in a small matter of this kind, we need data base. At a time India has a shortage of energy, 20 million marginal farmers in India have no animals at all. In most of Africa, 80 per cent of the energy is from manual labour, not animal. At a time like this, in India eight million male buffalo calves are being killed at birth. Ten days is the life span of a male buffalo calf. How do we know? There are 35 million females, only eight million males. Only technology of disposal is different—in Maharashtra, poisons are used, some kick them to death, some allow them to die by themselves. Even if you take Rs. 100 as the cost of a buffalo, that itself is Rs. 800 million. This is a staggering figure.

There are 40 million animals that are being slaughtered in India today. I want you to look at the animal, not just as an energy source alone, but as an integrated milk system, meat system, social forestry, lucerne tree, animal husbandry, rural transportation, bio-gas plant, all these systems are one single system. Today they are all segmented. Tractor people are advertising: 'We can replace 20 bullocks'. That is not the correct way to look at it. The way to look at it is how can you select your breeding, dual purpose animal, how can you combine the milk and draught power animal system, raise them and slaughter them in the end. The average cost of an animal is Rs. 200 at the time of slaughter. Whereas the modern slaughter house is paying Rs. 1000 per animal. Therefore, Rs. 800 is the loss per animal, per farmer, which is a very big figure.

I have counted and weighed 150 animals at the beginning of their journey to the slaughter house and weighed them at the end. They lose ten kg of weight on an average. That loss itself is enough to cover the cost to transport them by truck or train. The Indian government which subsidises every form of transportation, is not prepared to subsidise the transport of animals to the slaughter house.

Three to four years of their working life is lost due to bad harnesses. If we can improve the harness, which is easily possible, we can get four more animal-years, that is Rs. 400 more per farmer. China does a much better job. There they use the animals for 300 days in a year. They are feeding donkeys. They have got Professorships for Donkey Breeding. They are using the donkeys for carting 500 kg of weight. We are using the donkeys only to carry 50 pounds of clothing, 30 of which are dirt whereas their donkeys, even little ones, can carry 400 to 500 kg Donkeys can be used to provide energy instead of the women in Africa and many parts of India also.

There is enormous scope for breeding, health care, credit, marketing, utilization, better harnesses, hitting devices, methods of transportation, methods of slaughter, the sledge device, for example. We use the animal only at the age of three, some other countries use them at the age of six months. It is all due to a better harnessing device with a collar. We can have many devices likewise and utilization of animals can go up 500 per cent. I am dreaming of a day when animals do not have to work for men.

Things have changed. It has taken 12 years and the ILO, the UNIDO, the FAO, the IFAD, and the World Bank, now have projects on draught animals. Animals are not going to compete with any form of energy. They are complementary, and supplementary. We can withdraw the day man can find a different way. But there is no other way because fragmentation will continue. We have numerous small farms below two acres. We have to provide energy for cultivation. Combines can come. You can have *chakkies*, you can have any amount of mechanization. But you cannot avoid cultivation, the basic purpose is food. Therefore, do not give the lowest priority of all to animals.

DAYAL

Professor Ramaswamy has touched on an area which is really rather important. Of course he says there are a lot of doubts and cynicism in this area but the fact which I want to assure him is that in so far as some of us are concerned, we do look at all the various types of sources of energy. It is a fact that in the Third World, particularly in India, animal power is very important especially for rural areas. We had a very detailed discussion on the question of animal energy and animal power in Bangalore in July which led to a lot of data and information. I think, the proceedings of that seminar will be available to everybody. So I do not think we need go into that in detail. But I do think that the subject should be considered with the importance it deserves, as we all realize these things are in fact all complementary. There is no unique solution to the energy problem either in the rural areas or in the urban areas. We have to make use of all the various forms but use them in the most optimal fashion. Animals certainly have their role to play at the present as well as in the foreseeable future. Of course, listening to Professor Ramaswamy, it is quite clear that there is hope for the animals, he is the vocal advocate for them. I think this will be an encouraging evening for all those who love animal power.

REFERENCE

1. C.L. Gupta, Energy for rural development in *Renewable Energy Source for Developing Countries*, Heliotechnic Press, 1981.

TECHNICAL SESSION 7

ENERGY DATA SYSTEMS FOR POLICY AND PLANNING

Chairman : NITIN DESAI
Theme Paper : ASHOK DESAI
Discussants : K. MISHRA
KIRK SMITH

FIFTEEN

ENERGY POLICY AND DATA SYSTEMS

ASHOK V. DESAI

INTRODUCTION

POLICY may be defined as management on a national scale, and planning as policy-making with a time dimension. Investment in energy industries such as coal mining and power has such long gestation periods that planning for them becomes indispensable. The scope of energy policy will vary with the range of responsibilities that a government can or will undertake: in some centrally planned economics, for instance, governments assume total responsibility, while in some of the less developed countries energy planning, if it exists at all, extends only to limited urban areas. But wherever there is a policy, wherever there is planning, there must be information if the chances of making mistakes are to be minimized.

The information required by policy makers is generated by or comes from producers, consumers, the executive departments of government, and official and non-official statistical organisations. The bulk of it comes from the producers and official organisations, and the more information is produced by one, the less is required to be produced by the others. In other words, the better the data produced and supplied by the producer, the less the government has to do to collect it. The policy makers are lucky if the producers use sophisticated management techniques which require good statistical information. If they are unlucky, they might be faced with producers who do not require any statistical information to carry on their business, in which case the government will have to collect the information. We have all types of producers in the energy sector. For example, oil and power are industries whose management techniques are sufficiently sophisticated to generate a wide variety of detailed statistical information for production control, which can be used for policy-making as well. On the other hand, coal mining, or the type of coal mining we have in developing countries, is really a labour-intensive business, and a small-scale business which does not require any statistical information to run it. Rural energy is an area where no information at all is needed by the producers and the consumers and, therefore, whatever information is required for policy-making has to be collected by the Government.

Coal

I shall, to begin with, try to show how far the data which has been described in the papers on the energy scene are relevant to policy-making. Let me first take up coal. The coal industry started in a big way in this country because of the coal requirements of government-owned railways. The railways were an extremely powerful organisation; they controlled the development of the coal industry until recent times, and it was they who led to the organisation of the basic coal statistics from the late 19th century onwards. They were joined after the First World War by the steel companies. These were the major users of coal and they insisted that a certain amount of statistical information on coal should be collected and published. They also, from time to time saw to it that study groups and commissions were appointed to look into the problems of coal resources and reserves; this is how we have inherited the statistical systems that we have today.

This system was supplemented in the Second World War by the Coal Controller's Organisation. Coal came to be rationed and for this purpose the office of the Coal Controller was set up which started collecting information chiefly on the priority coal consumers, that is, the railways, steel, power and cement. This is how we came to have the first coal despatch or off-take statistics. The office of the Coal Controller itself lost its importance in the late sixties when a surplus of coal emerged, but the statistics collected by the Coal Controller have survived. Today we have two basic coal data systems.

In the early 1970s the coal industry was nationalized. Only then was it realized how doubtful were the statistics being collected by the two coal organisations. The output was being understated to a substantial degree, the reason being that there were two mechanisms for fixing the prices of coal and the wages in the coal industry which were quite unconnected. They worked in such a way that there were a great many high-cost mines which simply could not have made a profit if they had earned the prices they were supposed to under the controls and paid the wages they were supposed to pay under the wage regulations. So they used to pay wages below the minimum and to sell coal outside the control mechanism. In other words, there was a large proportion of unrecorded coal output which came to be recorded with nationalization, and a big statistical jump in coal production in the early 1970s.

So with nationalization, the quality of coal statistics improved, but certain problems still remain. In particular, the management system of the coal mines remains the same, especially in the smaller and more backward coal mines, as it was before nationalization, and the quality of the statistics received from the small coal mines remains unsatisfactory. It is also possible that the earlier operations with unrecorded output of coal which was sold outside the system still survive to some extent and the output is underestimated. We have also a new problem after nationalization, namely, that the coal mines are now owned by three or four separate organisations : most of the coal is produced by Coal India; Singareni Collieries produce some; and the Tata Iron and Steel Company and the Indian Iron and Steel Company have their own captive coal mines, and lignite is produced by the Neyveli Lignite Corporation. Today we have a great many statistics which refer only to Coal India which controls 90 per cent of the output but which nevertheless is not fully representative of the industry. It is a nuisance to have detailed statistics which are nevertheless incomplete.

The major problem with coal statistics is that the management of the coal industry still remains such that it does not require the sort of quantitative information which could be useful for overall planning. Therefore, the management itself requires and produces only a small range of statistics, especially statistics about output, manpower, accidents, etc. The rest of the statistics are really produced as a byproduct of a very large and diverse administrative machinery some of which is not necessarily concerned with the coal industry at all. Consequently, the rest of the statistics are quite out of date. There is a strong output-orientation in the management of Coal India, so we have fairly good output statistics, and no market orientation, so market statistics are virtually non-existent. For example, the only information we have about the types of coal is by grades according to which the prices are fixed. We have no idea what sort of coal is required by different industries, or would be required or demanded if they had free choice. We have no idea how the quality composition of coal has changed. While the coal output has been increasing, its average calorific value has been falling and one really does not know how far it has fallen, what is the average calorific value of coal, what sorts of coal are being sold in the market. There is an intermediary in the form of the Linkage Committee which allocates coal. In my view the coal statistics could be improved in a number of directions but they are unlikely to be improved until the management of Coal India in particular is modernized.

Oil

The first oil statistics began to be collected when a domestic crude oil industry emerged in the early 1960s. There was a little bit of oil being produced in Assam earlier, but major quantities of oil were discovered only in the late 1950s and began to be exploited in the early 1960s. This gave the Government an opportunity to regulate imports on the basis of domestic production. This is how statistics of production, imports and offtake came to be collected. They became more systematized by the late 1960s : a fairly easy task because most of the refined oil was produced by multinational companies with good data systems and the same data systems were adopted by the public sector oil refineries. So by the late 1960s there was a fairly standardized set of output and offtake statistics available for oil products, and this data system has survived to this day.

However, the late 1970s have brought some major changes in the oil industry. For one thing, oil exploration and oil drilling have become much more important, and much more capital-intensive with the discovery of offshore oil. Investments in oil drilling are today larger than investments in oil refining. But this is an area on which very little data are collected or published. The sector-wise details of offtake are still inadequate and until it is known how much of the oil products goes to agriculture, to small industry, and for road transport, the breakdown available for the offtake will remain inadequate.

Finally, very little information is available which would enable us to say anything about the efficiency of the oil production, refining and distribution systems. The only exercise I know that has been done was done by Ramesh Bhatia about seven years ago on inter-refinery planning, when he encountered enormous difficulties in collecting the data. Such data are still not available even for a few refineries.

Electric power

The electric power industry was nationalized in the early 1950s; at that time a data collection system was set up which has survived to this day. It is an extensive and comprehensive system and since the early 1960s we have statistics which cover capacity, output, inputs and distribution in some detail. The electricity boards in this country are not very consumer-oriented; hence they do not have much information about who the consumer are, or how much they consume. The quality of the offtake statistics has been deteriorating with the distribution of more and more electricity without meters as agriculture has emerged as a major consuming sector. A fuzziness has arisen at one end of the distribution spectrum which is difficult to remedy.

Here again, with the increase in investment costs, investments in the power sector have increased enormously. For the management of this investment, we require data which do not exist and whose lack is clearly felt in the long gestation lags and the cost over-runs. There are also no statistics on the quality of power supply which is highly variable and in some cases, especially in the rural areas, extremely poor. We have no idea how poor it is and, therefore, it is difficult to know how much scope it offers for improvement.

Biomass

There are so many producers and consumers of that it would be unrealistically costly to try and collect actual statistics for this sector. The best one can do is benchmark surveys, which have to be sample surveys to keep down the cost. There have been a number of benchmark surveys. There was an official survey, the National Sample Survey, in 1963-64. It was repeated in 1973-74, but its results are still not out. A number of surveys have been done by the National Council of Applied Economic Research, the latest being the one done in 1978-79 which was extremely detailed. These surveys have established broad orders of magnitude of consumption of various types of biomass, for example, in the case of fire-wood, they tell how much of it is in logs, stems, branches and so on. However, the policy problem that arises in this area is that of supply and demand patterns. In particular, it is a layman's impression that India like many other countries is running out of wood, which is leading to deforestation, and that this is somehow due to excessive consumption of firewood. Similarly, there are a great many books on energy published abroad which assert without question that India is short of fuel and, therefore, people burn dung and take it away from its alternative use as manure. But from whatever I have read, I am convinced that both these impressions are extremely misleading. There is growing deforestation undoubtedly, but it is very doubtful whether it has anything to do with firewood consumption. Similarly it is quite clear that only a certain proportion, three quarters at most, of the dung produced is being actually collected, and that burning dung as fuel is not a phenomenon which is in any way connected with the shortage of fuelwood. It is a cultural phenomenon confined to the Gangetic plain; one hardly finds it in the peninsula. So the questions of biomass supply and demand have not been looked at. I think it is possible to investigate them with the statistics we have, but no one has got down to doing it.

Other countries

Coming to the energy data in other countries, I am very unfamiliar with them and I can comment only on the basis of what has been presented at this workshop. I have some familiarity with the Thai data, which are extremely good. The range of data is, however, limited as compared to the Indian energy data. But whatever data are available are of good quality and are brought out fairly expeditiously. The Thai problem of energy planning is in some ways simpler than in a large country, for there are only two major forms of commercial energy, oil and electricity, and oil is entirely imported. So the question of planning for oil supply boils down to a question of management of the balance of payments, and Thailand is one of the countries which has managed this problem very well. So it has really not faced a serious problem.

The case of the Philippines is rather different. The Philippines is heavily dependent on imports of oil even for power production. It was hit hard by the crisis of 1973-74, and again by the crisis of 1979, and the only short-term solution for it was to import oil even by borrowing money, and to invest heavily in oil substitutes, especially in power. Again the investment could only be financed by borrowing from abroad. So the problem of the Philippines boils down to that of borrowing from abroad. It did not have the same options in this regard as, for example, India had. Quite a large proportion of its borrowings had to be commercial. So the incentive for the production and collection of energy data in the Philippines came from the need to present the energy picture to potential lenders and to develop the energy options which could be turned into projects for financing. The presentation of the Philippines statistics is consequently extremely good. They are the most beautifully brought out statistics I have seen in this region. I am told that they are not entirely accurate. I do not know of this first-hand, but some of the figures might need to be looked at again.

I am very unfamiliar with Brazil, but I have the impression that Brazil faces a situation which is similar to that of India in some ways. It also has an institutional structure which is similar to India, namely, it has a great public power company, Electrobras, and a big oil company, Petrobras, which have formed the basis of the energy data system. These are institutions which generate a lot of data for themselves and which provide the basis for the national energy data. The rest of the sectors are poorly served, and this is where Brazil faces difficulty in collecting data and in relating them to the rest of the energy picture. Efforts to do so really started only after the oil crisis because until then Brazil did not have a problem. Brazil has the institutional structure required to generate the kind of data it needs, and will probably advance further in this direction in the next five years. It has also tried some very imaginative solutions; alcohol is the most obvious one. But it is possible that in such a large country, an even larger number of options is available and should be considered together when making policy. It is in this area that I think Brazilian policy will advance further in the next few years.

Coming finally to China, I must say that I am even more ignorant about China than about all the other countries. I have always regarded the Chinese energy picture as full of puzzles. For example, China has been for quite some time the third biggest coal producer in the world. It produces over 600 million tonnes of coal. Today it is also perhaps the fourth or fifth largest producer of oil; it produces over 100 million tonnes of oil. Compared to most developing countries, it is extremely well-endowed in its energy sources, and is exploiting these sources with the technologies it has. What has always puzzled me is the relationship between this enormous commercial energy production and China's industrial base. Somehow I feel that this sort of energy base could support a very much larger industrial base. This raises questions regarding both the data and the technology. I mean the first impression would be that somehow energy resources are being utilized inefficiently, but it may well be that China's industrial base is larger than we think it is.

The other puzzle in the case of China concerns non-commercial energy. It is clear that a large proportion of the Chinese population lives in areas which are extremely densely settled where there is little surplus land, where the land that is cultivated is intensively irrigated and produces products which are not really energy products. In other words, there are no large areas of forest in the areas in which the bulk of the

Chinese population lives. There is not much scope for the production of non-agricultural bio-mass, and given the crop pattern of China, the agricultural bio-mass production is also not very high either. I have often wondered how the Chinese population has managed with such low supplies of rural energy. It is just possible that the large production of coal goes partially into rural energy consumption; but we do not really know. So there are a number of puzzles about the Chinese energy picture which need to be sorted out.

CONCLUSION

Let me conclude this impressionistic survey of national policies and data systems with the proposition that good information does not lead to good management or good problem-solving, but a good problem-solver will need good information. What has happened in India, and what is happening in more and more countries in the world, is that energy problems have become important. They have become energy problems, instead of just problems of coal or oil, and consequently a need has arisen for better information on all energy sources which is being met in various ways in various countries.

When a policy-maker perceives the need for information it is unlikely that the information will be available readymade but there is a likelihood of it being available. It is likely to be available if it is being used in the management information systems of the producers themselves. This is why there is a close connection between the management information systems of the producers and the information used by the policy-makers, and there is nothing unnatural about it. That is precisely how it should work. But there will be areas where producers do not collect information; there the policy-maker will have to get it collected for himself. Thus I think the policy maker's approach should be an entirely opportunistic one, namely to use information where it is available and to collect it where it is not available.

DISCUSSION

SENGUPTA

I have some comments regarding the power sector. Let us understand this issue clearly, that whereas on the one hand data acquisition is a very important aspect, data processing is an equally important aspect. Now at the moment whether the data are entirely authentic or not so, a huge mass of information is being compiled by the Central Electricity Authority and the regional electricity boards. Let us say it is there. We just cannot do everything overnight. At the moment the major on-line information are being collected at 66 kV level. We have a central computer housed in the National Informatic Centre; if all the electricity boards and the regional electricity boards acquire their computers, these computers could be inter-linked. Whatever data we collect are processed locally to the extent possible and then transferred through this computer hierarchy, which is well established in other countries. This is something which we can obtain in an organized manner at a fairly low cost. Having done that, or while doing it simultaneously we have to go down the line.

I think it is not easy to disaggregate data at the consumer level; it is easier said than done. In an industry you have motive power, induction motors, and so on, which are being driven. Also thermal ovens, for example, which have been set up. If you have to know how much energy is being consumed by each unit, then after a point the cabling has to branch out and we have to instal meters. As it is, in my surveys, I have found that even enlightened industries have just one meter where they monitor the amount of energy they purchase or draw every day. So, unless we have the cabling system broken down into separate lines with metering at every stage, it will be very difficult to obtain information on where the power flows in each respect. I think it is necessary to prune, there is no end to going down into details. I think it is extremely important for us to decide how much we will be satisfied with; what percentage accuracy is satisfactory. I think one can overdo the problem. In summary, I would say, let us not go overboard in looking for details, in that case we may lose sight of the forest for the trees. Let us get details to the extent possible which is cost effective but organise the computer hierarchy system, interlinked, so that the processing may become more effective.

BHATIA

Having worked with Indian data for the last 10 to 12 years, I have learnt two things: one is that data is invariably available, but it is scattered over different places. For example, in the case of oil the Indian Institute of Petroleum (IIP) does have fairly detailed data by depots and by uses and they do use this information for demand projection, and so on, and I did get detailed information when I did depot-wise projections. However, that does not get published as part of the Indian Petroleum and Petro-Chemical Statistics. It is not really widely available to the researchers. Some of this information, and also the demand projections that the IIP made, are also not available to the outsiders. There may be reasons for that. What I want to say is that information is available and it is generally not considered necessary to be given to the outside world. Perhaps because of some organizational arrangements.

Similarly, cost data of a large number of projects which have been completed or which have been sanctioned are available with the Planning Commission. Some of that information could be culled out and be made available in annual publications like the Indian Petroleum Petro-Chemical Statistics. Similarly for bio-mass, the livestock surveys are there; livestock census is done every five years. We have data on production of different crops. We can generate some coefficiency, grain to straw ratio if we made some effort. Some surveys can be conducted to refine these numbers that are available. The present availability is not very good. So one can try and develop a series on the animal residues, on bio-mass availability, as well as some utilization figures. The problem has been that nobody has really thought about it in a serious way, because there is no agency which has been really charged with handling of data or collecting or publishing the data for the energy sector as a whole. That brings me to the organizational part. Now that we have a unified Energy Ministry, perhaps one of the suggestions of the workshop could be that a separate cell on energy data could be created in the Ministry with the emphasis that the data should not only be just on supply and demand but also on costs. A large number of projects are available from which cost data can be given.

SHAH

Just some comments about the availability of information on coal as well as on electricity. Dr Desai mentioned that Coal India does not have information about the consumer end, of what the needs were, what the quality was, and that they did not have an arrangement whereby they could be sensitive to consumer needs. I believe they have a marketing system, a Chief Executive of Marketing who is the primary liaison officer with the consumers for their needs as well as for processing complaints about quality. One can always expect things could be better than what they are. But to say that they are virtually non-existent might not be quite fair.

Now the quality of coal, as received, is certainly very well-known. Unfortunately not very much has been done to improve this, but the knowledge is already there both at the sending as well as at the receiving end, not only on a gross basis but on a sample by sample basis of consignment. That is available not just coal field-wise, but also individual mine-wise and over a period of time. So data on how the quality of an individual mine varies is also available.

Similarly, I think Dr Desai mentioned about the off-take electricity statistics, as not being adequately available. I do not exactly know what he had in mind, but I believe electricity is one area, where because of the individual metering a lot of information is available. We should leave aside these relatively minor aberrations of non-meter supply, which might not form even a fraction of one per cent of the total electricity sold. We have a very close analysis of sectoral sales over a long period of time. One might wish to have it even more detailed. That is also available at specific points of time for specific needs. For example, the broad guidelines specified by the Central Electricity Authority are being followed and information is being collected.

I would also like to react to Dr Desai's feeling that 20,000 crores of investment is probably not giving the best result due to lack of information collection. I think wc ought to be a bit careful in this, in the sense that there are far greater factors which are responsible for whatever deficiencies we might find in our investment efficacies and utilization than mere information. Whatever information problems do exist are, I think, internal to organizations who do not really have very well defined management information system arrangements and these, to that extent, contribute to delays of projects. But I would likc to suggest that merely by good information we are not going to solve the problem of this magnitude. I again endorse what has been mentioned earlier repeatedly that it is a question not so much of availability of information but of management of the available information, and of generation of available information by approaches such as Dr Sengupta mentioned. There is an enormous amount of available information about the quality of supply. For instance there is a computerized system in one of the boards I know which regularly monitors every feeder, every interruption, what duration, and other factors which affect the quality. All the 14 or 15 hundred feeders are monitored monthly. I think our recommendations perhaps could be to what use we want to put all the information. Having decided that, most of the available information can be either re-classified or re-categorized or its periodicity changed and in order to support that effort, the collection, the acquisition efforts, ought to be strengthened a good deal more through use of computerization, as Dr Sengupta has been recommending.

MUBAYI

There is a programme at the National Energy Administration in Thailand to look for substitutes to imported oil whose cost is perceived to be a heavy burden on the economy and balance of payments. In this regard, the recently discovered natural gas in the Gulf of Thailand is going to be a major energy source over the next decade, although they have been having some problems recently determining the size of the resources, and reserves. Lignite is also going to come in a major way in electricity production at the mine-mouth.

On the subject that Dr Sengupta also touched—data acquisition in the power sector, we have had over the last two or three days a series of discussions on technology and fuel substitution in the various sectors of the economy and the data needs to be able to formulate sensible programmes both as regards choice of fuels and the investment and economic viability of such substitution. One plea is, that in the industrial sector, the use of electricity for various energy purposes should be monitored in some regard. Generally one can distinguish the non-substitutable uses of electricity, particularly for motor drive, which happens to form a large part of industrial electricity consumption, from the use of high grade electricity for producing low grade heat. This, because of ease, and because of logistics, has been on the rise. The simplest example is putting a resistance heater to heat your home and to use this high grade expensive fuel to generate low grade type of energy commodity. Perhaps there should be some aspect of metering, for what could be broadly distinguished, as substitutable and non-substitutable uses of electricity. The clearest example would be to meter differently the use for heat and the use for motor driving. I hope this can be done because otherwise for large intensive consumers, there is just no basis on which a good analysis can be done of what the various possibilities are.

S.S. SRIVASTAVA

I want to make a brief comment in continuation of

what Dr Ramesh Bhatia has said. Of course, in the last few days we have seen there is an enormity of data availability of certain kinds at least. The problem is that these data are not being processed well and the system of storage and retrieval has not been streamlined and needs quite substantial development. As of course most of us would be knowing, this is the situation in many sectors of the Indian economy in the sense that in many other sectors also we collect much information, but our system of storing and retrieval is not adequate. As a consequence, we do not get whatever we need for the purpose of policy making and planning.

There are areas where the data are simply not available, for example, as Mohan had mentioned, data on economic and non-technical variables. We know that traditional and non-commercial sources of energy in certain countries form a substantial part of their total energy consumed. The data system is also very inadequate. The question is what should we recommend about all this? I would have liked to see in those organization charts that Dr Sengupta showed the role of management scientists, information or data specialists, in the system.

I want to bring to the notice of this forum that the Indian statistical system is being continuously reviewed on a regular basis by a conference of Central and State Statistical Organizations. In the last session of this conference held in New Delhi in 1981, energy data systems were discussed but the papers that were presented there by the economists and the statisticians did not touch upon technical data. I will read out the recommendations. First, there was no nodal Ministry or Department of the Government which was looking after the data needed by the planning and policy makers, the data that should be collected, stored and retrieved. Our recommendation there was that the Ministry of Energy should be declared as a nodal Ministry which should undertake the coordination of activities pertaining to the development of energy statistics. We had gone on to say that the Ministry of Energy should take steps to develop a proper management information system and perhaps a statistical cell for this purpose. Amongst data gaps that were identified by this group was the data base for the non-commercial fuels. Also knowledge of renewable energy sources, particularly solar energy, etc., were identified as gaps. Then this group also found that there were problems about uniformity in definitions and concepts of various things. Even the definition of 'villages electrified' for example, is not standardized and, therefore, the data you get in that regard may not be comparable. Another recommendation was that all data collected on energy through surveys should be tabulated in detail. Of course, this is also a very serious problem. Even in the National Sample Survey and some other surveys, we collect quite a lot of information, but sometimes as much as 90 per cent of the information simply lies somewhere and then gets lost in the process.

The last recommendation was that efforts should be made to compile energy statistics for rural and urban areas respectively disaggregated at appropriate levels.

SARMA

I would like to stress two points. One, about definitions, is similar to what has been just made. The other is about access to information that is being collected. There are areas where information is yet to be collected but in the Indian government today, we have got an enormous amount of information in many fields, with the Central Electricity Authority or petroleum statistics, or coal statistics. But the difficulty is that this information is not accessible. If a research scholar or some University or institution wants the information, it is not readily available. Even for a person like me in the government, when I want information from a different department either for official work or academic work, I find that it is almost impossible to get. Dr Ramesh Bhatia will corroborate my statement. For the last $1\frac{1}{2}$ months we are trying to get information on fertilizer cost statistics for completed projects. But we find that the other departments do not part with the information. When the Rajyadhaksha Committee on Power, as an official high level committee, wanted information from other departments, from other state governments, from state electricity boards, this information was not forthcoming. As a community of energy researchers, for the people who are concerned with energy, should it not be a recommendation of this seminar to do something about it? In the US, for example, there is a Freedom of Information Act. As a lobby should we not do something?

I find that on a number of government publications is written 'for official use only'. Now I can tell you, many of these documents can be released to the public without any difficulty. Their information content is not of a secret nature.

The second thing is about comparability. For example, in the power sector, take the definition of a concept like plant load factor, that is capacity utilization of a power station: different State Electricity Boards look at it from different points of view. They have got their own definitions. In fact, in the Power Ministry when we tried to compare the performance of different power stations, the data we got were not comparable. We still compare, nevertheless. This is one of the important problems a seminar of this type should really tackle to evolve a proper definition.

When we were studying the financial performance of State Electricity Boards in India, we asked people for the report on profits and losses. The word 'profit' is defined in different ways by different State Electricity Boards. There is a lot of public debate on it without any comparability. We encountered this problem with the Rajyadhaksha Committee also. As far as the power sector is concerned, we are trying to evolve some common definitions. But in the energy sector as a whole we should try to evolve a set of definitions.

NITIN DESAI

On access to information, I can give you one simple bureaucratic trick. Instead of asking for information, you make your own estimate of fertilizer cost and send it to the Fertilizer Ministry for comment. And they will comment on it and give you the figure.

SENGUPTA

Can I endorse this view of Mr Sarma's? Over my last six month's research, I found an enormous amount of data published, stored away and "all for official use only" and you really have to smooth your way to get at it. That is how I got hold of a large number of files meant for official use only but didn't find on going through these the kind of information which really could be held against those people or that which could endanger anything.

NITIN DESAI

I wanted to request Prof. Mena, Mr. Limsumalee and Prof. Levy to react to some of the observations.

MENA

I think the information that Dr Ashok Desai has given about the Philippines is right. We are too much dependent on oil and the present thrust of the government now is precisely to develop other alternative sources of energy. Of course, we are also developing our hydro-power, because it is not yet fully tapped. Another source which is very promising is geothermal. Besides that I would also like to say a few words on three relevant issues.

The complexity and importance of national data collection processing and analysis in the energy sector have been brought into focus during this workshop. The data have come from various sectors of the economy and have to be processed and analysed to precise needs which include among others the following:

Data needs for assessing the technical potential of a given technology; data needs for assessing the economic and financial viability of the technology; data needs for assessing the social and political viability of the technology; data needs for investment priorities; and it was also mentioned, the data related to revolutionary break-throughs in energy technology.

There is need to have all these data collated and analysed by a central body. In the case of China I noted yesterday, they have adopted the dual statistical system for cross-checking and supplementing additional information in cases where there are information gaps. The information gaps can be minimized if the various agencies involved in data collection are made aware of the functions and responsibilities that they have in the national data network system. I therefore suggest that in defining the extent of the national and the regional networks the functions and responsibilities of each agency in the network be emphasized.

The second issue I would like to comment upon is that one of the tendencies everywhere is to react to problems, as for example, the energy crisis, environmental degradation, etc. We have a tendency to react to problems when they have already reached the crisis proportion. Right at the very start, the very concept of planning is missing. In other words, under such circumstances we plan to correct what we should not have done in the first place. It is for these reasons that there is a wide information gap between what is available and what is needed in some instances resulting in a similar wide gap between what is going on and what should be done. Therefore, a good monitoring programme for continuing data collection, processing and analysis should be emphasized if we want our short, medium as well as long-term plans to be economically, financially, technically, socially, and politically feasible.

The third point that I wish to raise is on the practical level of sophistication in both data collection and analysis that developing countries may reasonably aim for. It may be desirable to aim for the level of sophistication that exists in developed countries, although in some developing countries this may not be possible for a number of reasons. I wonder if this workshop can provide some guidelines on this issue.

NITIN DESAI

Thank you Prof. Mena for your comments. May I request Mr. Limsumalee to say something?

LIMSUMALEE

I think much of the data are available in different countries. In the case of my country the purpose of data utilization is first to go into the policy and planning, and secondly, to go into the overall picture of the energy situation. In our country, the NEA has the function to respond to the policy and planning formulations. How do we get this detailed information? We know that the different data are kept by different agencies, and the NEA makes the effort to collect this from different agencies. But the non-technical data, we try to collect ourselves and we try to publish it in our annual report.

LEVY

I would like to make an observation at this stage which is perhaps obvious to many of you. I think it should be stressed. One should not make it look as if that there was a domestic relationship between the data and the policy. In other words, it is not enough to have sufficient data in order to formulate policy. There is an intermediate stage which is essential, which is to have a model of the situation. This intermediate stage, actually always reflects, an external set of constraints which are either implicit or explicit. This means that data needs change with constraints and change with the model which we are using either implicitly or explicitly. I think a very good example, as Dr Mubayi was insisting several times in this Workshop and I agree with him, is the fact that we need now a user oriented data, instead of production oriented data. We are looking now for substitution, for fuel substitution and energy substitution and the data we need for this is not available because for a long time people are gathering data as if the energy crises did not exist and they continue with their old methods. Now, of course, our needs of data are not the same as they were some years ago. Therefore, the system must be changed. I think one should stress in any data system for a given country, the need for flexibility so that the data can be re-arranged or re-obtained depending on the needs. I think, of course, in this respect a good computer system will be extremely useful.

MISRA

I will take this opportunity to speak a few words, based on what I have heard this morning. As we look at the energy data system as it obtains today, we are really struck by a few things. I entirely agree with Dr. Ramesh Bhatia and Dr. Srivastava that it has become necessary in India now to have a nodal agency. Right now, a lot of energy data is being generated by a number of agencies in the country, but there is no nodal agency. Because the Ministry of Energy has been newly constituted, consisting of the Departments of Coal, Power, Petroleum and Non-conventional Sources of Energy, perhaps it would be appropriate if this type of coordinating function is carried out in the Energy Ministry. There is some degree of coordination, but it is not quite up to the mark; it should be much more intensive.

At present in the system we have a Committee of Secretaries in the Government of India and most of the matters pertaining to energy policy, energy problems, are referred to this high level Committee. This Committee of Secretaries presided over by the Cabinet Secretary, normally takes an overall view of the entire spectrum of energy policy. Above it, there is the Cabinet Committee on Energy, and most of the problems pertaining to the area of energy are finally referred to it. This Cabinet Committee on Energy takes a view of the entire situation and lays down policy guidelines. But the time has come when a division or a cell should be set up, may be under the Energy Ministry, may be outside, which should be institutionally responsible, not merely for collecting and compiling the data, but also for integrating and interpreting it and drawing from this the relevant conclusions.

There are certainly gaps in the area of energy statistics. There is a lot of room for further refinement in the definitions and concepts, for further standardization of the data, for using uniform conversion factors, so that when we speak of something, we all mean the same. There is also a wide area in our economy where perhaps the data has not been collected in a very detailed manner. I am referring to the area of non-commercial sources of energy. India in that way is a very fascinating country, where the sources of energy range from atomic fuel to cow dung and vegetable wastes. Non-commercial sources of energy account for almost 45 to 50 per cent of the total energy consumed in the country in terms of heat value. From time to time certain surveys and some studies have been conducted, but perhaps there is a need to refine and revise this data and to make it a continuing exercise.

Recently in connection with the hosting of the World Energy Conference, we are trying our best to collect the relevant energy data of the last ten years in the entire country. I was apalled by the fact that certain data are just not available. For instance the data relating to the process loss, the stock changes, the loss in transit, are not available. We have also looked into the data relating to the production of coal and my strong hunch is that we have not got the data regarding the various grades of coal. At least not in Delhi. I am thinking of sending somebody to Calcutta with the objective of procuring this data.

Similarly in the area of petroleum also, there are certain gaps in the data of consumption pattern. We have to request data for the various end-products. What are the various distillates, and end-products which are being produced, which end-product is going to which category of consumers and in what proportion, these aspects are not readily available.

In the Department of Power, I am associated with power planning. This is a very complex exercise which depends upon the interaction between various utilities of the state governments, Department of Power, Central Electricity Authority and the Planning Commission. I would, therefore, like to present a few points based on my observations in the Ministry of Energy.

I hardly need mention that energy is a very valuable input in any economic activity and its balanced development and growth is vital if we have to meet adequately the aspirations of our society. Because construction of power stations take a lot of time, demand forecasts have to be done years in advance. In fact it takes five to six years to construct, erect and to commission a thermal power station, following the investment decision. The investment decision itself is preceded by a lot of detailed investigation, preparation of the project report, and also techno-economical appraisal by the Central Electricity Authority. Similarly it takes about eight to ten years to construct and commission a hydroelectric station and the investment decision in this area also is preceded by a very detailed investigation in the areas of hydrology, geology water discharge, and a whole host of other disciplines. It takes even longer to commission a nuclear power project. Because of the long gestation period involved in the construction of these projects, it is required that we should have the demand forecast years—10, 15, 20 years in advance—so that we avoid to the maximum extent possible the discrepancies between the demand and supply in the later years.

There is a danger if the demand forecast is not very reliable. If the demand forecast is over-pitched, this might result in the over-building of capacity. Now for a country like ours, where capital is very scarce, where the opportunity cost of the capital is very high, we just cannot afford to over-build the capacity. If the demand forecast is underpitched, then it might lead to underplanning of the capacity and the result might be that certain areas of economy will be deprived of this vital input. So what is necessary is that there should be a very balanced forecast of the demand. It is a very delicate exercise. We have also to achieve a balance between hydro power and thermal power. Thermal power is supposed to meet the base load of the system; hydro power, because of its inherent flexibility, is supposed to cater to the peak requirements of the system. Now the reliability of the forecast obviously depends on the quality of the data; on the adequacy of the entire data system, and on the comprehensiveness of the survey; it would also depend upon the soundness of the methodology that we follow and also on the assumptions that are made in calculating these things.

Some time back, the Working Group on Energy Policy had laid down the likely requirements of energy in the 1980s and the 1990s. Right now, it takes Rs. 10,000 to instal one kW of capacity in the system. It takes exactly the same amount, almost Rs. 10,000 to build up the matching transmission and distribution lines to take that 1 kW of power to the consumer at the end of the line. The Working Group on Energy Policy had recommended that by the turn of the century, by A.D. 2000, perhaps on the most optimal level forecast, we might need 465 m. tonnes of coal, 69 m. tonnes of oil, and 465 TW hours of electrical energy. For supplying the energy, they had estimated that we would have to build up a capacity of the magnitude of 36,000 MW in the 1980s and 50,000 MW in the 1990s. For building up a capacity of this magnitude they had estimated that they might be required to invest

Rs. 50,000 crores in the 1980s and Rs. 70,000 crores in the 1990s and this forecast was based on the assumption that one kW of capacity would cost only Rs. 5,500. Now it is costing Rs. 10,000 and there will be further escalations in the years to come. So you can well imagine how colossal is the size of the investment in the power sector. Right now in the Sixth Plan, out of an outlay of Rs. 97,500 crores, almost 30 per cent has been allocated to the area of energy, of which 20 per cent belongs exclusively to the area of power. When we are dealing with such massive investments, naturally we have to pay a lot of attention to the demand forecast, to the adequacy of the data, and also how meticulously the data is collected, how thoroughly the data is collated, how dispassionately the data is examined, integrated and interpreted in the normal course because the investment decision of the Government of India is going to depend on this data.

I will take a couple of minutes more to indicate how energy data would influence policy decisions. Most of the surveys that we have done in the past, have revealed that amongst fossil fuels, coal is the only fuel which we have in reasonably good proportion. For a number of decades to come, coal is going to be the primary source of energy in India and coal would progressively substitute other fuels wherever possible and technologically achievable. Now all the thermal power stations in our country are coal based. We use some furnace oil to start up the boiler, or occasionally to ensure flame stability in the boiler, if the quality of coal happens to be poor or when the coal is somewhat wet. Over the last few years, systematic endeavours have been made almost in all the thermal power stations in the country to bring down the consumption of oil.

Apart from the primacy of coal, the second most important tenet of our energy policy would be that we have to manage the demand of oil, and have to still depend very heavily on imported sources. A lot of efforts have been made, and promotional support is being given to contain demand. The Cabinet Committee on Energy has already taken a decision that the demand for oil would be pegged to 40 m. tonnes by 1984-85, and by then we should meet almost two-thirds of our requirement indigenously, so that we can reduce our dependence on imported oil. Now to promote this energy policy, a number of steps have been taken by the Government of India. For instance, the entire rural electrification programme of railway traction has been accelerated. Now a programme has been undertaken that by 1984-85, the Ministry of Railways would be undertaking almost 1000 km of electrification of railway traction every year. About 5,300 km have already been electrified. Work is currently going on for a stretch of about 3,000 km and by 1984-85 the programme is such that every year, they would be electrifying almost 1,000 km of the traction.

To promote fuel efficiency in thc road transport sector, decisions have been taken to manufacture multi-axle vehicles. We have got currently in our country eight manufacturers of commercial vehicles, and decisions have been taken to book almost 20 per cent of this capacity for the manufacture of multi-axle vehicles, which will have a much higher carrying capacity, higher almost by 30 per cent. To promote fuel efficiency a decision has been taken that all of the eight manufacturers of multi-axle vehicles, would be developing turbo charge engines. TELCO and Ashok Leyland have already designed, developed, and manufactured turbo charged engines. The other six manufacturers are also developing that type of engine which will bring down the consumption of oil in the system. Numerous other policy measures have been taken.

SMITH

I want to emphasize the last statement of Mr Dayal in the rural energy section of this meeting that there is no unique solution to the energy problem. I think there is no unique data itself to examine the energy problems well. In all fields there is some micro-macro demon, that is, you do not know what questions to ask, if you do not understand anything about what is out there, and you do not know what data to collect, unless you know what questions to ask. We tended, as one might expect in this meeting, to concentrate basically on what is there, how do you understand, how do you categorize, how do you make the definition uniform, and so forth. But as we have seen, particularly in the rural energy area, you could end up collecting so many data that completely overwhelm, unless one thinks about the macro questions that one is trying to ask. If I had more time, I would try to convince you that actually there is no universal energy data itself, that can answer all your questions. Indeed it is not possible to create universal energy data itself because of some of the indexing problems, the aggregation problems, the use of gross energy, the basic physical and economic terms. One should not even try.

The kinds of questions I am talking about are fairly obvious, how do you determine oil imports? How do you justify various project developments? How do you determine technology feasibility? How to determine energy constraints on economic development? What are the environmental impacts of energy use? What are the opportunities for economic investment in the energy sector? Indeed, hiding information is a very important use of data itself, as the US government has found in dealing with international oil companies, and as many other governments have. We provide information to the consumer so that the consumer can make the decision himself about how to use energy, how to consume it, and so on. So hiding of information is surprising. I would now like to try to summarize a few of the major points in the rural energy section that were discussed yesterday.

Rural energy is a very complicated area. We should try to give some sort of structure to it. I think basically rural energy today is a question of biomass fuel, including animal energy. Basically what I am trying to address myself to, is the post biomass transition, not the post-oil transition. People in the world depend for most of their energy on biomass, not oil. Oil is most of the energy but not for most of the people. Biomass is in transition from collecting and gathering to the management of supplies of various kinds. For example, fuel wood plantations. Biomass fuels tend to be used in a fairly low quality state, solid fuel basically needs to be converted into higher grade liquid on gas. That will go along for not only meeting the basic needs but supplying the needs of economic development.

What was common in our discussion on rural energy was how, in the main, location-specific things were. There are too many villages in the world and in some sense every village is different and indeed every household is different. There is a huge variety in the cultural, geographical, climatic and ethnic variations among rural villages and consequently in the end-use of energy. An implication of this is that in all these sectors, very careful attention must be paid to socio-economic information in addition to energy data.

There are severe conceptual and methodological problems in collecting data on biomass and rural energy use in general. There are such problems in all energy forms, including gas and coal. Perhaps they are most difficult in the case of rural energy: how do you deal with animal energy, solar energy, and human energy, how do you add a joule of solar energy available at 100° to electricity which is being of a very high energy quality. It is a very difficult problem conceptually, not well dealt with in the existing energy data system. There is also question of how to do surveys, in what combinations of intensive or very detailed surveys with a small number of the villages and households, or extensive with fairly small data from a large number of households.

We are all trying to answer in these discussions a set of generalistic questions, why, where, who, how, and when the survey should be done. One thing that could be added to that list is the caveat, that in rural energy survey unlike other types of surveys that are being considered, one is dealing with individual households. There is an ethical question of how much each local household life is worth. It is not a similar problem as when dealing with firms and government corporations.

A very important issue raised by Dr. Dayal was the question of whether one cannot only consider the basic needs, basically for cooking fuel or are you going to provide high grade fuel needed to increase agricultural production, for rural industry, etc. The important area developed by Dr Gupta was primarily about preparation of data about rural energy demand. One has naturally to match that rural energy demand with the preparation of data about the possible technology to meet that demand. There has to be knowledge about the cultural, environmental, economic and engineering implications of various technologies.

Other points that were made are also critical. They

need monitoring afterwards. One cannot just put the gadgets into the villages and then walk away. It is very important to maintain a monitoring network so that feed-back occurs. In addition there is need for marketing information. What may seem to make the system viable from an economic analysis, a host of paper analyses, need not actually attract the people to buy something. If in the US everyone bought automobiles based on the basis of cost per passenger mile, we would all be driving Toyotas. If the country has a wide range of automobiles it is because of a wide range of needs.

Particularly very little is known about how much straw is produced for different types of grains; or about what the stoves can do in improving rural energy use; very little is known about how they work and also potentials for upgrading the fuel so that you can provide high power fuel needed for rural development. One way of looking at the distinction between the rural energy system and some of the other systems is that in the industrial sector including the urban domestic sector most of the energy is consumed in intermediate production. About two-thirds, for example, is typical in a country like the US. This is in intermediate production, not in final end-use, not by consumers buying gasoline and electricity for the home. One can expect that the criteria that drive people to make decisions on the energy intermediate production sector can be very well modelled by engineering and economic criteria. They are going to react to price mechanism, if they have the information. They are going to try to find the least cost solution to their energy problems. You cannot expect that if you are going directly into the final end-use sector. People are very complicated. They do things for a variety of reasons. A good example in the US, as I mentioned before is automobile choice for reasons other than energy efficiency. Similarly in rural areas in developing countries, the stoves are needed for other than cooking. They have cultural and religious significance in some places. They also cater to their other needs. You cannot expect people to make a decision on stoves based solely on economic and engineering criteria. Indeed the distinction in the rural area in developing countries is about 90 per cent or more of the energy users is in the final end-use sector and not in any intermediate production.

Therefore, you can expect to do much more pooling and modelling using external economic and engineering criteria.

Finally, I would like to say that we should think about fuel data system. Although there are certain problems in the fuel data system in all countries, there is not much mystery about what you need in the case of oil, coal, and gas and even in some degree in biomass. The concept of energy is different from the concept of fuel and if you can narrow down these things, you will get the final picture of what is happening. This is what is called the energy world view. With the energy world view it is possible to understand what is happening in an economy and to the society during its energy flows in addition to fuel flow. It is possible to make adjustments in energy systems to make things better for society. We have to rationalize the oil system, the coal system, and others. I would say that it still remains to be seen exactly what it means to use energy as a tool to modify society as opposed to rationalizing the individual fuel system. There is much work to be done on conceptual and practical aspects as to how exactly one would gather data, in particularly rural areas in developing countries where it is quite clear that energy is the problem. It is not clear how much the lack of energy is a constraint on development.

NITIN DESAI

Listening to the various comments that have been made, one thing I find is that there have been several complaints about the lack of clarity in definitions, the fact that data is available but is not readily accessible, the fact that it is all scattered in different agencies, etc. The focus has been on a central organization for collecting this data. I don't think it is enough. Because all that may happen is that the central organization will just compile and bind together the statistics produced by the different agencies.

I think it is time we started thinking in terms of energy accounting matrix on lines analogous to social accounting matrices which are used in national accounting statistics. It will allow you to integrate the sort of data which comes from producing organizations, with a user-oriented data system that you require; it will also ensure compatibility in sector classification by different agencies, it will ensure certain compatibility of concepts. If you have such an energy accounting matrix as a base, and then you clearly identify that so and so agency, say, the CEA is responsible for these cells in these matrices; then the central agency will put this out in the form of an organized set of energy statistics, rather than just a compilation of what is being collected by different agencies as part of the system. I feel that this is a concept to which the seminar can devote some attention in the summing up which is to follow, and it will require a great deal of work. But I think this is a feasible proposition now. I have a sort of framework in mind and it is something like what Professor Lu Yingzhong had outlined in his paper on energy balances for China. One can build on that sort of framework into energy accounting matrix for data collection.

The second point that I would wish to make after listening to Dr Desai and others is that our whole data system is too dependent on what I describe as administrative statistics, that flow from the management and administration of energy producing enterprises. Now Dr. Desai, for instance, mentioned one thing: the fact that certain elements of consumption data for electricity are becoming less certain because we have moved away from metered supply. We cannot change the management system and insist on meter supply for every body just because we want reliable data. Apart from this, it is very inefficient to have metered supply for small consumers. If we are going to, then there will have to be compromises in actual administration and management. We should not become totally dependent on the flow of data in the processes of energy production and distribution by the various producing enterprises. We have to supplement it with independent information on what is happening in this sector, particularly at the user end. There is a need for independent sample surveys for collecting data on some items which are not generated in the official system, for cross-checking the official data. In this connection, I want to make a suggestion that the conference may recommend that all administrative agencies should examine the possibility of utilizing the various surveys that they do right now. For instance, almost all countries have a census of manufacturing industries. We in India have an Annual Survey of Industries, and also an Annual Survey of Small Scale Industries. We have annual, very detailed surveys, which we call cost of cultivation studies. There is a field man who visits each farmer in the sample survey, every second day, and collects data on what the farmer did over the previous two days. What I feel is that instead of trying to superimpose a fresh survey of users, one must investigate the possibility of elaborating or expanding the relevant blocks in the schedules used in the census of manufacturing industries or in the farm management surveys, which I am sure will be there in most countries. They are certainly there in India. I think that will generate information, more efficiently than trying to superimpose a fresh energy survey. I think you may still need field level surveys just on energy use. There is a lot of work which has been done in this regard.

I find much of the energy survey work is really what are called surveys-by-recall that the NCAER conducts. They go to each house and ask how much firewood have you used over the past week, month, and year. This is subject to tremendous error, particularly when you are talking in terms of the utilization of collected, rather than purchased, forms of energy. We should think in terms of observation based surveys on energy use. Station a man there, and let him observe what happens exactly as we do in the case of farm management studies. We do not go and ask the farmer what did you do over the previous season. We go every second day and ask him what did you do over the previous two days thus covering the whole year and we compile this data. Incidentally, I may suggest that somebody who is interested should pursue and retrieve the data from the schedules and use it for some rural energy study. I feel that the time has come to call for such detailed observation based study at least in the case of rural energy.

Dr Sengupta had mentioned about the difficulty in disaggregation of data for energy use. There is a point beyond which you cannot go. An enterprise has just one meter. One may really not know where the energy is being used. I personally think this need not be the case. Certainly as far as large items of equipment are concerned, which utilize a substantial amount of energy (you can fix a cut off—it can be 500 kW perhaps) we must insist on metering as any equipment of that size is sufficiently expensive. It is worthwhile to just instal a meter at the point at which power is spread. Certainly within a company, or an enterprise, there will be cost centres. The accounts of an enterprise are not being maintained just for the enterprise alone. Often individual machines will become cost centres in an engineering enterprise. There is no reason why you cannot preserve and obtain energy use data at the cost centre level at least. Of course, there is no need that this has to be reported at a higher level. If this data is maintained at the enterprise level, the sort of thing Mubayi was mentioning, segregating energy use by the type of application will become feasible, you need not ask for equipmentwise data but certainly you can ask that it be segregated by type of use.

SENGUPTA

In some recent experiences in Karnataka State, the difficulty that we faced is this. We go to a fairly large industry and find they have just one meter. Whereas I totally agree with you and Dr Mubayi as to the need of obtaining data, particularly for thermal use, for motive power use, the question is how do you enforce this on the users, and some of the institutions. We have got to look at the incentives. If an industry is energy intensive, then they, on their own, because of their own needs, try to conserve energy and try to do some metering. This is what I have seen in INDAL in Belgaum. But the difficulty is this: for one particular process industry, for Rs. 100 product cost—hold your breath—the cost of electricity is 48 paise. You give me one reason why they should bother. Whereas I take your point, that for fuel substitutions and other things, it is absolutely essential and worthwhile to obtain this disaggregated information, I think one has to think of some methodology by which these bulk consumers can be induced to having separate metering and supplying this information. So one can think of disincentive/incentive schemes to be offered to the bulk consumers so as to enforce them to provide this information.

NITIN DESAI

I agree with you. It is true. It is not as though it will come automatically. You will have to do something to ensure that this comes. We are trying to bring in energy audits now. That will generate this sort of information. I will now request Ashok Desai to respond.

ASHOK DESAI

I find Professor Levy's observations extremely just, namely, that first of all there must be a perception of a policy problem only then will one start wondering how to solve this problem. For that, one must have modelling. Only when one defines one's model will one know what data one needs. That is when one will start looking for data, and eventually might find it. This is really the process and this is what I would say in reply to Mr Shah as well. It is perfectly true that good information does not lead to good management or good problem solving, but a good problem solver, will look for good information and he cannot do without good information. In no way can you solve a problem well without good information. What has happened in India and what is happening in more and more countries in the world, is that energy problems have become important. They have become energy problems, instead of just problems of coal, or oil, and consequently a need has arisen for information which is being met in various ways in various countries.

When a policy maker perceives the need for information it is very unlikely that the information will be available readymade but there is a likelihood of it being available. If it is already being collected, it is likely to be available, especially if it is being used in the management information system of the producers themselves. This is why, not just in India, but in all countries, there is a close connection between the management information system of the producers and the information used by the policy makers. I would say in reply to Nitin Desai's point that there is nothing unnatural about it. This is precisely how it should work. But there will be areas where producers do not collect information and hence the policy maker will have to collect it. In this case I think the policy maker's approach should be an entirely opportunistic one, namely to use the information from where it is available and to collect information from where it is not available.

Coming to the Indian scene, I agree with the points made by a number of speakers that plenty of information is available. What is lacking is the processing, summarizing, and the packaging of information in such a way that it can be used to solve problems. I do not think that this problem can be solved by centralizing the information collection because the risks are very great. You might set up a first class organization and it might do a first class job and get you first class information. On the other hand you might set up a second rate organization and all the information might be wasted. You might get no result. Also as I said the incentive for collecting and processing of information, arise out of the problems. What happens in any agency which collects statistics regularly is that it loses sight of the objective for collecting information. For example, the CSO, and the NSS, collect millions and millions of statistics every year but they have no idea why the statistics are being collected and consequently you get surveys which are routinized and which cannot help us solve any problem. So I think centralizing of information is not the solution in this case. The solution really lies in improving the information system that exists, in increasing the processing of information and this, I would again say, can be achieved best if the managements of the particular energy sectors are themselves improved. It was in this respect that I was talking of Coal India. It is really when the management of Coal India improves, that the management information system of Coal India will improve.

VALEDICTORY SESSION

Chairman : K.S. HINGE

Valedictory Address : K.C. PANT

VALEDICTORY ADDRESS

HINGE

I believe this exercise that we have undertaken in collaboration with several institutions has been a unique one, since a large number of related and interested institutions have taken part, and the topics covered also are quite extensive, almost covering all aspects of energy. I do not find the transport sector covered, but I hope that the discussions would have had some reference to the transport sector too. I only hope that this concentrated attention focused on the data systems has been useful and that we are now wiser about how to harness information and put it to use for researchers, planners, policy makers, and so on.

With these few initial remarks, I would request Mr. Pant to deliver his address.

PANT

I have had something to do with the problems of energy, but you have probably covered all aspects of the problem. From the few concluding remarks I heard, I wish I had been present at the other sessions. I take this opportunity to welcome all the participants who have come from abroad. The crux of the usefulness of such a Workshop of course, is, the follow-up action and I see that already Dr Pachauri is taking steps to make sure that there is follow-up action.

I got the impression from listening to Dr Pachauri's summing up, the Chairman's and Dr Mubayi's remarks, that perhaps the availability of energy data in India is not as widely known as it should be, but I would not say it is not available. There are gaps of course. But the energy committee's report for instance, the Annual Reports of the various Ministries to Parliament, contain a fair amount of data and the possibility of making use of that data is there. It is not made use of properly in most cases. That is a different matter.

If you allow me to make a personal reference, when I came into the Energy Ministry for the first time, we were in the midst of a power crisis, I think in the year 1973-74. I tried to find out from my Ministry officials as to why exactly this was so, and why there was a gap between generation and demand in the various states. To my surprise, I learnt that since power was a concurrent subject dealt with by the States and the Centre, the Centre did not call for data regularly from the States. This was, therefore, left to the States. They were hesitant to ask for this data from the States but within one month we organised a system of information under which every day I could get figures such as, which particular unit in the country was generating how much, which was out of commission, which outage was because of maintenance, which outage was forced, when was it likely to come back, and so on. In other words, the data was all there. Within one month the system could be organised, centralised to the point that I could send questions to them the next day. And similarly, I used to get figures about production and transport and stocks of each coal company in the nationalised Coal India Ltd. every morning.

The fact is that data is available, perhaps scattered, and it is a question of organising the data system in at least the commercial energy area. The problem of long term planning is somewhat different but because energy planning is so essential to economic development, and to human welfare, that there is a relationship between them. Indeed I would make a distinction here which is important for developmental purposes. The rural energy in a country like India, is important for survival. Ever since the discovery of fire by man there has never been any real danger that hearths would not burn, that he would not be able to cook; but one may come to a point unless this problem is tackled, there will be adequate food but not the energy to cook it. It may happen in some parts and, therefore, it is a question of survival.

I, for one, am glad that the shock of the oil price rise has jolted the world out of a sense of complacency, with which it was sliding into an era of dwindling fossil fuel resources which would have given a much worse jolt later on. In a way, this was very timely. I think it is just as well, that all of us have woken up to this crisis that is creeping upon the world and are taking stock of the crisis. In this context the developing countries which are also oil importing, have been extremely hard hit. They have been hit not only because the price of oil has gone up but also because the price of machinery and other items which they import from developed countries have also gone up. Developed countries on the other hand have got the surpluses of the oil exporting countries, which have certainly acted as a buffer to the oil price rise shock for them; but we who are in the least position to really bear that kind of shock have been doubly hit.

I am very glad that the UNDP is associated with this seminar and you will be glad to know that the UN has very recently passed a resolution, which we have all been hoping for some time, on this question of development of energy resources of developing countries. This was on 26th November 1982. It refers to Nairobi's programme of action, it refers to the work that is being done by the UNCTAD, it emphasises the importance of intensifying the capabilities of the UN in the collection, analysis, and dissemination of information in the field of development of energy resources in the developing countries. This has a direct bearing on this Workshop, and therefore, I mention it. After the preamble the resolution says that the General Assembly requests the Secretary General to prepare, within the context of the international development strategy for the Third UN Development Decade, a comprehensive report on the development of the energy resources of the developing countries to be submitted to the Economic and Social Council at its second regular session in 1983. Then it goes on to give the details of the kind of report which would be required by the General Assembly. I mention this to you because it is a very fortunate coincidence that this Workshop is meeting at a time when the UN has asked the Secretary

General to prepare such a report. It is obvious that while we need the energy data for our own national purposes, here, there is an additional sense of urgency that this data will be needed in order to enable the UN to take action on this particular resolution. But I do feel that the wheels have been set in motion in the United Nations and perhaps the international financing agencies will willy-nilly be brought more into the picture of developing energy resources in the Third World countries. Once these efforts start bearing fruit, the developing countries will have to be ready with the blue prints of programmes; they will have to be ready with estimates of funds that are required with projections of their energy needs and they will have, if possible, to do it in a way which can be acted upon quickly.

In this context, if this Workshop could set up a small body which could standardise a scheme of collection and utilisation of these data, go further and even think of economic models in which it could be used, I think this standardisation would go a very long way in helping the cause of developing countries when the United Nations finally gets down to analysing the data as well as give assistance for energy purposes to Third World countries. Therefore, I would very much wish that we would go one step beyond and increase the interaction between the developing countries in this field, increase co-operation, the sharing of knowledge, of technology and prepare concrete plans and schemes in close co-ordination. I think this is completely possible, I think it is necessary that this Workshop can lay a foundation for this purpose.

Now the position of Brazil which has a National Energy Documentation Centre, we were told, is perhaps better than ours; we do not have such a Centre. I find this idea very useful and think we should adopt this.

You must have gone through the general energy picture in India, the use of energy in industry, in transport, in agriculture, in commercial activity, and so on. This is a field which is better known, better covered, better documented statistically, than the rural energy sector. The household sector is mostly non-commercial and almost 40 per cent of the total energy in India is non-commercial. But the household sector in the rural area, which means 70-80 per cent of people, almost invariably uses non-commercial energy for cooking. It may use kerosene for lighting, but for cooking purposes it uses non-commercial energy, in most cases either firewood or animal or vegetable waste. This is an area of darkness so far as statistical information is concerned. It is not that we do not have some information but it is not information upon which we can rely completely. It is one of the challenges of the Third World economic planning in energy sector.

So far as the commercial energy sector is concerned, when we come to current data, it is more easily available as it passes through the market, and the market mechanism makes it very easy to at least broadly follow the energy movement. But when it comes to long term planning, when it comes to the question of reserves, then we come across difficulties. Energy production is very capital intensive: I think 25 per cent of the plan investment is about the figure that goes into the power sector, and it is a long gestation period activity.

About oil, all I can say is, you take a long term view and take shale and tar sands into account, which are available in some parts of the world. There, of course, the reserves are very large but if you take the economics into account, today you cannot count them amongst the reserves which can be used at least in the near future. So which are the real reserves? It depends on the point of time and on the cost. These reserves are very variable figures. It is not only a question of certain margins but they are very variable, depending on the cost and time. If other types of fuel disappear, and no alternatives are available, then certain kinds of fuel which are impossibly expensive in other conditions may become acceptable.

And, regarding the question of coal. We in India have estimated our coal to a certain depth. Now we have gone deeper. So in our last Energy Survey Report the reserves of coal have gone up because we have gone deeper. We have gone up to 1200′, as Dr Marwah was telling me. Now in Gujarat we have come across a huge deposit of coal, about which very few people know. It is below the oil level. But because the coal experts have drawn a line at 1200′, this coal is not shown in the reserve. It depends on how deep you go; it depends on the quality of coal you are prepared to use;

some of the coal is very bad but you can beneficiate it and use it; it depends on the techniques of extracting the coal. You recover only 50 per cent of the coal today by the normal deep mining method. Suppose you can gasify it underground tomorrow, I know efforts have been made but have not succeeded so far to the required extent, suppose you scan through hydraulic means which will succeed to a certain extent, recover more of the coal, then the entire picture of coal is going to change in a few years. And I am sure that the present practice of leaving 50 per cent of coal unmined cannot be tolerated when fossil fuels are dwindling so fast.

Similarly, the question of the efficiency of coal use is pertinent. For the old boiler in the train, the steam locomotive, the efficiency is very low. With the magneto-hydro-dynamic system, you reach an efficiency of 60 per cent. It changes the entire efficiency picture of coal usage, and hence indirectly it changes the picture of coal reserves.

Similarly, in the hydro-electric sector, I was asked the other day about India's hydro-electricity capacity. Well, figures vary, and one reason for that is that in the north-east of our country, there is a certain place where if our friends from China and we co-operate, a lot of energy can be generated. If we do not co-operate, it cannot be done. And similarly, in the Brahmaputra in the north-east, a large amount of energy is locked up but harnessing it means a tremendous amount of capital investment and other problems. I will not go into that. Then there is the question of how many micro sets you put up, the fall is there, you can put up any number of micro sets, you can use bulb turbines, you can use the gradients of the canals. So what is the measure of reserves of hydro-electric potentials?

Today in India we have used up most of the easy sites. Now if we go up in the Himalayas, we have more sites, but the costs are very great. Again, it is a question of cost. It is a question of other competitive sources of energy and of course, the environmental problems. So, all told, while we have to go by a set of figures, we have to refine the entire process of assessment of reserves to a point where they take into account the cost and time factor. Only then, can one have realistic picture.

Take nuclear energy. Take uranium. If you go in for a fast breeder reactor and use the uranium in it, the reserves of uranium will go up immediately. And so whichever source you look at, these are factors which face one when dealing with questions of long term planning in the energy sector.

I will tell you something about load forecasting, which is of great interest. Every 5 years in India we have a drought, that is, rain falls short of requirements. Consequently, catchment areas do not have enough water. Therefore, our hydro-electric stations tend to have much less water in the dams. So the potential goes down in the very year that we need additional energy because more energy now goes to our agricultural sector for working the pumps. So precisely when we need it most, we have an energy shortage because almost 35 to 40 per cent of our power comes from the hydro-electric sector. To what extent can we build this into our load forecasting? This will occur at least once, sometimes twice in 5 years. Can we afford the luxury of developed countries of having a large spinning reserve which remains idle for 3-4 years? Here is a cruel choice. It has to be made. My own view is that we cannot have a large spinning reserve because as a poor country we just don't have the money to put into such a large spinning reserve. We should instead concentrate on having a national grid as quickly as we can and use thermal and hydro-electric power generation units in a complementary manner, so that we can use the dams to store the water and use hydro-electric energy when we need it. This kind of planning on a national basis is what we should aim for in India.

I now come to the most important question, in my view, of the non-commercial sources of energy—the mass fuel, used by 90 per cent of the families, even in the city. It would be a mistake to assume that everyone in a city uses commercial fuel; they do not. I know that in some cities they have started using animal waste as fuel. One of the reasons is dairy farms, which used to serve the country side, have come within the ambit of the enlarged cities and since there is no use for animal waste in agriculture, they tend to become fuel in the city. In this manner the problem is becoming more complex. Figures for agricultural waste are very uncertain. The agricultural wastes are also used to feed cattle and it is very difficult to know precisely how much is fed to the cattle and how much is available for the purpose of generation of energy.

Then there is the question of bagasse which is often taken as an agricultural waste. However, bagasse is not available for any other purpose, except as fuel in the sugar plants themselves. They use all the bagasse. None is available although it appears in statistics of agricultural waste.

Similarly, there are questions about biogas. I know that China has done remarkable work in the field of biogas. But one of the problems which I have seen in India is that when you really get to the micro level, the number of families in a village who have 5 heads of cattle or more, is perhaps 12½ per cent. Only they can have their own biogas plant. Suppose one says that the whole village should have a biogas plant, then the problem of having adequate biogas plants comes in. But suppose we decentralise this, only 12 per cent of the families have biogas plants, then the social problems arise. Because the poorest of the village also depend upon animal waste for their fuel. If all the animal waste goes into these few biogas plants, where would they get their fuel from? So you see, these are peculiar social problems which can only come to light at the micro level, although in principle it is obvious that the biogas revolution is something which all of us would welcome.

So the planning has to be in terms of the realities on the ground. The reality on the ground is that most people in the villages get free fuel, they get wood from the forest, they get the animal waste free, they get the agricultural waste free, they do not have the purchasing power today to go in for alternative fuel.

I will tell you the danger of going by macro statistics. You will find that 'x' number of villages have been electrified. When you go to a village, in some places, the electricity has reached one pole, and that village is recorded as having been electrified. Even in those villages where there are sufficient arrangements for supplying electricity to individual households, I think on an average only 12 to 13 per cent have taken electricity connection for lighting—not more than that. It is not that they do not want it. But the low purchasing power inhibits them. So mere supply of electricity is no guarantee that electricity will be used. The purchasing power of the people is the most important factor.

Some goes for coal. Many people give the advice, "Why don't you in this country save forests and supply coal. You have plentiful coal. Supply coal to the people". Yes, we can. But they will have to buy it, whereas today they do not buy their fuel. To expect that section of the population which is already in a difficult economic situation, below the poverty line, to find resources to switch over to and to purchase some other fuel is, I think, a pipe dream. Therefore, in our energy planning for rural areas, we have got to concentrate on fuel wood for the time being. We have got to concentrate on renewable quick growing forests. We have got to devise methods of using vast tracts of land which cannot be used for agriculture like the arid zones, which have been used in Brazil, I understand, even to produce oil. So we have to use this land and we have to have quick maturing trees on a large scale. I think that is a realistic approach to the rural problem. Otherwise to expect people to purchase commercial fuel is not a very realistic proposition.

I come to the question of forests. I have seen statistics of forests in India. I am sorry to say that forests are being depleted at a faster rate than what we think. Where you think you have large areas which are supposed to be under forests, when you do go into the forest, a little inside, you will find large bald patches. Therefore that is another piece of statistics that I would like you to consider very carefully. I would like you to take this into account before the destruction of forests actually reduces the extent of firewood that is really available.

Now we talk of biomass. I do not think we have any accurate data, and we cannot have that accurate data, because it is highly decentralised. We talk of wind energy, we have yet to develop wind maps. We talk of solar energy, insolation. Well, there is some data, and again not enough, sometimes one has broad data. You go by the latitude, and you say that this is the insolation at this latitude, without taking into account the cloud cover, which may change the entire ability to make use of that insolation. So there cannot be a broad generalisation on these questions. You will have to go to the micro level, and collect data at that level, if you are really going to make use of it.

Then similarly there are other questions of wave energy, ocean energy; the question of data on the transport of energy sources to the consumer. They are generally available at the source. But as somebody said what the consumers receive is a different matter. There is the question of T&D losses. Somebody said that China has the same order of T&D losses as we have. It came as a great source of satisfaction to me because I was beginning to wonder, if this was something which was peculiar only to us in the world. It is true that our T&D losses are very high. We have so many step down transformers before we reach the ultimate tube-well, but part of it is because of pilferage of power. The suggestion of having meters, low tension meters, on a wide scale is also most welcome. It is also a challenge to the pilferers to find a way to get round the meters and believe me, they have found very ingenious methods of doing so. But that is, of course, a continuing struggle.

Similarly there is diversion of coal. There is undoubtedly diversion of coal from one set of consumers to another. There is a certain amount of pilferage also. But pilferage or diversion does not mean that coal is not used within the system.

Now the question of the unit for planning has often been raised. My own feeling is that the village would possibly be too small, and a district would be too large. I have been convinced during my study of this problem, that it would be begging the question, if one only went in for macro planning at the national level, or at the state level. It has got to be a question of balancing the need against the supply at the micro level. You can have a unit somewhere between the village and the district, perhaps the block can be the unit. There we have to have detailed data and answers have to be based on that data. You have to have a proper mix of various energy sources, commercial, non-commercial, traditional, new sources, and also renewable sources. So the most important element of energy planning today is a proper interlinking of micro planning and macro planning and if we can achieve that interlinking then we will get at least a clear idea of where we want to go. Otherwise we may be able to take care of developmental needs only, but we may not be able to take care of the needs of the people and that is my real concern.

I have mentioned earlier, cultural factors and social factors along with economic factors, of purchasing power and so on. The question of 'social power' is important. In China all the human waste is re-cycled. In India human waste is not re-cycled, which in my view is very wasteful. Similarly solar cookers can be used today. There is no problem, except that the housewife is accustomed to cook inside her kitchen. This is a very relevant question. It is related to the cultural tradition and habits of the people. This in short is the point I want to make. I do not want to go into the industry or other aspects.

I do feel that the Workshop is very timely. I think we in this country have got to do something which nobody else can do for us, that is, we have to have much more interaction between researchers, policy makers, and those directly involved in the operation of data and information system. This is lacking. Unless there is a spread of knowledge, policy decisions cannot be taken quickly, they cannot be implemented effectively in a manner which affects the targeted consumer. So this is the most important immediate problem and we are lucky that the world has entered the phase of an information explosion, and the hardware and the software are available. Micro processers and mini computers have come in a big way. There is the fibre glass technology; there is the cable system. The entire information; industry has changed; the technology has changed; we have to take advantage of it. I think we in the developing countries could find no better area to begin to take advantage of this revolution in the information technology than energy, and if this Workshop can give a fillip to it, then I would say it has served a very major purpose.

In the end I would like to hope once again that this Workshop will lead to follow-up action, which will lead to very concrete results in this country. The presence of so many eminent people in this country who are dealing with energy, leads me to hope that this will happen and that the results will be forthcoming within a matter of perhaps a couple of years. Thank you very much.

HINGE

Thank you very much Mr. Pant. Thank you for giving us some excellent suggestions and also for your exposition of the energy scene in this country. Most of our people live in villages and 70 to 80 per cent of their energy requirements is for cooking. That poses a very serious problem, apart from the problem of exhaustion of the conventional sources, in our country. We need to do something very urgent about how to stall the rapid deforestation which is posing the threat of food being available but not the cooking medium. Thank you very much.

Subject Index

Aggregation, problems of, 382
Agro-climatic regions, data for, 381
Alcogas, use of, 232
Alcohol products, 321; subsidy on, 329; use of, 322-24
Animate energy, 448-52; contribution of, 411-12; sources of, 384, 481
Ankleshwar oil field, 42, 45
Annual Survey of Industries (ASI), 200-1, 365
Annual Survey of Small Scale Industries, 485
Argentina, 276, 281; classification of energy information, 280; data transmission network, Arpac, 306; energy planning, 276; Energy Bureau (FB), 276, 283, 287, 304; energy data systems, 276; electrical energy bulletin, 286-89; feasible improvements in information system, 302; Fuel Bulletin, 285, 287-88; gaps in information, 281; National Statistic System, 283; organisation of energy planning entities, 277; regional information, 290; non-energy information, 297-301; petroleum data bank, 253
Assam oil fields, 42

Bagasse, 498; use of, 321, 439
Biogas, 430, 434, 498; fertilizers, 438; plants, 273, 440, 498; plants maintenance problems, 274; successful adoption of, 387
Biomass, 460; sources, 384; transition, 480
Bombay High oil fields, 44
Brazil, 461; energy information system, genesis of, 311ff; Inter-industry matrix, 310; Inter-ministerial collaborations, 316; Inter-sector matrices, 314; National Energy Documentation Centre, 484; National Energy Balance, 310
Brazilian Energy Information, 310ff; energy matrix, 310-14; energy model, 321, 328; firewood consumption, 313

Cabinet Committee on Energy, 475, 478
Capital output ratios, 85
Centrais Electricas Brasileiras (ELECTROBRAS), 311, 313, 316, 319
The Central Electricity Authority (CEA), 70, 77, 83, 103-5, 129-30, 150, 362, 464, 467, 470, 476
Central Fuel Research Institute (CFRI), 161, 171, 187
Central Mine Planning Design Institute (CMPDI), 171-72, 187
Central Mining Research Institute, 185
Central Mining Research Station (CMRS), 171
Central Statistical Organization, 488
Charcoal and firewood, use in Thailand, 244
Climatic data, 431

Coal, calorific value of, 187, 488; pilferage of, 500; wholesale price of, 170

Coal, consumption data, 169, 180, 184; demand assessment, 177-78; forecast of, 172-73

Coal Controller, 162-63, 174, 177, 182 83, 186, 362, 456-57

Coal equivalent units, 52, 54; exploration, criteria for, 159

Coal fires, 170; fuel cycle, 179; gasification, 181

Coal industry, employment in the, 186; historical development of, 456; management, 457-58; material handling in the, 175; production data, 162-63, 175-77; projects investment, 171

Coal mines manpower data, 164; statistics, 162

Coal mining, changing conditions, 187; output, under estimation of, 457

Coal plants, emissions from, 149

Coal replacement units, 52, 54

Coal resources in India, 160

Coal transportation data system, 166-67

Coal use, efficiency of, 496; environment aspects of, 181

Cobb-Douglas production function, 218

Co-generation, 209, 214

Commercial energy, 441; sources, 272

Commission for Additional Sources of Energy (CASE), 380, 429

Commission Nacional de Politica Energetica (CNPE), 333-36, 338, 357-58

Committee for the National Energy Balance, 312, 316-18, 322

Computers, use of, 6, 26, 31, 64, 97, 103, 135, 152, 161, 164, 172, 306, 368, 413, 464

Connected load block, 147

Conselho Nacional de Petroleo (CNP), 313, 318

Consumption of energy, norms of, 202, 207

Conversion technologies, 344

Cooking energy, 492; and deforestation, 502; diversity in requirements, 428; non-commercial energy for, 495

Cooking stoves, improved, 387, 437, 438

Cost of cultivation studies, 485

Cost of capacity additions in the power sector, 477

Cost overruns of power stations, 114-28

Daily load profiles, 429

Data, absence of, 14; attributes of, 175; availability, 97; collection, sophistication in, 473; communication, 32; compatibility of, 174, 193; conflicting requirements, 38, 129; consistency of, 174, 181, 252; cross checking of, 186; excess of, 130, 135; gaps and deficiences in, 366; processing, 96; product-oriented, 40, 48; retrievability of 40; sources of, 362; standardization of, 176, 216; unavailability of, 368; utilization, 83ff; use of, 130; withholding, 130, 205-6, 209, 215, 471, 480

Data for management, 129, 134-35,

Decision-making, at the firm level, 211; at the national level, 212; uncertainty in, 33

Deforestation, 460, 499, 502

Demand assessment, 183; data requirement, 347; forecasting, 19, 476-77; seasonal variations in, 31; management, 136, 145, 151

Demographic system, 129

Depot consumption, 31, 465

Developing countries, energy planing in, 414
Diesel oils use in agriculture, 25
Director-General of Mines Safety, 162-64, 179
Director-General of Technical Development (DGTD), 153, 205, 207, 219
Distribution losses, 274-75; planning, 32, 64; models, 29-30
Domestic energy consumption, 241-42
Dominican Republic, 324, 333ff, 382; alternative scenarios, 357-58; National Energy Assessment Project, 333, 336
Down-stream facilities, 35
Droughts, effects of, 497
Dungcakes, 398, 407; burning as a cultural phenomenon, 460

Economic, efficiency, 222; resource, 42; system, linkages with, 10; viability, 448
Econometrics, use of, 322
Efficiency improvement, 24; changes in, 50
Energy Development Information System (EDIS), 338-39, 341, 344-47, 356-59; problems in implementation, 360
Election Commission, 129
Electricity, cost of, 486
Electrobras, 461
End-use, analysis, 22, 30; approach, 181; categories, 251; energy demand, 251; method, 85, 131-32; structure, 325; technique, 208
Energy accounting matrix, 483-84
Energy analysis, 191-92, 368; assessment, 13; balance, 262, 290, 293, 295, 305, 310-14; conservation, 212, 219, 223, 229; forms of, 361; indexing, 12, 480; information, classification of, 280; intensities, 175, 204-6; legislation, 202, 206, 213, 219
Energy consumption, norms of, 208, 221
Energy cost of coal transport, 188
Energy data, availability of, 491; centralized agency for, 10, 367, 466, 475, 488; knowledge of, 10
Energy demand, 293, 343; elasticity, 201, 203, 209; quality of, 15; renewable sources of, 144; resources cell, 392; sources of, 362; substitution, 252; supply and demand, local control of, 380
Energy economy, planning, 330; interaction, 330
Energy, non-commercial sources, 390, 475, 497
Energy pricing, 229; productivity, 218, 220
Energy planning, organizational apparatus for, 247ff; overview of, 334
Energy statistics, comparability, 14-15; use of, 5
Energy surveys, 242
Energy technology, break-throughs in, 331; data, 327
Enhanced recovery process, 45
Environmental aspect, 172, 180, 214; of coal use, 149, 181, 184, 186-87; data variability of, 216; deterioration, 185; monitoring, 215; problems with alcohol production, 322-23
Eucalyptus plantations, 442
Evergreen forest reserves, 244
Exploration investment, risk in, 43

Farm management surveys, 485
Fast breeder reactor, 497
Feedback, monitoring network for, 482
Financial and commercial data, 134
Financial analysis, 327
Final energy, 54, 55
Firewood consumption, 313; sup-

ply, 386
Flexibility in data systems, 474
Forecasting, interpretative skill, 33; reconciliation, 29
Freedom of Information Act, 471
Fuel Policy Committee (FPC), 363, 388-89
Future rural energy needs, 426, 440

Gas del Estado bulletin, 292
Generation outages, 86; losses, 86; planning, 132-33
Geo-data base, 38; system, 158-59
Geological data, 41
Gestation periods, 477
Ghazipur windmill, 435-36

Heat recovery system, 327
Household sector, 53
Hydraulic efficiencies, 437
Hydro-carbon prone rocks, 52
Hydro-electric capacity, 329, 496; production, 272

Igneous rocks, 51
Income elasticity, 136; international comparisons, 203
India, coal data system, 157ff; energy policy in, 475; rural electrification in, 94
Indian Bureau of Mines, 179
Indian coal formation, 159, 495
Indian society, poverty aspect of, 3
Indo-US Joint Working Group, 48
Information borings, 41; explosion, 501; reverse flow of, 447
"Immersion-perception" technique, 409
Input-output models, 85; tables, 263, 345
In situ oil, 44
Installed capacity, 101, 133; generating capacity, 107, 234-36
Integrated rural energy system, 14
Interactions amongst energy sectors, 96; between supply and demand, 136
Inter-fuel substitution, 25, 41, 49, 396, 468, 474
International comparisons, 53
Inter-refinery planning, 459
Internal rate of return, 211, 213
Interview methods, 408-9
Investment analysis, 133; in technology, 328; planning, 330
Irrigation efficiency, 437

Krayvis dollar, 204

Life cycle costing, 220, 432-33
Load forecasting, 84, 101, 131-32, 144
Load control through micro-processors, 145; management, 145, 208; shedding, 151
Long run marginal cost, 138-39, 151, 187
Losses commercial, 89; technical, 89; transmission and distribution, 89, 90-91; of crude oil, 32; optimal, 138-40

Management information systems, 463, 487-88
Marginal cost of production, 180
MARKAL Model, 322, 331
Market segment data, 440
Matching of current demand with technologies, 430, 437, 481
Metering, 133-34, 148, 205, 459, 465-68, 484-86, 502
Mineral Exploration Corporation (MEC), 161
Mineral Resources Data Bank, 37
Mine equipment data, 166-68; safety data, 164
Mines, environmental aspects of, 172
Mining systems, economic data for, 180
Modelling, 29, 30, 219-20, 308-9, 322, 331, 337, 474, 387
Motor spirit consumption, 41

Naharkatia oil field, 42
National Council of Applied Economic Research (NCAER), 191, 194, 206, 410, 460
National Informatics Centre (NIC), 77, 103-5, 464
National Information System for Science and Technology (NISST), 37
National Productivity Council (NPC), 102, 207-8, 210
National Sample Survey, 381, 470, 488; Organisation (NSSO), 364, 366, 388, 397
New energy technology, 385
Non-commercial fuels, 384; energy, 495
Non-conventional energy sources, 396-98
Non-energy information, 296-97
Nuclear power, 141, 144, 150, 153, 379

Oficina Nacional de Estadistica (ONE), 335
Oil Coordination Committee (OCC), 58, 61, 63
Oil economy budget, 61; in place, 44, 46; prices committee, 63; production, efficiency of, 459; traps, 52
Oil exploration and foreign companies, 37, 459
Oil Industry Development Board, 58
OLADE, 288, 290, 308
Organic Rankine Cycles, 433
Outages, 138-39

Pakistan, results of survey of non-commercial fuels, 392
Parametric wells, 41
Parana, State of, 325
Pay back periods, 433
Peak load, 142
People's Republic of China, 153, 260-675, 498, 500; coal production, 272; distribution, losses, 275; disposal of nuclear waste, 150; energy system network, 263, 267, 287; energy situation in, 114; dual statistical system, 260, 472; energy balance table, 262, 268-9; input-output table, 263; maintenance problems with biogas plants, 274; National Energy Data System in, 260ff; non-commercial energy in, 462
Performance comparisons, 130
Petroleo Brasileiro S.A. (PETROBRAS), 311, 313, 319, 461
Petroleum Conservation Research Association (PCRA), 33, 58
Philippines, alternative source of energy, 472
Photovoltaics, 332, 433
Pilferage of power and coal, 500
Plant Load Factor (PLF), 112, 417
Pollution control, 230
Power crisis, 1973-74, 492
Power sector, 69; commercial data, 76, 95; construction data, 76, 92; costs of capacity additions, 477; decision-making, 136-37; investment in, 477-78; operations data, 76; organizational structure, 69; plan investment into, 495; system data, 76, 101
Power supply, quality of, 459
Price elasticity, 136
Price as a signal to users, 138, 145, 150
Process market share, 350
Production, marginal cost of, 183
Projections, probabilistic nature of, 34
Purchasing power, 4, 498

Rail transportation, 53, 176, 184
Railway, electrification of, 479
Rajadhyaksha Committee on Power, 89, 100, 101-2, 132, 363, 470-71
Recoverable reserves, 46

Recovery factor, 34, 45
Reference Energy System (RES), 325-26, 339, 341, 345-47, 358-59
Regional Electricity Board (REB), 73, 104, 133
Regional Load Survey Group, 131
Regional Power Survey, 142-43, 150; office, 77
Reliability of electric supply, 138-40, 144-45
Renewable energy, 379, 382, 427ff; future contribution of, 442-45; sources, 324
Renewable Non-conventional Energy Project, 244
Renewable systems, feasibility of, 428-29
Reserves, categorization of, 34, 43; defining, norms of, 39; prognosticated, 34, 36, 39
Resources, classification of, 180, 183; production, 344; sub-economic, 42
Road transportation, environmental problems from, 184
Rural electrification, 93, 102, 152, 230, 233, 498; energy, 381, 463, 480, 492; problems, 273
Rural Electrification Corporation (REC), 93, 94, 147
Rural Energy Data Systems, 371ff
Rural energy studies, variability in, 371, 372; energy survey, 398, 410, 481; comprehensive, 399; determinants of, 400; methods of data collection, 407; selection villages and households, 401, 446; short duration, 399, 402: standardization of, 373; types of data to be collected, 405, 418-9

Sedimentary basins, 37, 39; rocks, 51
Seismic data, 41; surveys, 52
Shale and tar sands, 495
Short-term planning, 41
Silicon Valley, 144, 145
Site specific data, 435, 480
Small scale industries, census of, 366
Social acceptability of new energy technologies, 379, 385
Social forestry, 442
Socio-economic activity, 293; data, 136, 139; surveys, 397, 398; information, 481
Solar cookers, 501: crop drying, 325; market segments for, 439
Solar energy, 220, 499; data, 327; pumping of water by, 381, 437; systems, 429, 431; ponds, 433
Solar water heating system, 438, 442
Source inventory, 210
Spatial distribution models, 29, 30
Spinning reserve, 100, 497
Standard survey procedures, 383
State Electricity Board (SEB), 75, 85, 89, 93-95, 99, 100, 103-4, 150, 471
Storage of electricity, 144
State Load Despatching Centre (SLDC), 77, 100
Stoves, non-cooking use of, 482
Subsidies, in supply of power, 379
Supply management, 136; oriented policies, 49
Surveys, limitations of large scale ones, 391; methodological issues, 396; by recall, 407, 485
System boundaries, 192

Taiwan, 144
Tandon Committee's Report, 221
Technology market share, 350; efficiency, 351
Thailand, 461; balance of payment, 228, 229, 255; coal handling port, 241; energy data system, 246ff; energy demand, 230ff; energy policy, 228ff; fifth social and development plan, 228; inland water transportation, 232; lignite for industrial uses, 241; lignite power plants, 240-41,

468; National Energy Data Systems, 227; National Energy Administration (NEA), 228, 240, 247, 249, 257-59, 292, 296, 468, 473; natural gas deposits, 234; National Energy Authority, 244; oil shale deposits, 238; options for use of natural gas, 235, 237-38; resources, 240; use in electricity production, 468; National Energy Information and Documentation Centre (NEIDCT), 247, 257

Thailand Energy Master Plan (EMP), 250

Theft, 146-47; of electricity, 148; of materials, 92

Thermal power station, 478; station performance, 130, 134; units, planned maintenance for, 60ff

Thermodynamics, efficiency, 222; first law of, 211; 213; second law of, 210, 211, 213, 221

Time lags, 31, 201

Traditional energy, 416; sources, 256, 272, 386, 388, 396; fuels, 384; sources, role of, 384ff

Transmission planning, 132

Transport sector, 53, 491

Transportation data, 180, 183

Turbo-charge engines, 479

Useful energy consumption, 54, 55

User oriented data, 474

Village level surveys, 394

Volcanic rocks, 51

Waste heat, recovery of, 214

Weibull distribution, 431, 436

Wind energy, 436, 499; mill, 436; resources data, 326; systems, 429, 430, 431

Wood fuels consumption, 386

Working Group on Energy Policy (WGEP), 48, 101, 159, 172, 183, 203, 263, 388, 389, 477

Name Index

Aditya, S., 37, 42, 44, 45, 57
Agarwal, S.K., 150
Ahuja Dilip, 29, 51, 150, 214, 256
Altomonte, Hugo A., 276, 308
Araujo, Joao Lizardo R.H. De., 310, 321, 322, 323, 325, 328, 329

Banerjee, A.N., 183, 186
Bhatia, Ramesh, 427, 446, 448, 459, 465, 469, 470, 474
Bhattacharya, D., 191, 200-2, 205-8, 217
Boonyubol, Charuay, 227
Bose, S.K., 157, 174-76, 178, 179, 182, 183, 186, 187
Bowonder, B., 182, 371, 380-83
Brar, 440

Darmstadter, Joel, 202, 203
Dayal, Maheshwar, 379, 426, 436-38, 440, 442, 443, 445, 452, 479, 481
D'Errico, Antonio J., 333
Desai, Ashok, 44, 47, 48, 54, 55, 202-5, 465, 466, 467, 473, 484, 487
Desai, Nikhil, 30, 51, 54, 55, 146, 147, 148, 149, 204, 205, 258, 259, 271-73, 308
Desai, Nitin, 37, 41, 42, 44, 46, 49, 53, 55, 471-73, 483, 487
Doernberg, Andres, 333

Fazal, Mohammad, 3

Garcia, Elis, 333
Ghate, P.B., 273, 435
Gopalakrishnan, N.K., 9
Gulati, I.B., 19
Gupta, C.L., 213, 427, 437-37, 442, 443, 446, 447, 481

Hinge, K.S., 491, 502

Jagannathan, C.R., 33, 41, 45, 47-51
Jayashankar, R., 31, 56
Jhirad, David, 32, 221, 300, 310, 329, 330, 331

Khurana, J.C., 31, 46, 52, 209, 442, 445

Levy, Maurice, 13, 256, 271, 368, 487, 474
Limsumalee, Suchai, 255, 256, 259, 321, 472, 473

Marwah, G.S., 178, 182, 186, 188, 415
Mena, Dean Marino, 142, 472, 473
Mishra, K., 474
Moulik, T.K., 380, 438, 446, 450
Mubayi, Vinod, 13, 132, 179, 182-84, 203, 204, 309, 321, 322, 329, 330, 437, 468, 474, 480, 491
Mulchandani, H.K., 19, 28, 30, 31, 33, 255, 323
Munasinghe, Mohan, 28, 135, 145, 151, 152, 211, 213, 221, 469

Nangia, Rita, 30, 151
Nathans, Robert, 333
Nyati, K.P., 208

Owens, Michael, 333

Pachauri, R.K., 10, 12, 13, 28, 48, 49, 52, 53, 791
Pant, K.C., 491
Prakasa Rao, S.V.S., 257

Rajan, K.S., 205, 209, 210, 213, 217, 221, 223
Ramani, S., 217, 221
Ramaswamy, N.S., 448, 452
Ramesh, S., 133, 138, 152
Rao, S. Subba, 174

Sarma, E.A.S., 470, 471
Seal, K.C., 361
Sengupta, D.P., 31, 69, 129, 132, 133, 135, 136, 142, 146-49, 152, 157, 205, 446-47, 464, 467, 469, 471, 485-86
Shah, J.C., 11, 129, 132, 142, 146-48, 150-51, 201-4, 296, 207-8, 255, 274, 275, 435, 438, 466, 487
Smith, Kirk, R., 12, 54-55, 145, 436, 447, 479
Srivastava, S.S., 200, 201, 408, 479
Srivastava, R.N., 329

Yingzhong, Lu, 150, 260, 271-75, 489